Tooth Development in Human Evolution and Bioarchaeology

Humans grow at a uniquely slow pace compared with other mammals. When and where did this schedule evolve? Have technological advances, farming and cities had any effect upon it?

Addressing these and other key questions in palaeoanthropology and bioarchaeology, Simon Hillson examines the unique role of teeth in preserving detailed microscopic records of development throughout childhood and into adulthood. The text critically reviews theory, assumptions, methods and literature, providing the dental histology background to anthropological studies of both growth rate and growth disruption. Chapters also examine existing studies of growth rate in the context of human evolution and primate development more generally, together with implications for life history. The final chapters consider how defects in the tooth development sequence shed light on the consequences of biological and social transitions, contributing to our understanding of the evolution of modern human development and cognition.

Simon Hillson is Professor of Bioarchaeology at the Institute of Archaeology, University College London. He has over thirty-five years of experience in teaching and research in dental anthropology, with research focussing on the development and diseases of teeth and the ways in which these can shed light on the way of life of people in the past. His previous books include *Teeth* (Cambridge University Press, second edition, 2005) and *Dental Anthropology* (Cambridge University Press, 1996).

Tooth Development in Human Evolution and Bioarchaeology

SIMON HILLSON
University College London, UK

CAMBRIDGE
UNIVERSITY PRESS

Shaftesbury Road, Cambridge CB2 8EA, United Kingdom

One Liberty Plaza, 20th Floor, New York, NY 10006, USA

477 Williamstown Road, Port Melbourne, VIC 3207, Australia

314–321, 3rd Floor, Plot 3, Splendor Forum, Jasola District Centre, New Delhi – 110025, India

103 Penang Road, #05–06/07, Visioncrest Commercial, Singapore 238467

Cambridge University Press is part of Cambridge University Press & Assessment, a department of the University of Cambridge.

We share the University's mission to contribute to society through the pursuit of education, learning and research at the highest international levels of excellence.

www.cambridge.org
Information on this title: www.cambridge.org/9781107011335

© S. Hillson 2014

This publication is in copyright. Subject to statutory exception and to the provisions of relevant collective licensing agreements, no reproduction of any part may take place without the written permission of Cambridge University Press & Assessment.

First published 2014

A catalogue record for this publication is available from the British Library

Library of Congress Cataloging-in-Publication data
Hillson, Simon.
Tooth development in human evolution and bioarchaeology / Simon Hillson.
  pages cm.
Includes bibliographical references and index.
ISBN 978-1-107-01133-5 (hardback)
1. Dental anthropology.   2. Teeth – Evolution.   3. Human evolution.
4. Human remains (Archaeology)   I. Title.
GN209.H58 2014
599.9′43–dc23
2013044151

ISBN   978-1-107-01133-5   Hardback

Cambridge University Press & Assessment has no responsibility for the persistence or accuracy of URLs for external or third-party internet websites referred to in this publication and does not guarantee that any content on such websites is, or will remain, accurate or appropriate.

Contents

	Acknowledgements	*page* vii
1	**Why development and why teeth?**	1
2	**Development schedule, body size and brain size**	5
	How development is studied	5
	Human growth in body size	6
	Body size growth in non-human primates	18
	Growth in different systems of the body	24
	Summary	25
3	**How teeth grow in living primates**	28
	Process of dental development	28
	Issues in studying dental development	32
	Dental eruption	42
	Tooth formation	49
	Summary	68
4	**Microscopic markers of growth in dental tissues**	70
	The tooth surface	70
	Microscopy of the crown surface	76
	Structures seen in sections of teeth	88
	Summary	110
5	**Building dental development sequences**	112
	Underlying principles	112
	Methodological issues	123
	Development chronologies for living and fossil primates	128
	Summary	147

6	**Human evolution, pace of development and life history**	149
	Life history	149
	Characteristic features of human life history	150
	Life history of fossil primates	153
	Weaning, giving birth and the expansion of post-canine teeth	155
	Fast and slow mammals and Schultz's rule of eruption	157
	Life history, development and cognition in primates	158
	Summary	159
7	**Dental markers of disease and malnutrition**	162
	Hypoplastic defects	162
	Wilson bands, pathological striae or accentuated lines	174
	Recording enamel hypoplasia by simple surface observation	176
	Building sequences of defects	181
	Causes of enamel hypoplasia	184
	Summary	195
8	**Health, stress and evolution: case studies in bioarchaeology and palaeoanthropology**	198
	Health, stress and prevalence	198
	Case studies in bioarchaeology and palaeoanthropology	205
	Summary	225
9	**Conclusions**	228
	Appendix A: Tables	231
	Appendix B: Technical information	261
	References	273
	Index	302

Acknowledgements

I wish to acknowledge my debt to the teaching of Alan Boyde and Sheila Jones at University College London (UCL) at the start of my career. Their extraordinary dental and skeletal anatomy course has been the foundation for many other researchers in this field as well as myself. The original idea for using microscopic incremental structures in dental enamel to build a schedule of development goes back to Alan Boyde's work published in 1963. I have also had the benefit of working at UCL alongside other former students of the course, including Chris Dean and Daniel Antoine. This has provided not only inspiration, but also a reality check for my ideas. I have further benefitted from discussions with other members of the extended family of histologists and anthropologists working in this area, in particular Don Reid at Newcastle University, Louise Humphrey at the Natural History Museum in London and Charles FitzGerald of McMaster University. My UCL colleague Tony Waldron has been generous with advice on epidemiology and I wish in particular to remember Phil Walker, who very sadly died in 2009. I greatly miss my conversations with him and the approach I have taken in my review of enamel hypoplasia in Chapters 7 and 8 grew out of a brief chat at a conference. Over the years I have learned much from similar conversations, especially with the international group of dental histology researchers such as Debbie Guatelli Steinberg, Tania Smith, Rebecca Ferrell and Paul Tafforeau. I am grateful for all this advice, but of course, I take full responsibility for the ideas and opinions expressed in this book. As always, I am also grateful to my much-loved and long-suffering family at home for all their support.

1 Why development and why teeth?

In comparison with other mammals, humans grow up slowly. This is integral to the way we learn the complexities of our language, social organisation and material culture. It is part of what defines us as human. Slow development is also one of the distinctive features of primates as an order of mammals and there is a relationship with size. The larger species within any mammalian order tend to develop more slowly than the smaller species. Humans are among the largest of primates but, even taking that into account, we are still by far the slowest primates to reach full maturity. In any development graph, we are an outlier, in a different category to the rest. One crucial question for archaeologists and palaeoanthropologists is therefore to determine when our slow rate of development evolved. If the earliest stone tools date to 2.6 million years ago, were they made and used by hominins (the subfamily of living and fossil primates to which we belong) for which the pace of development was as slow as ours, or was the pace relatively faster, as in other living primates? This is surely central to understanding the meaning of these tools, the social and behavioural context within which they fitted and the cognitive abilities of their makers.

The evidence to answer this question can only come from the fossil remains of hominins. Most fossils are teeth and jaws, not only for hominins, but for all mammals throughout the geological record. Teeth have evolved for daily wear and tear in the testing conditions of the mouth, so it is not surprising that they survive well in the ground. This means there are more dental fossils which are preserved in a state in which they can be identified and studied than, say, the bones of the limbs. Not only that but, as explained in the following chapters, teeth are the *only* fossils in which growth *rate* can be studied. This is because they have a daily clock built into the layered structure of their tissues against which the growth rate can be measured. That is not to say bones are of no use. Bones of young primates have indeed been found as fossils, frozen at the stage of development reached at death, but on their own they do not have an inbuilt clock to estimate the age at which this stage was reached. For this reason, the debate about the evolution of the human development sequence centres on teeth.

Another special feature of teeth is that, once formed, dental tissues remain little changed throughout life unless they are modified by the effects of tooth wear and dental disease. Bone, by contrast, is constantly being replaced at a slow pace, so all traces of its initial development gradually disappear. This is known as tissue turnover. The lack of turnover in dental tissues preserves a layered structure that records a detailed chronology of development which is preserved throughout adult life. This makes it possible to determine the ages at which different stages of tooth development were achieved, to reconstruct the sequence in which the teeth were formed and to measure the rate of crown and root growth. If the individual was still growing at the time of death, it is possible to estimate its age on that day. Also preserved in the layered sequence is a record of disruptions to growth caused by fevers, nutritional deficiencies or other factors related to health. On the one hand, the presence or absence of these disruptions might indicate whether or not the individual being studied is a good subject for demonstrating normal development. On the other hand, it leads to other questions. The establishment of a slow pace of development in hominins, similar to that of living humans, might imply cognitive, social and behavioural changes. Could these have led to an increased level of care for the young or greater security of food supply? What was the effect on growing children of increased technological complexity, the adoption of farming or industrialisation? Many studies have attempted to address these questions by examining the record of disruptions to dental development preserved in fossil and archaeological teeth.

The preferential preservation of teeth and jaws is one reason for their importance as fossils. Another is their detailed variation in shape and form. Their morphology has always been a major element in the definition of living mammal species. Traditionally, zoologists collect a skin and a skull with its teeth; the great natural history museums of the world have enormous collections of this kind, including the holotypes or defining examples of each species, against which all other specimens must be compared before they can be identified. Amongst the primates, the order of mammals to which living humans belong, there is considerable variation in the dentition. The prosimians, or lemurs, lorises, bushbabies and the like, show a variety of distinctive dental features related to their diet and behaviour (Swindler, 2002). Overall in the primates, there is a correlation between tooth and body size and the prosimians in general are small on both counts. New World monkeys such as the howler or spider monkey are distinguished from Old World monkeys such as macaques or baboons by the possession of three permanent premolar teeth in each quadrant of the dentition. Old World monkeys, great apes and humans all have just two permanent premolars in each quadrant. The most obvious difference between human and chimpanzee, bonobo, gorilla and orangutan dentitions is in the much larger canine teeth of the non-human primates, which vary considerably between males and females. This changes the whole shape of the dental arcade and the way in which the teeth interlock when the jaw is closed. Canine reduction and the greatly reduced level of so-called sexual dimorphism are additional important questions in human evolution. Further back in the dentition, the pattern of cusps on the molars does not differ greatly between, say, chimpanzees and humans, but the dental enamel which coats the tooth crown surface is substantially thicker in humans. The earliest hominins in the fossil record had much larger post-canine (that is, premolars and molars) teeth than any living hominins, with very thick enamel indeed. Important trends in the evolution of humans include upright bipedal locomotion, manual dexterity and large brains, but no less important is the evolution of smaller teeth and jaws. Within our own species, *Homo sapiens*, this is the most marked change seen in either the skeleton or the dentition over the past 26 000 years.

Care is needed when using words like hominin, hominine, hominid, hominoid, pongid, pongine or ape. In the past, humans were separated as the family Hominidae, containing just one living genus *Homo* and a number of extinct genera. Chimpanzees, bonobos, gorillas and orangutans were placed in the separate family Pongidae. More recently, genetic evidence has shown how closely related chimpanzees and bonobos are to humans, so it has become common to include them along with gorillas, orangutans, humans and related fossils in the family Hominidae (Appendix A, Tables 1 and 2). To describe a living or fossil species as a 'homin<u>id</u>' means it is a member of the family Hominidae. Usage of other names starting in 'hom-' depends on the classification system favoured. Homin<u>ine</u> means the subfamily Homininae, which most would take to include all the living hominids except orangutans, which are separated off with similar fossil species into the subfamily Ponginae (or family Pongidae). Homin<u>in</u> means a member of the tribe Hominini, a subdivision of the Homininae. Some include living chimpanzees, bonobos and gorillas along with humans and related fossils. Others confine it to the latter, separating chimpanzees, bonobos and gorillas as the tribe Panini. Homin<u>oid</u> means a member of the superfamily Hominoidea, which includes both the hominids and the hylobatids, or gibbons and siamangs. 'Ape' is an old English word, originally implying monkeys as well as chimpanzees, gorillas, orangutans and perhaps gibbons too, but never humans. Until recently, most anthropologists confined it to chimpanzees, bonobos, gorillas and orangutans, but now some include humans as well (Robson and Wood, 2008). Not everyone agrees, so the word is avoided in this book which seeks only to be clear and does not deal with issues of taxonomy. Common names are used rather than the binomials, or official names for species, except for fossil taxa which have no common name.

This book therefore aims to address two questions of current interest in palaeoanthropology and archaeology using the evidence provided by microscope study of the dental tissues; particularly enamel and dentine. The first is the point in the fossil record at which the slow pace of modern human development appeared. The second is whether or not the major biological and cultural transitions shown in the fossil and archaeological record affected the health and well-being of the growing young. The book starts with an examination of the special place of humans in the world of mammalian development. Chapter 2 explores the particular features of the human body size growth curve and contrasts them with other living primates. It is not so much the overall shape of the human curve that is unique, but more the timing and combination of features. Chapter 3 continues with the evidence for a unique pattern of development in the human dentition, based on direct observation or radiographs of living children and other young primates. It is supported by data presented in Appendix A. By far the most straightforward aspect of dental development to observe in the living is the emergence of the teeth through the gums into the mouth. This, however, is the end of a long chain of developmental stages hidden inside the jaws, for which a variety of evidence has to be drawn together. For these reasons, they are not as well known as might be expected. The core of the book is Chapter 4, introducing the microstructure of enamel and dentine together with the tiny incremental features which are the basis for everything that follows. Techniques for microscopy are outlined in Appendix B. Chapter 5 goes on to describe the ways in which developmental sequences have been built from counts of the incremental structures in dental tissues. There is a tension here between the quality of information provided and the necessity to cut microscope sections from valuable specimens. The few sections which have been made from fossils are an important resource, but fortunately some non-destructive options for microscopy have been made available over the past decade. The resulting sequences make it possible to compare the rate of development in extinct hominins with living species. This comparison fits into a wider debate on primate life history and cognitive development which is outlined in Chapter 6. Disruptions to growth are first described in Chapter 4 along with normal structure, but Chapter 7 returns to the issue of defects and discusses the clinical and experimental evidence for the factors involved in their formation. Finally, Chapter 8 provides a critical review of the theoretical framework within which the defects are interpreted and presents examples from palaeoanthropology and archaeology.

2 Development schedule, body size and brain size

In comparison with other primates, the sequence of human development lasts a long time and has an unusual pattern of peaks in growth rate. But is it unique? Long development might be a function of size and humans are among the largest of primates. Similarly, development patterns vary between primates and it could be that humans form part of a continuum in the sequence of growth stages. Then there is the large human brain. Do humans follow a unique growth trajectory to achieve it?

How development is studied

Development is studied by measuring growing individuals at different ages. The study may be *cross-sectional*, with just one examination per individual in a group which includes a variety of ages, or *longitudinal*, with several examinations at different ages for each individual. The measurements taken vary. Some relate to general body size. In humans this is usually *stature*, but in young children the equivalent is *supine length*, where the child lies on its back between parallel head and foot boards. There are two alternatives for measuring body size in a foetus or young baby. *Crown-rump length*, from the top of the head to the most prominent part of the buttocks, is approximately equivalent to sitting or trunk height in older children and adults. *Crown-heel length* is measured with the baby lying on its back and one leg stretched out, and is the nearest equivalent to stature. The upright bipedal stance of the human body makes it difficult to find directly comparable stature measurements for non-human primates. For them, crown-rump length (trunk length) is commonly measured, along with lengths of segments of the limbs. Simpler than any of these, however, is to measure *body mass* (or weight in everyday language). This represents all the different systems and tissues of the body and shows different patterns of growth from stature, which reflects largely the size of the skeleton.

Whilst growth studies can be distinguished as cross-sectional or longitudinal, there are grey areas. For a fully longitudinal study, it is necessary to follow every member of a group of children throughout the range of ages included, but constant effort over many years is needed to maintain this. Some children leave, miss examinations, or join part way through. This situation is known as a *mixed-longitudinal study* and requires statistical manipulation to make sense of the results. A pure longitudinal study thus tends to include a small group of children but, because it follows individual children over time, it is particularly useful for showing how stature or mass change with increasing age. By contrast, cross-sectional studies are much larger and are more useful for understanding variation in stature or mass for a given age.

Human growth in body size

Growth rate, or velocity, is the gain in stature, mass or other measurement per year (or per month or week, depending on the animal). A *velocity curve* is the graph showing the change in growth rate from year to year. Each individual child has its own curve and the most famous is that based on the regular measurements made by Philibert Guéneau de Montbeillard between 1759 and 1777 of the height of his son François from birth to 18 years of age. This was done at the request of his friend Georges-Louis Leclerc, Comte de Buffon, who published the measurements in his enormous many volumed *Histoire Naturelle* (de Buffon, 1777, pp. 376–83). de Buffon's table of measurements is still regarded as one of the most detailed individual records (Tanner, 1981) and has been the basis of several published velocity curves (Figure 2.1). To those expecting a smooth curve it looks surprisingly irregular. The jagged line shows considerable variation from one age of measurement to the next, but children grow like that. Short bursts are interspersed by slower periods and de Buffon noted that the boy grew faster in summer than in winter. There is, however, a strong general trend shown in the curve. Growth *in utero* is many times faster than at any time after birth and this is apparent from the near vertical fall in the line during the first three years. François then grew at an irregular rate, fluctuating between 6 and 8 cm per year, until around 7 years of age, after which growth slowed still further, reaching a minimum rate of about 4 cm per year at 11 years of age. After this, there was a sustained rise in rate, slow at first, but increasing to a sharp peak of 12.5 cm per year at 14.3 years. This is the growth spurt accompanying puberty and de Montbeillard's son was not far from the average age in boys today. His growth rate then fell as rapidly as he approached adult size, ending up a rather tall 1.89 m or 6.2 feet. François Guéneau de Montbeillard became a cavalry officer, fathered three children and died at 88 years of age.

Growth curves are presented (Figures 2.2 to 2.5) for height (supine length from birth to 2 years and stature thereafter) and body mass, both in terms of height or mass attained for age, and growth velocity. They are intended to represent an idealised norm and all the minor variations have been smoothed out. The height velocity curve just after birth is falling rapidly from the overall height velocity peak, which is prenatal (see below). The gradient becomes less steep between 3 and 6 years of age, and the velocity continues to fall to a minimum around 10 years in girls and 12 years in boys. This is followed by the marked peak of the pubertal (adolescent) growth spurt and a final fall in rate as adult size is approached. In some children there is a slight velocity increase at around 6 years of age, the mid-growth spurt, but

Human growth in body size

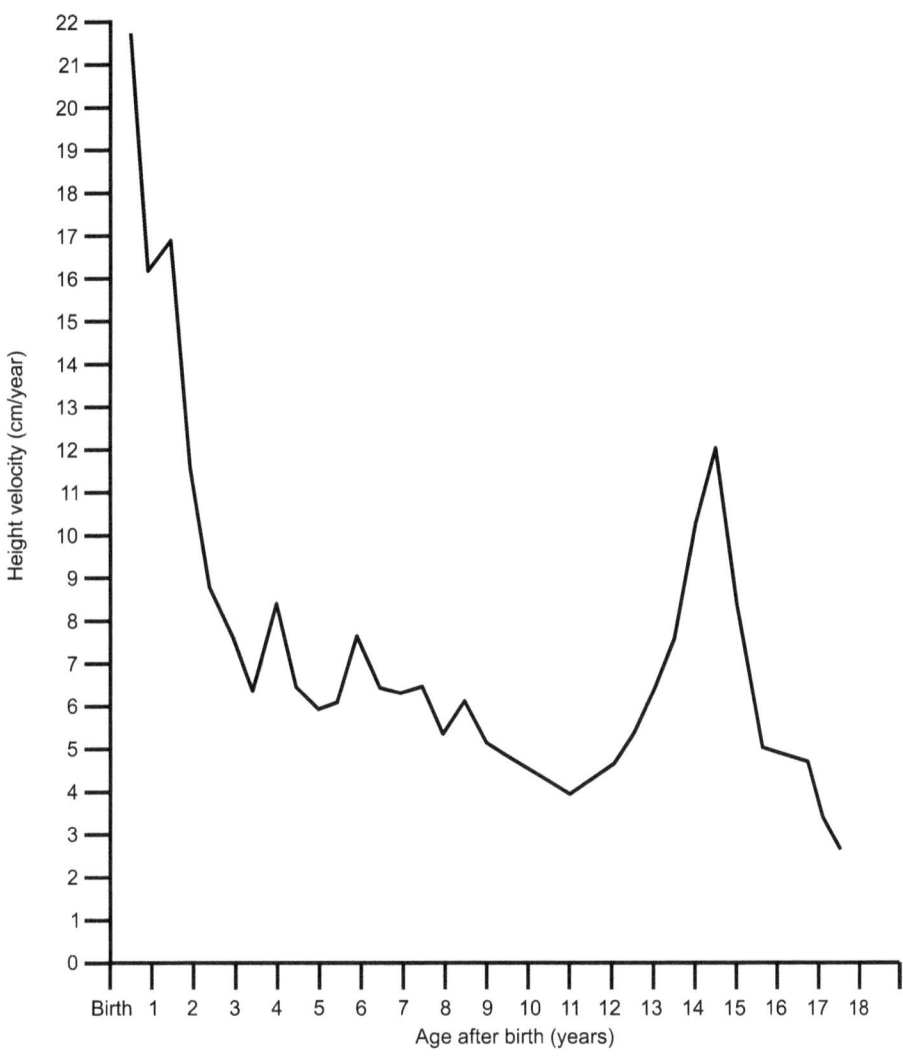

Figure 2.1 Body height growth velocity curve for de Montbeillard's son, plotted from measurements taken between 1759 and 1777. Re-plotted from Tanner (1981), figure 5.3, p. 105.

this varies and some studies show it more prominently than others. It is not shown in the norms presented here. The velocity curve for body mass in humans is broadly similar to the curve for stature, particularly in the timing of the adolescent growth spurt. Just after birth, however, body mass velocity rises to a sharp peak (see below) and then falls rapidly to a childhood minimum between 2 and 5 years of age, after which it climbs again gradually.

The timing and height of the main peaks and troughs (particularly the adolescent growth spurt; Figure 2.6) vary between different children, who also show individual patterns of departures from the main trend. This is partly due to a consistent difference between boys and girls, partly to other inherited variations and partly

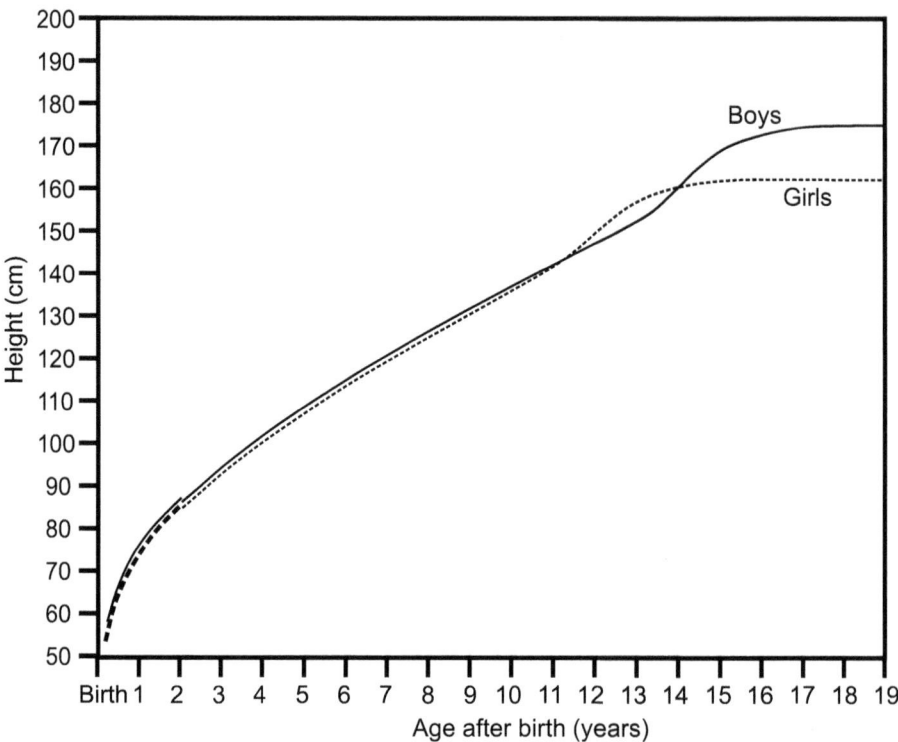

Figure 2.2 Body height curve from birth to 19 years of age measured in London boys and girls. Re-plotted from Tanner et al. (1966), figure 10, p. 467.

due to variations in the diet, health and social conditions within which the children are growing up. It does, however, need to be taken into account when trying to draw a curve representing the typical pattern of variation in growth velocity for a given group. This is one of the difficulties with cross-sectional studies which, for each year of age, produce a range of (say) statures attained by different children. From this range an average, or some other best estimate value, can be used to represent the stature usually attained by that age. The increase between this value and the mean or estimate for the next year gives the figure for growth rate in centimetres per year. If rates calculated in this way are used to plot a growth velocity curve, the variation included in the rate estimates at each age tends to smooth out the curve so it does not look at all like the curve produced by measurements for any one child. For example, it makes the adolescent growth spurt peak too wide, too low and not pointed enough. In order to draw a curve that represents a typical child (Figures 2.2 to 2.5), it is necessary to use the shapes of curves derived from following individual children in a longitudinal study, combined with values for the start and end of the curve derived from cross-sectional studies (Tanner, 1989). This is a particular difficulty with studies of growth in non-human primates, where most studies are of necessity cross sectional.

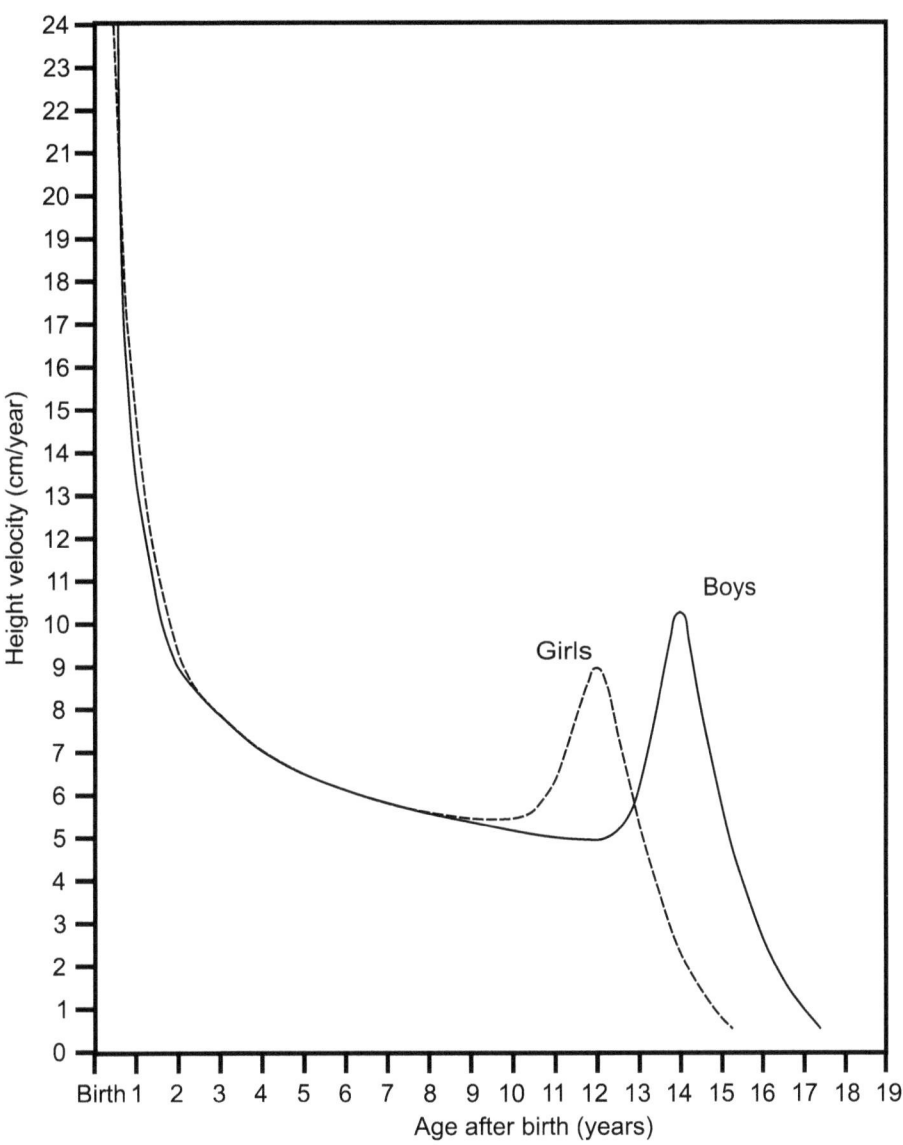

Figure 2.3 Body height velocity curve from birth to 19 years of age measured in London boys and girls. Re-plotted from Tanner *et al.* (1966), figure 8, p. 466.

Prenatal development

Modern studies of growth *in utero* are mainly based on measurements of the foetus in ultrasound scans of the mother, but older studies employed direct measurements on children born prematurely. The point reached in gestation is expressed as a *gestational age* in weeks (see Definition Box 1). For 5–15 weeks gestational age, the usual measurement is crown-rump length (as discussed earlier), but for 15–22 weeks it is the circumference of the head or femur length. Modern tables to convert these

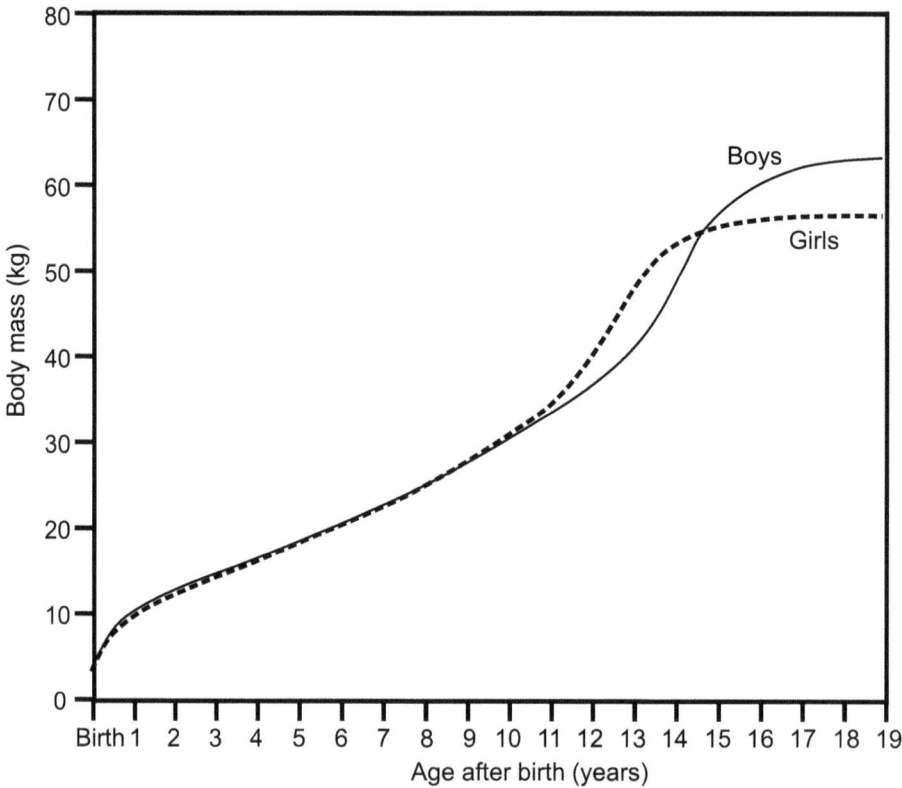

Figure 2.4 Body mass curves from birth to 19 years of age for London boys and girls. Re-plotted from Tanner *et al.* (1966), figure 11, p. 467.

measurements to gestational age are available, but are confined to those periods of development (Loughna *et al.*, 2009). Crown-rump length tables covering the full period of prenatal development are found in older textbooks (Arey, 1974; Patten, 1976) and were the basis for gestational age estimation in most studies of deciduous dental development (page 54). It needs to be borne in mind that there is an element of circularity in argument here because one developmental measure (body size) is being used to calibrate another.

Definition Box 1. Gestational age, chronological age, embryo, foetus, neonate and infant

Obstetricians count *gestational age* from the first day of the last menstrual period of the mother. This typically pre-dates ovulation by 2 weeks. Fertilisation of the ovum takes place within 24 hours of ovulation and implantation of the embryo in the uterus wall occurs around 1 week later. Variability of these events contributes 4 to 6 days to the overall variation in gestational age at birth (Engle,

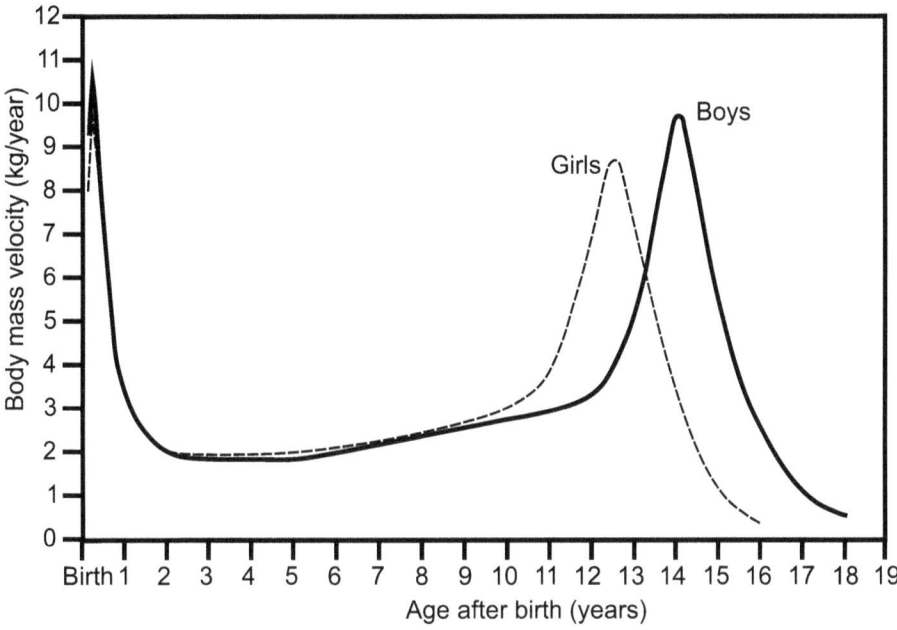

Figure 2.5 Body mass velocity curves from birth to 19 years of age for London boys and girls. Re-plotted from Tanner et al. (1966), figure 9, p. 466.

2004). By definition (Engle, 2006), in a normal *full-term* pregnancy, the child is born between the first day of the 38th week after the first day of the last menstrual period and the last day of the 42nd week. Babies born at a younger gestational age are known as *preterm* and those born later, *post-term*. In chimpanzees, normal full-term birth takes place between 29 and 37 weeks gestational age (Thompson and Wrangham, 2008), whereas for gorillas it is around 36.4 weeks (Robbins *et al.*, 2006). Gestational age is commonly used for development studies of the deciduous teeth. The age of the child as we would normally give it, counted in years from birth, is distinguished as *chronological age*. It is important to establish which age is being used in any study.

The word *embryo* is used in zoology to mean the offspring of an animal up to the point when it is born. In human embryology, the developing child is regarded as an embryo until 8 weeks after fertilisation, when it has most of the human features including limbs, head and heart, and then it is called a *foetus*. That term is used up until the time of birth, after which the child is known as a *neonate* for the first 28 days and then as an *infant* until weaning.

Such measurements show that foetuses develop faster than a child ever grows after birth. Crown-heel length velocity may peak at 10 cm per month near the middle of intra-uterine development (Tanner, 1989). If the length of the foetus is, say, 30 cm at that stage, this represents a 33% increase per month. By comparison, peak

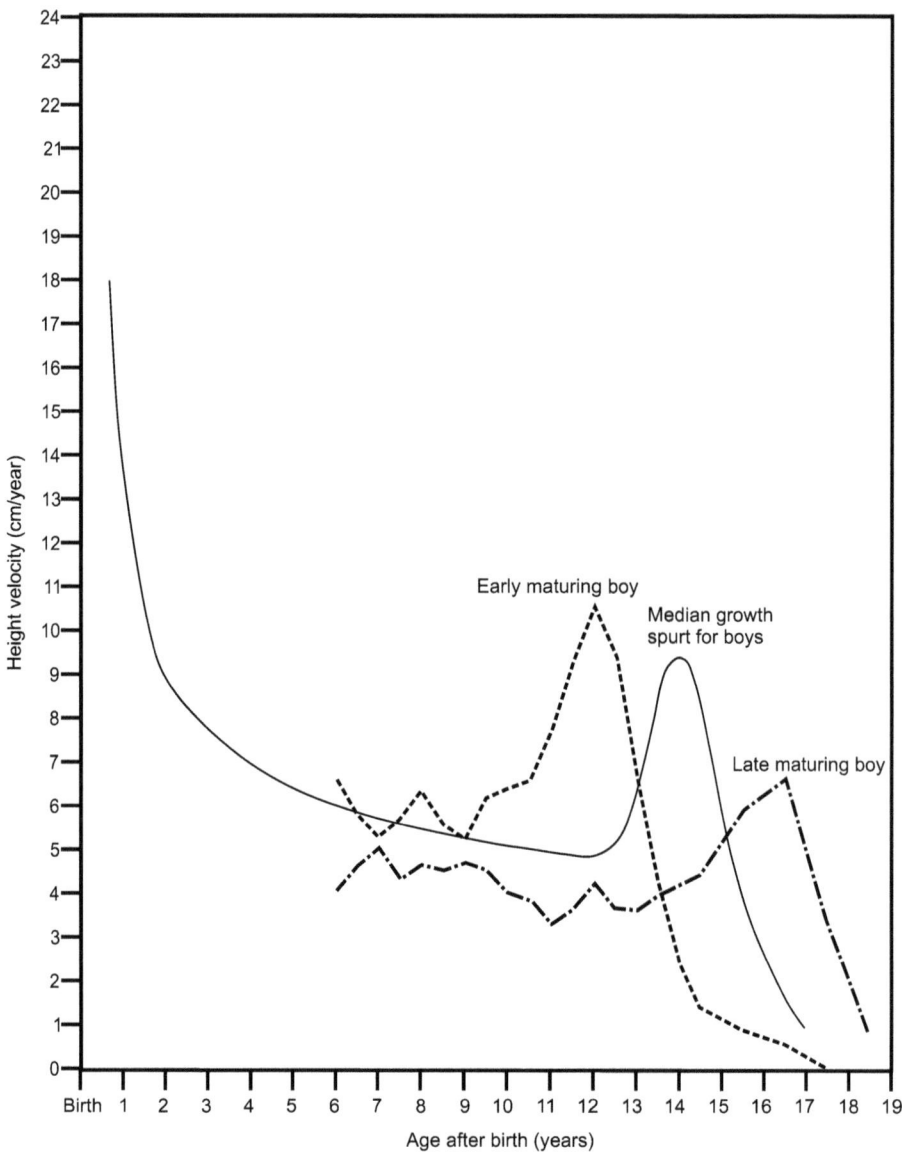

Figure 2.6 Variation in timing of the pubertal growth spurt in London boys. Median body height velocity for the whole group is shown by the dashed line, with curves for an early maturing and late maturing boy superimposed. Re-plotted from Tanner and Whitehouse (1976), figure 5, p. 176.

height velocity at the growth spurt of puberty (see below) might be 10 cm per year for a boy (Iuliano Burns et al., 2001), which is equivalent to just 8 mm per month (0.05% of mean height). Head or abdominal circumference, foot length and limb bone length growth velocities (Chitty et al., 1993; Altman and Chitty, 1994; 1997; Chitty et al., 1994a; 1994b; 1994c; Chitty and Altman, 2002) fall off from about

20 weeks gestational age. Older tables, based on direct measurements of stillborn foetuses, suggest that the peak velocity for crown-rump length is at 14 to 17 weeks (Patten, 1976), whereas Tanner's (1989) approximate graph suggests about 18 weeks for the crown-heel length (perhaps up to 6 weeks later). Tanner also observed that the peak growth velocity for body mass occurs later, at about 34 weeks. After this point, it slows rapidly until birth at 40 weeks, after which there is a short burst of catch-up growth lasting about three months (see above, Figure 2.5). Babies with a smaller birth mass tend to show a more pronounced increase during this catch-up phase than babies which are larger at birth. In addition to inherited factors, nutrition, disease and antenatal care of the mother all play a strong role in the size of babies at birth: presumably they affect the rate and pattern of prenatal growth.

Birth requires a number of adaptations for the neonate. *In utero*, the foetus relies on the placenta for the delivery of oxygen to its bloodstream and elimination of the carbon dioxide that results from its metabolism. Immediately after birth, the baby's own lungs must start to function and be provided with a blood supply (Bhutani, 1997). Any failure may cause brain damage and other disruptions. Another change is in calcium metabolism. Calcium is essential for the development of bones (and teeth), which are also the principal store of the element and, *in utero*, calcium is transferred to the foetus across the placenta. At birth, the baby is abruptly disconnected from this supply and there is a sharp dip in its levels of serum calcium over the next two days (Hsu and Levine, 2004). It must provide its own supply by mobilising calcium from the skeleton and by absorption from the digestive tract. Even in normal circumstances this may well affect the growing dental tissues, but, in some cases, there is an exaggerated fall known as hypocalcaemia and this has been associated with defects of dental enamel (page 192). Calcium metabolism after birth is regulated by parathyroid hormone (PTH), but vitamin D is also important. The foetus obtains all its supplies of vitamin D from the mother and the neonate needs to build its own supply from the action of sunlight on the skin or by absorption from the diet. If the mother is vitamin D deficient, this may lead to early-onset hypocalcaemia (during the first 4 days after birth) in the neonate. This condition is also associated with breathing difficulties or trauma at birth, premature birth and maternal diabetes. Late-onset hypocalcaemia (5–10 days after birth) is associated with vitamin D deficiency and a failure of the immature kidneys to respond to PTH. Hypocalcaemia is self-limiting and the child usually recovers rapidly, but it may leave its mark in the dental tissues.

Postnatal growth

Clinically, the period between birth and 28 days after birth is known as the *neonatal period*, and a child at this stage is called a *neonate*. It is a period of high risk (Lawn *et al.*, 2005). Body height velocity, already falling before birth, continues to fall rapidly during this period, even though it is still faster than at any stage that follows. Body mass velocity rises sharply immediately after birth to recover from the restriction which characterises the final weeks before birth. For the archaeologist and palaeoanthropologist, the neonatal period is best followed by the increasing size of the deciduous teeth as the first molar and first incisor, the first permanent teeth to initiate formation, are as yet too small to survive well. The long bones of the limbs are increasing rapidly in length and the key to skull development is the changing size and shape of the basilar part of the occipital bone.

Infancy follows the neonatal period and lasts until weaning which, for most children in non-industrialised countries, is between 2 and 4 years after birth (Humphrey, 2010). Dependency on milk is the defining feature of infancy. Body height growth velocity continues its rapid fall throughout infancy but, as described earlier, weight velocity rises to a peak at 2 to 3 months after birth. After this point weight velocity also falls rapidly to reach a minimum at 2 to 3 years chronological age. From a dental point of view, infancy is the period during which the deciduous (milk) teeth emerge into the mouth. The last deciduous tooth emerges on average a little after 2 years of age in modern humans (page 45). An archaeologist or palaeoanthropologist can monitor the development of infancy through the rapid succession of changes in the dentition with the initiation of most permanent teeth. The limb bones continue to grow and a whole variety of additional bony centres starts to develop, including the epiphyses at the ends of the limb bones. In the skull, the crucial changes are in the fusion of different parts of the occipital bone and the development of the tympanic plate (in the ear region) of the temporal bone.

Bogin (1999) has discussed at length the characteristic phases of development in humans and has distinguished childhood and juvenile stages. By his definition, *childhood* lasts between weaning and 7 years of age. A *child* in this particular usage of the word is no longer dependent on milk for survival, but nevertheless still requires food and protection from parents and community. It is a period during which the decline in growth rate slows or ceases and its end may be marked in some children by a small increase in growth velocity, which has been termed the *mid-growth spurt*. In London children, Tanner and Cameron (1980) found a slight increase in mean body mass velocity between 7 and 7.5 years in girls, and between 7 and 8 years in boys. Body height velocity was much less affected, with the boys showing a 'diminution of deceleration' at 7 years and girls no sign at all. Childhood also marks an important stage in the growth of the brain (page 25), which grows much more rapidly than body size throughout the prenatal, neonatal and infancy stages, but slows during childhood. The brain reaches almost adult size around 4 or 5 years of age. Most importantly from the point of view of the discussion in this book, the first permanent teeth appear in the mouth, starting with the first molars at about 6 years of age. Bogin's *juvenile* stage lasts from 7 years until the onset of puberty. By his definition, a juvenile is no longer dependent on its parents to provide food and can survive without their protection. It is a period of minimum growth rate in height and a relatively constant or slowly rising growth rate in mass. Dentally it is characterised mostly by a mixed deciduous and permanent dentition; the last deciduous tooth is not usually lost until after 10 years and the bulk of the permanent dentition (with the exception of the wisdom teeth or third molars) is established by about 12 years with the eruption of the second molars. For an archaeologist or palaeoanthropologist, both childhood and the juvenile stage are best monitored through the development of the dentition, which still shows a rapid succession of changes. By contrast, the skeleton displays far fewer changes, with the exception again of the occipital bone, which shows further fusion of its different parts.

Puberty, adolescent growth spurt and variability

Puberty is defined by the appearance of pubic hair and the sudden rapid growth of sexual organs. It is marked in humans by the pubertal or adolescent growth spurt – a rapid increase in both stature and body mass growth velocity to a brief peak, followed by a decline as full adult size is approached. In boys, puberty usually starts with the rapid enlargement of the testes between 9.5 and 13.5 years chronological age (Tanner, 1989). This is followed about a year later by the start of the growth spurt, which peaks after one further year. In girls, the earliest event of puberty is often the start of the growth spurt, followed by breast enlargement at 8–13 years. Menarche, the onset of the first menstrual period, occurs relatively late in comparison (perhaps two or more years later). In humans this does not necessarily mark the arrival of full reproductive capability in all girls, which often takes a further one or more years. The growth spurt peak occurs on average around two years earlier in girls than it does in boys. For a few years, girls are on average larger than boys of the same age (Figures 2.2 and 2.4), but the male growth spurt is more pronounced and it is this that results in adult men being on average larger than women. As with other aspects of growth there is considerable variation between individual boys and girls from one population. An early maturing boy might reach the peak of his pubertal growth spurt five years before a late maturing boy (Figure 2.6) from the same population and socio-economic group (Marshall and Tanner, 1970) and the age of menarche can similarly be a good four years earlier in some girls than others. This variation has both genetic and environmental components as discussed below.

The end of adolescence is marked by the attainment of full adult size. In the skeleton, this is associated with the loss of the cartilaginous growth plates at the end of the limb bones (fusion of the epiphyses), a process which is highly variable and often lasts well into the twenties. In the dentition, the start of adult life is marked by the eruption of the third molar, once more with considerable variation. The differences in body shape that distinguish men and women are established during adolescence, and only afterwards do sufficiently clear diagnostic features in the skull and pelvis allow identification of the sexes from the skeleton.

Adolescence is difficult for archaeologists because, for age estimation and studies of growth, all they have to go on are the skeleton and dentition. In girls, epiphyseal fusion for the main limb bones is some two or three years in advance of boys (Scheuer and Black, 2004). As will be discussed in Chapter 3, the difference in dental development is generally less, but girls still reach development stages on average about six months in advance of boys. It is difficult to distinguish between the skeletons of boys and girls and most children's remains are classified simply as 'sex not determined'. This makes it impossible to choose the appropriate standards. Epiphyseal fusion shows considerable variation, which leads to very broad age estimates, but there are problems even with dental development. The most rapid period of change in the dentition is before 12 years of age and there are few stages of development that could be used between that age and the eruption of the third molars in the late teens and early twenties. Adolescence is equally troubling for palaeoanthropologists, for whom the main issue is the timing and pattern of growth in fossil primates. Compared with most mammals, many living primates show high levels of sexual dimorphism, both in body size and in development schedule, and with extinct primates the level of dimorphism is often controversial. If it is not possible to be sure which fossils are males and which are females in the fully mature adults, then their young present a real challenge.

Effects of environment on the pattern of growth in children

The fully developed adult form of an individual mammal – its shape, size, physiology, pigmentation, hair and so on – is defined as its *phenotype*. For that same individual, the total collection of genetic material in its cells is known as the *genotype*. If the genotype defines the potential size, shape and other features of the phenotype, it is the process of growth that brings about the realisation of this potential. External factors, such as nutrition, disease, exercise and socio-economic conditions, affect development. These factors are together termed the *environment*. Thus the final form of the phenotype is produced by a combination of genotype and environment through the processes of growth. The environment provides the resources from the diet that are available for development and also initiates the things that have a competing claim on those resources, such as disease or level of physical activity. A large proportion of the resources available is taken up by basic body maintenance. Growth requires a surplus. If the resources available do not exceed those required for maintenance, then growth stops, to be resumed when the level of resources increases. The phase of rapid growth that follows is known as *catch-up* and it can return a child almost completely to the normal position on the growth curve for a child of their age. This is what happens with an acute or short-lived interruption to growth. If the interruption goes on for a long period of time, then the effect may be permanent and the child ends up smaller as an adult than the potential defined by its genotype. For this reason, adult stature can be a useful measure of childhood development. The figures for men need to be separated from those of women but, within one population, the children of affluent parents grow up on average to be taller adults than the children of less well-off parents. In addition, the affluent children are taller on average at each year of development. From the end of the nineteenth century in the industrialised countries of Europe and America, there was a consistent rise both in mean adult stature and mean stature for children at different ages. In Europe, for example, the mean stature of a 5–7-year-old child increased by 1–2 cm per decade between 1900 and 1990 (Tanner, 1989). This is known as the *secular trend* in stature. Other developmental factors also showed this trend such as, for example, the decreasing mean age of menarche in girls. For most of those countries it lasted until the final decades of the twentieth century and then slowed greatly. In effect, children had reached the potential for growth defined by their genotype. In other countries, less well provided with resources and health care for families, this secular rise started later and has continued.

The term *tempo of growth* was coined by the anthropologist Franz Boas (1935) to describe the variation seen between children in their attainment of different stages, or their rate of progress in size and other features. These are coordinated, so that a child with a fast tempo of growth will, for example, achieve its skeletal and dental development stages at a younger age than a child with a slow tempo, as well as being larger at a given age, entering puberty earlier and maturing younger. This integration of development stages is an important concept to grasp when considering individual variation, sexual dimorphism, population and species differences. A fast tempo child does not necessarily end up larger, because a slow tempo child with which it is compared may catch up later. A large part of this variation in tempo seems to be inherited. There are correlations between non-twin siblings, and between parents and siblings in the ages at which they attain growth stages. Clearly the shared family environment may have some effect, but the strength of genetic control is evident in the greater degree of similarity in tempo between identical twins than in non-identical twins. There is also variation in tempo between different populations. North American children whose families originated in sub-Saharan Africa have, on average, a faster growth tempo (Eveleth and Tanner, 1990) than those whose families originated in Europe when their socio-economic circumstances are similar. When individual children are compared, the difference in tempo can amount to several years and care needs to be taken in comparing hominin fossils, in which there is often only one individual, or just a small group. There will certainly have been variation in growth tempo within the living species. Who is to know from which point in the spectrum that particular individual came? It makes sense to base comparisons between species on the differences between stages rather than absolute ages, because variations in tempo shift all the development stages earlier or later. For one population, the spacing of different stages should stay relatively constant in both slow and fast tempo children.

The main environmental factors that affect growth are diet and disease. The two are interlinked because disease not only itself uses a proportion of the surplus resources over and above the requirement for maintenance, but decreases the child's ability to eat. The combined impact of disease and malnutrition are greatest just after weaning. Psychosocial stress (page 203) also seems to have an effect, although this is more difficult to measure because emotional deprivation of this kind is often associated with nutritional deprivation. On the other hand, psychosocial stimulation through play and caring behaviour seems to cause a more rapid catch-up growth in undernourished children (Nahar *et al.*, 2008). The difference in tempo of growth between the children of affluent and less well-off families must be due to a whole mix of factors, including the attention paid to care of the child in the home, the effects of unemployment, crowding and substance abuse, as well as diet and disease. Given this environmentally caused variation in tempo of growth, it is also important to consider the environmental background when non-human primates are studied. Most animals for which growth data have been recorded lived in captivity. There is some evidence that captive animals have a faster tempo of growth than their wild counterparts and there has been considerable discussion of their use in comparison with extinct primates (page 39).

Body size growth in non-human primates

Growth in rhesus macaques

Rhesus macaques (*Macaca mulatta*) have been kept as laboratory colonies for many years, with meticulous records maintained, so they make a good place to start the discussion of growth in non-human primates. One particularly closely studied colony was kept at Yale University School of Medicine between the 1930s and 1950s. It was very carefully monitored, with measurements and observations of the babies every day until 4–6 months, weekly measurements until 1 year of age and then monthly. This provided the data for a longitudinal study of height and weight (van Wagenen and Catchpole, 1956). As matings within the colony were planned and the date of insemination recorded, true gestational ages were known for the babies at birth and for foetuses removed during hysterectomy, thus providing cross-sectional data for prenatal development (van Wagenen et al., 1965). In addition, very detailed records were kept for tooth eruption (page 33), providing a useful link between general growth and dental growth. Altogether, it is one of the most complete primate growth studies ever carried out. Clearly it was a very managed group, not only in terms of mating, but also diet and health care, so it is not in any sense to be regarded as a representative of wild monkeys. However, it provides a good case study for understanding how growth in primates works.

Gestational age at birth varied between 140 and 195 days, with a mean of 168.9 days. The study included records of body mass at birth for a variety of gestational ages (van Wagenen et al., 1965). These cross-sectional data can be used to calculate mass velocities, which suggest that growth rate peaked *in utero* and then fell to a low just before birth. van Wagenen and Catchpole (1956) plotted growth velocity curves from birth onwards based on their detailed longitudinal data for both body mass and crown-rump length (Figures 2.7 and 2.8). Mass velocity, but not length, shows a sharp rise after birth, rising to a peak occupying approximately the first few months. In this, it is fairly similar to the human growth pattern. The peak is followed by a gradual fall in mass velocity to a minimum between 1 and 2 years chronological age. There is a pubertal growth spurt, seen strongly in the mass velocity curves, but far less for the crown-rump length velocity curves. As will be seen, this is a common pattern in growth studies of primates. In females, the growth spurt peaks at a lower velocity and earlier than in males. After that, the body mass velocity reduces steadily, but final adult size is not reached in males

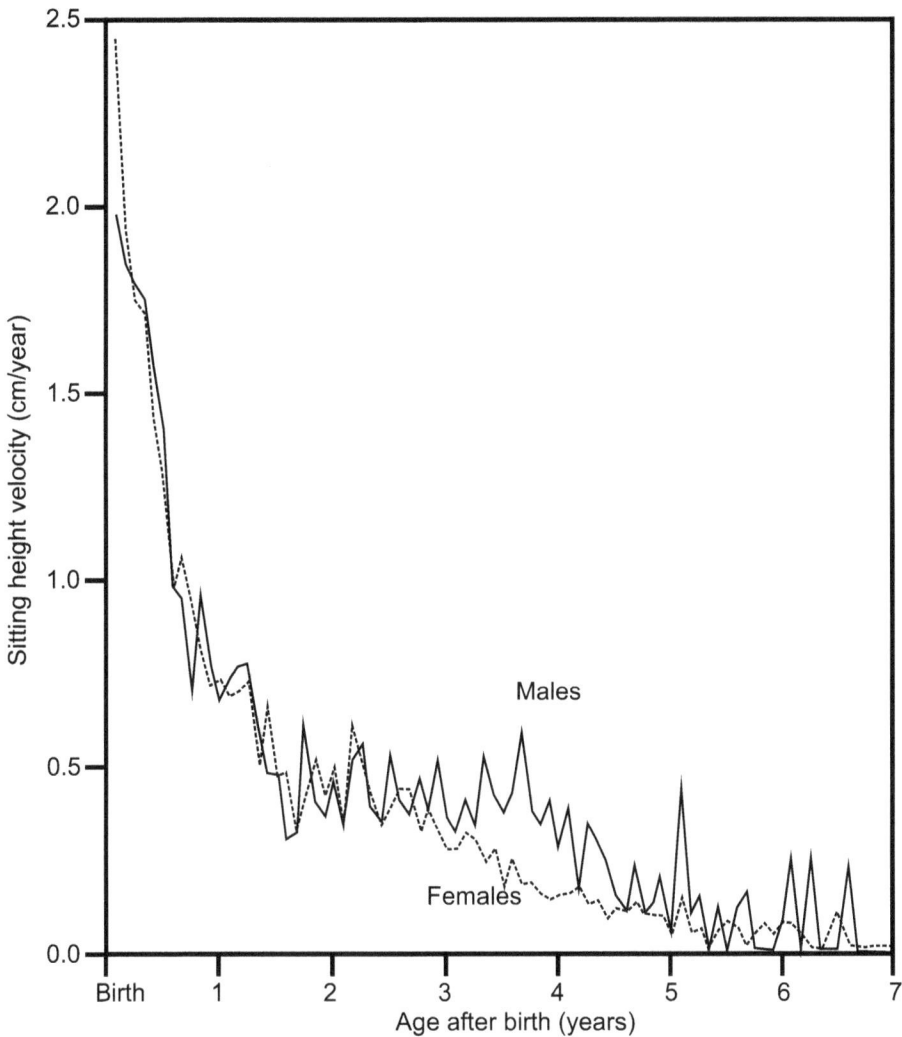

Figure 2.7 Sitting height velocity curves from birth to 7 years of age measured in male and female rhesus macaques. Re-plotted from van Wagenen and Catchpole (1956), figure 8, p. 258.

during the period represented in the graphs. The females became sexually mature during the second to third years after birth and passed through several pregnancies well before the end of their development. van Wagenen and co-workers defined two groups of females: one with an early menarche (mean 1.7 years) and one with late menarche (mean 2.4 years). The body mass velocity curves for these groups (Figure 2.9) show that this was accompanied by an early and a late growth spurt, respectively, not dissimilar to the variation seen in humans.

Crown-rump length (sitting height) velocity showed a different pattern, with a strong decrease during the first and second years after birth, gradually levelling off for the remainder of the curve. In both sexes a slight reduction in the rate of fall,

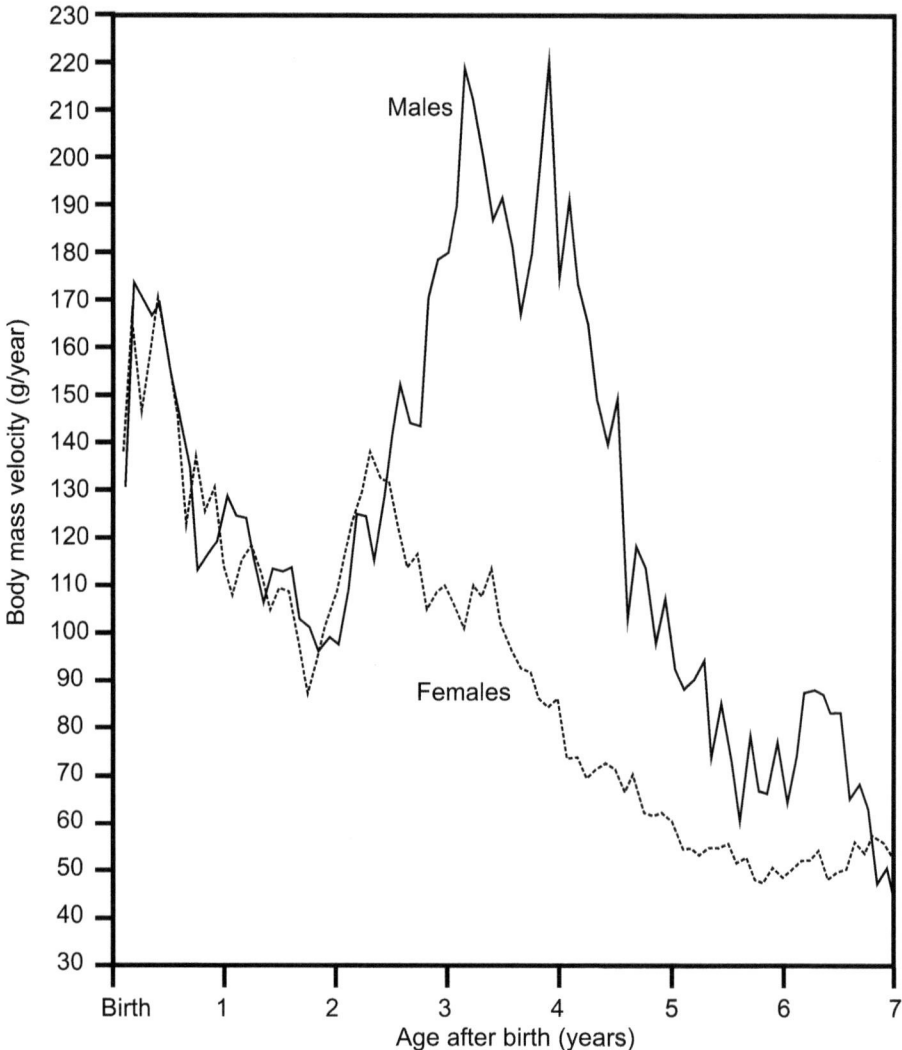

Figure 2.8 Body mass velocity curves from birth to 7 years of age measured in male and female rhesus macaques. Re-plotted from van Wagenen and Catchpole (1956), figures 5 and 6, p. 255.

to make a plateau, corresponded to the period of the body mass growth spurt. This was more marked in males than in females. Tanner *et al.* (1990) carried out a detailed longitudinal study of two small groups of rhesus macaques living in colonies. For more than two years they were measured monthly for crown-rump length, mass and tibia length. In one group, housed indoors, there were clear growth spurts associated with puberty in all three measurements. In the other group, which was housed outdoors, the crown-rump growth spurt was less marked.

The macaque body mass growth velocity curve from a detailed longitudinal study thus shows similarities to the equivalent curve in humans. There is a restriction in

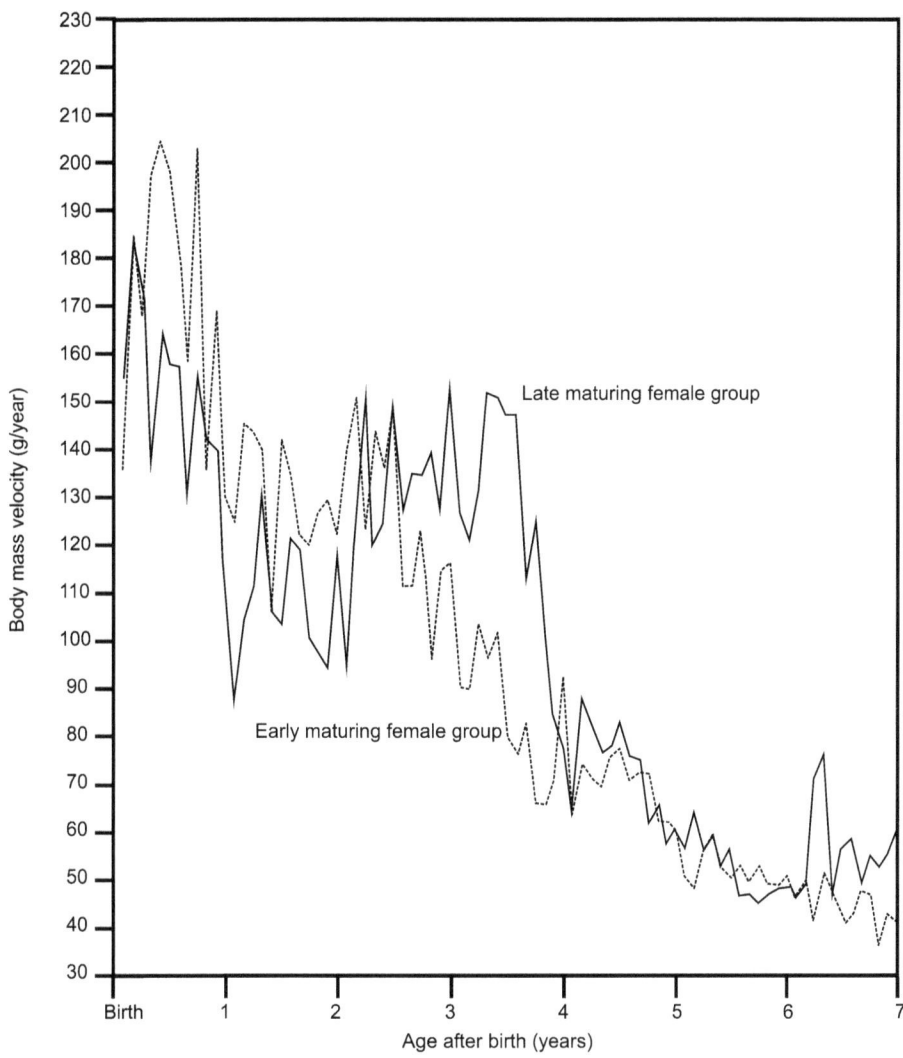

Figure 2.9 Mass velocity curves from birth to 7 years of age measured in early and late maturing groups of female rhesus macaques. Re-plotted from van Wagenen and Catchpole (1956), figure 7, p. 256.

growth rate just before birth, followed by a rapid rise and peak just after birth. In turn, this peak is followed by a period of low growth rate. In comparison with humans, this interval is relatively short at one to two years. Its end is marked by a prominent pubertal growth spurt which is more marked in males than females but, as in humans, occurs at a younger age in females than in males. It is initiated at a much younger age than in humans, but macaques are much smaller animals and, as is discussed in Chapter 6, body size in mammals is strongly correlated with the length of the growth period. One marked difference from humans is that in macaques the pubertal growth spurt is only shown strongly in body mass, not in height.

Growth in other primates

Are the growth velocity curves for macaques representative of the majority of living primate species? Leigh (1996) analysed body mass records for nearly 2400 individual primates from 24 different species, all living in laboratories or outdoor colonies. He treated the age versus mass data in a cross-sectional way (although at least some of it was longitudinal) and fitted a line to it by LOWESS (locally weighted scatterplot smoothing) methods. The weights predicted by this line for given age intervals were used to estimate body mass growth velocities. As he pointed out, cross-sectional studies would be likely to minimise the height of any growth spurt peak (as discussed earlier), so his approach was conservative. By this method he found that some primate species showed no evidence of a pubertal growth spurt in either sex. Only one of the 12 New World monkey species studied, *Cebus apella* (black-capped capuchin), had such a spurt and, even for this species, only in the males. All 16 Old World monkey species had growth spurts – seven in males only and nine in both sexes. Neither the gibbon nor the siamang showed growth spurts in either sex, but they were present in male chimpanzee (*Pan troglodytes*) and perhaps orangutan (*Pongo pygmaeus*), and for both male and female gorilla (*Gorilla gorilla*) and bonobo (*Pan paniscus*). The male growth spurts in his graphs are large and, as in the rhesus macaque described earlier in this chapter, the lower female peak is at a younger age than the male:

Gorilla female peak 5 years, male peak 7–8 years
Bonobo female peak 5 years, male peak 6–7 years
Chimpanzee male peak 9 years.

Given the difficulties with cross-sectional studies, it is important to back these up with evidence from longitudinal studies and studies have centred on the chimpanzee. Grether and Yerkes (1940; redrawn in Tanner, 1962) reported a study of 10 male and 17 female animals from the Yale Laboratories of Primate Biology, measured at monthly intervals between 1925 and 1940. Body mass velocity was calculated, as in the rhesus macaque study described earlier, from the mean of the increments in weight between each period of measurement (Figure 2.10). There was a pronounced male pubertal growth spurt, peaking at 8 years of age. Females showed a much less marked spurt, with a lower peak at 7 years. This corresponded only moderately well with the onset of menarche at 7 to 10 years. Hamada and Udono (2002) carried out a longitudinal study of sitting height (effectively crown-rump length) in five male and seven female chimpanzees from Kumamoto Primate Park and the Primate Research Institute of Kyoto University in Japan. Measurements were taken every three or six months between birth and 14.5 years of age. They were not able to find consistent evidence of a pubertal growth spurt in sitting height.

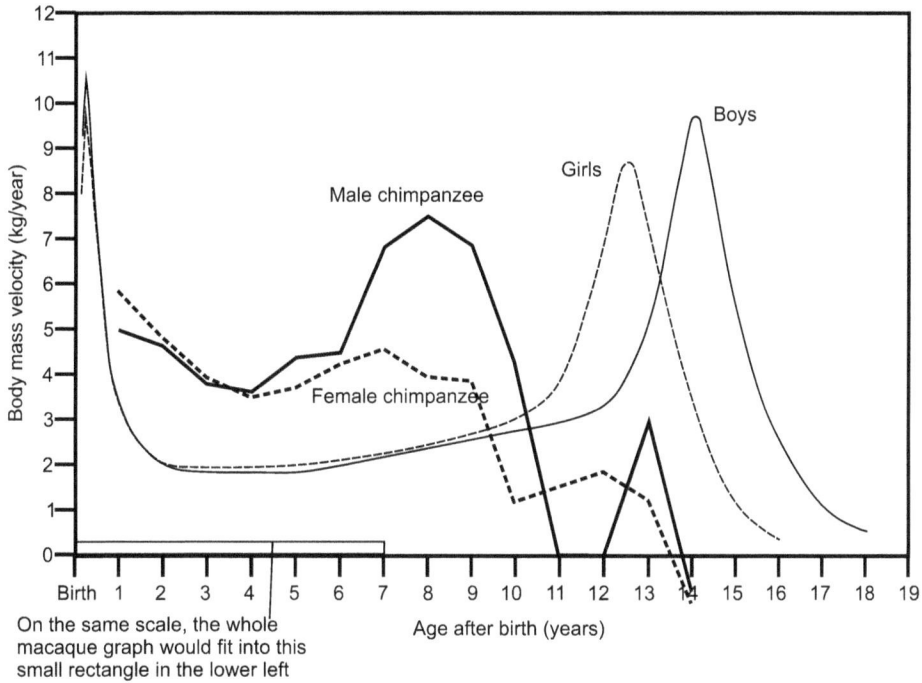

Figure 2.10 Mass velocity curves from birth to 14 years of age measured in male and female chimpanzees, plotted on the same axes as the human curve from Figure 2.5. The velocity curves were calculated from mean body mass in a small sample of chimpanzees; this is the reason for the negative values after age 13 years. Re-plotted from Grether and Yerkes (1940), figure 4, p. 189.

Growth in other mammals

Body mass or height/length velocity curves for most mammals differ from the higher primates described in the preceding sections. They do not show a peak in growth rate during the prenatal and/or neonatal periods; nor do they show a pubertal growth spurt (Bogin, 2003). Tanner (1962) illustrated examples from several orders of mammals, including guinea pigs, mice, rats, rabbits, sheep and cattle. For all these mammals, growth velocity rises rapidly through the prenatal stage and continues to rise after birth to a single peak (Figure 2.11). After this point, the velocity falls rather more slowly until the final body size is reached. Birth is well before the peak of maximum growth velocity and puberty is typically after it. Growth velocity curves have been published for only a few mammal species, but this seems to be the most common pattern of development.

Pubertal growth spurts, at the time of writing, have only been described in primates. Some non-primate mammals, however, do appear to show a juvenile period; that is, a substantial period between weaning and puberty when the young animal is independent of its parents. It is found not only in humans and other primates, but also in strongly social carnivores such as lions, wolves and hyenas, in elephants

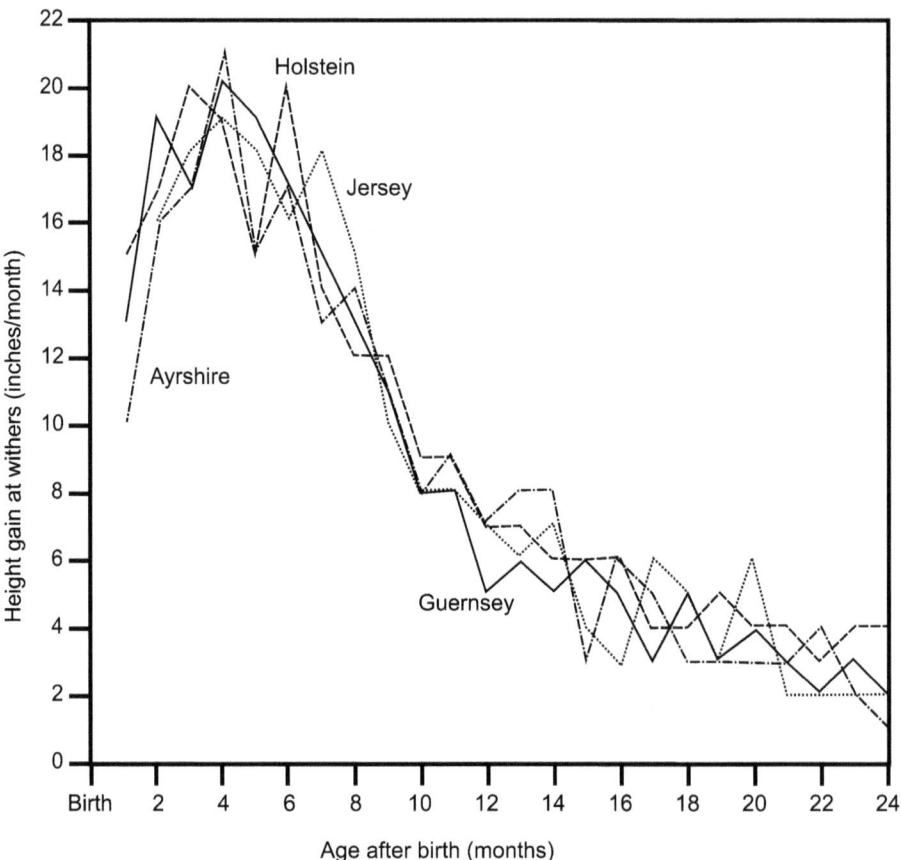

Figure 2.11 Withers height velocity curves for four breeds of cattle. Re-plotted from Brody (1945), figure 16.59d, p. 554.

and in whales and porpoises (Pereira and Leigh, 2003). Bogin (1999; 2003) has suggested that a well-defined childhood stage between infancy and the juvenile period is a unique feature of human development (page 14). Childhood follows weaning, but the child is still dependent on its parents for the provision of specially prepared foods. The young of some social carnivores and primates are also dependent on their parents for food and training in hunting and gathering. Human children, however, are dependent in this way for about four years and they also require specially prepared foods to suit their still small digestive tract and rapidly changing mixed deciduous and permanent dentition.

Growth in different systems of the body

In a pioneering description of growth which formed part of the book *The Measurement of Man*, Scammon (1930) published a diagram of the relative development of different body systems in human children which has been reproduced in various versions in almost all subsequent books on growth. The present book is no exception (Figure 2.12). Scammon's original diagram summarises very clearly and simply the different development schedules of the body as four superimposed curves. One curve is for growth in the body and shows the rapid increases in infancy followed by the slower development of childhood and the juvenile stage, and then the pubertal growth spurt which slows as adult size is reached. This is often known as the somatic growth pattern. Limb bone lengths follow this trend. The second curve is for the neural pattern of growth, with a very rapid development during infancy which slows during childhood to reach adult size well before puberty. This is followed by the brain, eyes and the skull bones that surround them. The third curve is for the growth of reproductive organs, with little development throughout the childhood and juvenile periods and then a very rapid rise during puberty. The features that distinguish the skeletons of men and women develop along this schedule, including the size of the main limb joint surfaces, the shape of the pelvis and the features of the skull. The skull is thus a highly complex mixture of these development patterns.

A fourth curve in the original diagram, the lymphoid type, was drawn from Scammon's own observations on the thymus and lymph glands and shows yet another pattern, but has been omitted here because it is not associated with features of the skeleton. The dentition follows a schedule of its own, which is not really possible to measure in the same way, but if the average ages at which teeth are erupted into the mouth are added to the graph (Figure 2.12), it is possible to see that there are two phases: the relatively rapid establishment of the deciduous (or milk) dentition and the slower establishment of the permanent dentition. Growth in the bones of the jaws also follows this schedule.

Few comparative figures are available for other mammals. One particular issue is the rapid development of the large human brain, which reaches almost full adult size during childhood. One estimate suggests that 95% of adult size is attained in humans between 3 and 4 years of age, whereas another suggests 90% by 5 years (Robson and Wood, 2008). The limited data for chimpanzees suggest that the growth rate is slower than in humans during the first year after birth, but is similar thereafter (Leigh, 2004). The smaller chimpanzee brain seems to achieve 90% of its full adult size by 4 years (Robson and Wood, 2008), which is not too dissimilar from humans.

Summary

Is human growth unique? The schedule is certainly a long one, particularly in relation to body size. In humans, full adult body size is reached, on average, at around 19 years after birth for boys and at about 17 years for girls. Considerable variation either side of these figures is shown and, in general, all aspects of development are linked, so that early developers are not only relatively large for their age,

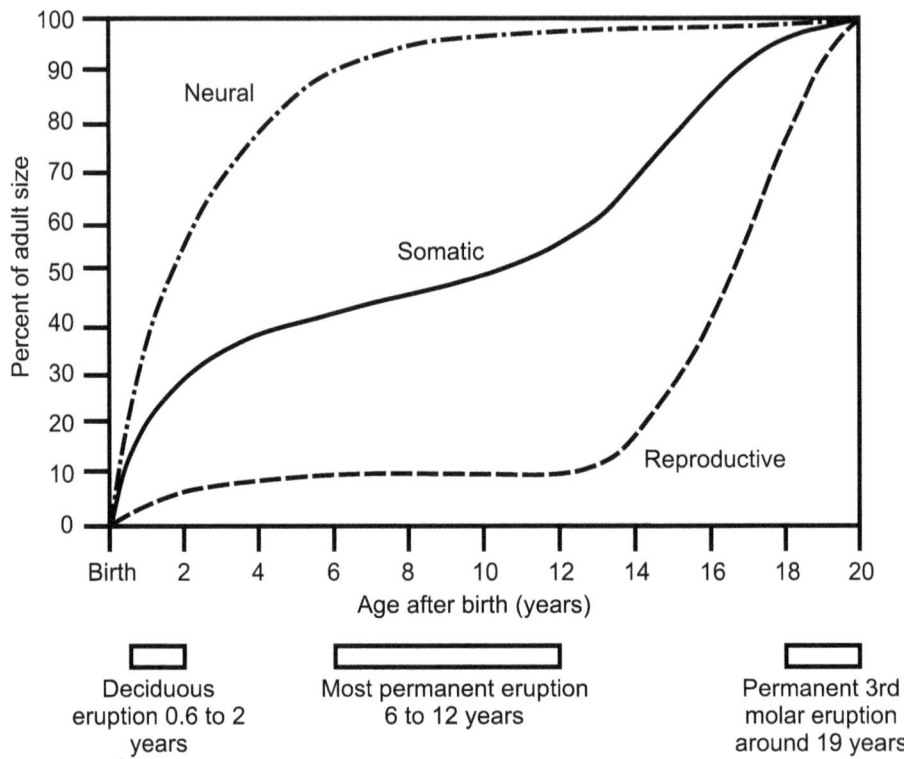

Figure 2.12 Different patterns of growth in humans. The curves represent the percentage of adult size attained at different ages. Ages for eruption of deciduous and permanent teeth are shown below the axis. Re-plotted from Scammon (1930), figure 7.3, p. 193. Eruption data from Chapter 3 below.

but they attain growth stages for the skeleton and dentition at younger ages. Late developers are relatively small for their age and attain growth stages at older ages. This seems likely to be true of many mammals, but far more is known about it in humans because there have been more and larger growth studies. A male chimpanzee on average might reach a fully mature size at 15 years and a female at 16.5 years, whereas a male macaque would do the same at 12.9 years and a female at 12.7 (Hamada et al., 2004). This difference is partly related to size, as the larger members of any order of mammals tend to develop more slowly, but primates in general tend to have long development schedules for their size. To put this in context, a very much larger mammal such as the cow might take three to four years to reach mature body size (Nadarajah et al., 1984).

What about the pattern of growth? In humans, body height and mass grow fastest *in utero*. A child born at a normal length of gestation is past its peak growth rate and is slowing rapidly. The rate levels off to a minimum during infancy, childhood and the juvenile period, and then with puberty there is a marked growth spurt in both height and body mass in both sexes. In girls, both puberty and the growth

Summary

spurt occur at a younger age than in boys, but the spurt has a lower peak growth rate. This is certainly an unusual pattern in mammals as a whole, where the peak rate of growth is generally after birth, followed by a gradual decline in rate, without any sign of a pubertal growth spurt (Tanner, 1962; Bogin, 1999). A similar pattern, if not schedule, of development to the human condition has only been described in some primates. The most detailed studies are of rhesus macaques and chimpanzees, both of which do show a pubertal growth spurt, although it is much more pronounced in body weight than in height. As in humans, females start their growth spurt a little earlier than males, but there is a much greater difference in the peak growth rate. As might be expected, the timing is later in humans: menarche occurs in girls between 10.5 and 15.5 years, whereas menarche in the macaque varies from 1.6 to 2.4 years, and in the chimpanzee from 7.3 to 10.2 years. Thus the human pubertal growth spurt is not entirely unique, but it is certainly uniquely pronounced and occurs relatively late, even in relation to body size.

3 How teeth grow in living primates

To what extent does dental development reflect the slow tempo of human growth? What is known about the schedules of dental development in living primates? Reliable figures are surprisingly difficult to find. Even in humans, by far the most numerous primates and very much the most studied, the sheer variety of different approaches makes it difficult to arrive at a consensus view. In comparison, non-human primates are rare mammals with a restricted habitat, whose numbers can never have been very large. There are few studies of dental development, the sample sizes are small and conditions for data collection not ideal. Most come from captive animals and there is debate about whether or not these are representative of their wild counterparts.

Furthermore, the great majority of studies, human and non-human, record the eruption of the teeth through the gums. This is difficult to compare with fossil jaws in which the soft tissue is not preserved. The stages of tooth formation which are more clearly seen in fossils are necessarily assessed by radiographs in the living and it is not straightforward to match these with direct observation of the teeth themselves. Collections of exposed jaws and teeth from individuals whose age-at-death is independently known are rare and precious. They provide observations which are directly comparable with the fossils, but the number of specimens is small. Nevertheless, it is possible, through a combination of evidence, to address the question and to provide a background to the discussion of the following chapters.

Process of dental development

Dentitions

Almost all mammals have two sets of teeth (Figure 3.1; Appendix A, Table 3): a deciduous set which generally starts to form *in utero* and a permanent set which replaces it and is then retained throughout adult life. The *deciduous dentition* erupts into the mouth soon after birth (before birth in some mammals) and is then gradually replaced by the *permanent dentition*, with the first permanent teeth often, but not always, arriving in the mouth around weaning. The eruption of the last permanent tooth to be formed is an event that marks the attainment of full adult size and proportions. There are thus three clear stages:

Process of dental development

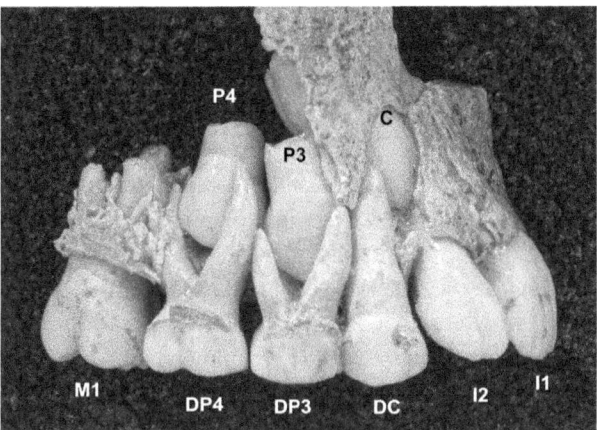

Figure 3.1   Mixed permanent and deciduous human upper dentition. This state of development would be reached by a modern human child at approximately 8 years of age. At the time of death, the deciduous canines, third and fourth premolars were fully erupted and in occlusion. Their roots were completed and resorption had just started at the apices to make way for the successional permanent teeth which were developing below. The deciduous incisors had already been exfoliated and the permanent first incisor was almost fully erupted. The second incisor was not yet erupted into the occlusal plane. At the back of the dentition, the permanent first molar was fully erupted and starting to wear, behind the deciduous fourth premolar. Its roots were still growing, although much of their length was already completed. The third and fourth permanent premolar crowns were completed but their roots were still growing. Just visible through a gap in the specimen is the permanent canine, in which the crown was also completed, with about one-third of the root length. Only a few small fragments of bone survive and this specimen gives a good idea of how much of a child's face is composed of teeth. It is an archaeological specimen from the teaching collection of the Institute of Archaeology, University College London.

1. establishment of the deciduous dentition;
2. the period of mixed dentition in which deciduous teeth are replaced by permanent teeth;
3. full establishment of the permanent dentition.

At all stages, the dentition can be divided into four *quadrants*: upper right, upper left, lower left and lower right. For the primates considered in detail here, there are normally five teeth in each deciduous quadrant and eight in each permanent quadrant. Deciduous quadrants contain two incisors, one canine and two premolars. Permanent quadrants contain two incisors, one canine, two premolars and three molars. The teeth in each series are always numbered outwards from the centre front of the dental arcade. Thus the incisors are labelled first and second, or central and lateral. By convention in anthropology, for both deciduous and permanent teeth, the two premolars are labelled third and fourth. This is because the 'generalised' mammal dentition, including some prosimian primates (Swindler, 2002), includes four premolars and by comparison it is clear that those species with only two represent a reduction from the front of the premolar row. It is therefore the first and second premolars that have been lost through evolution. Note that many dentists describe the deciduous third and fourth premolars as deciduous first and second molars, respectively. They also describe the permanent third and fourth premolars as first and second premolars. This is only one of many tooth labelling problems and the anthropologist's terms will be used in this book.

In humans and in all living monkeys and apes, the deciduous teeth erupt into the mouth in a sequence starting with the first incisor and ending with the deciduous fourth premolar (the timing of the canines varies). The permanent dentition can be divided into *'successional teeth'* that replace deciduous teeth and *'additional teeth'* that are added behind the deciduous fourth premolar. These additional teeth are the three molars. The successional teeth are the permanent incisors, canines and premolars, which erupt in a highly variable order among the primates considered here.

Initial development of teeth

In humans, the mouth cavity starts to develop *in utero* from around 4 weeks gestational age (page 10). It is lined with a layer of tissue known as *epithelium*, underlain by a tissue called *ectomesenchyme*. At 5 weeks gestational age, the epithelial layer becomes thickened along horseshoe-shaped zones of the maxillary and mandibular processes, mapping out the line of the future dental arcades (Osborn and Ten Cate, 1983; Ten Cate, 1998). This is called the *primary epithelial band.* Soon afterwards it grows into the underlying ectomesenchyme to form a strip of epithelium, the *dental lamina*. At intervals along the dental lamina, a series of local epithelial thickenings develops, starting from nearest to the midline of the oral cavity and spreading around the arch. These are the *tooth germs* and, by 8 weeks gestational age, there is one for each of the deciduous teeth. The permanent molar tooth germs are added when the jaws have grown more, from around 14 weeks gestational age, by extension of the dental lamina to posterior. Permanent incisor, canine and premolar tooth germs arise as proliferations at the point where the deciduous tooth germs are connected to the dental lamina. Each deciduous tooth is normally followed by a permanent successor in the same location. The permanent molars have no deciduous precursors and develop immediately behind the deciduous tooth row. They are very like the deciduous fourth premolars in form and the small tooth germ of a potential successor tooth also forms at an early stage (Ooë, 1979). These 'abortive successors' fail to develop further, but clearly show the developmental place of the permanent molars as part of the deciduous series.

This is part of a much broader discussion about the evolution of mammalian dentitions and control of the pattern of dental development. Edmund (1960) suggested that mammals had two waves of dental development, or *zahnreihen,* marshalled by some external controlling factor. Furthermore, Ziegler (1971) proposed that the first zahnreihe consisted of the deciduous teeth and permanent molars and the second zahnreihe consisted of the permanent incisors, canines and premolars. Several alternative developmental mechanisms have been proposed (Smith, 2003; Jernvall and Thesleff, 2012), but the pattern is well established in living and fossil vertebrates. Developmentally speaking therefore, the first, second and third permanent molars could be considered to be the fifth, sixth and seventh deciduous premolars but, although important for understanding the developmental series, it would greatly complicate tooth nomenclature to insist on changing their names.

Development of the dentition can thus be seen as two series. The first series consists of deciduous incisors, canines and premolars, followed by the *additional teeth*, the permanent molars. In humans the first deciduous teeth start to be formed around 14 weeks gestational age. The permanent first molars are the last teeth to be initiated *in utero*, just before normal full-term birth. Permanent second molars are initiated after birth, followed by third molars. The second series involves the *successional teeth*, starting just after birth with the permanent incisors, canines and premolars. In humans, incisors largely overlap permanent first molars in formation, but this varies in non-human primates. Permanent canines and premolars similarly overlap second molars. Initiation of third molars varies between taxa. Those primates that have large permanent canines initiate these teeth as early as the first incisors and complete them late in the sequence. As will be seen, the degree of overlap in development timing between teeth is one of the central issues addressed in this book.

The first stage of each tooth germ is a simple epithelial swelling from the dental lamina, known as the *bud stage*. The swelling develops an indentation on one side and so passes into the *cap stage*. Finally, the edges of the cap grow and the indentation deepens, developing a point at its apex, and the germ enters its *bell stage*. Each bell stage tooth germ contains four elements:

1. the *enamel organ* (or dental organ) which is derived from epithelium folded in on itself to make the bell shape;
2. the *cervical loop* which makes the edge of the bell;
3. the *dental papilla* which is derived from ectomesenchyme and fills the bell;
4. the *dental follicle* which is also derived from ectomesenchyme and encloses the whole tooth germ in a little bag.

The inside layer of the enamel organ is known as the *internal enamel epithelium* (or internal dental epithelium) and it grows by dividing, at first evenly throughout the layer. Early in the bell stage, a little group of cells in the internal enamel epithelium at its apex stops dividing, stopping growth at this point, whilst cells in the rest continue to divide. This is what causes the enamel organ to buckle into a bell shape, with a point at the apex. A short time later, a ring of additional cells just outside the apex region stops dividing, followed by another ring and another, spreading like a wave down the inside of the bell to map out the eventual form of the tooth. In a tooth with multiple cusps, there are initially several points at which cells cease dividing so, as the cells in between continue to divide, the internal enamel epithelium buckles into several lobes, one for each cusp. When all the cells have stopped dividing, the form of the internal enamel epithelium corresponds very approximately to the shape of the future enamel-dentine junction (page 90).

It is late in the bell stage that enamel and dentine start to form. Once they have stopped dividing, the cells of the internal enamel epithelium begin to differentiate into *ameloblasts*. This signals to the surface of the dental papilla underneath, where cells start to differentiate into *odontoblasts* which begin to form *dentine matrix*. In turn, this triggers the overlying ameloblasts into forming *enamel matrix* on top of the dentine matrix. In both cases, the matrix is laid down in layers and each succeeding layer expands as the group of fully differentiated ameloblasts and odontoblasts grows. The details of layered growth can be followed by microscopic examination of tooth sections, as explained in Chapter 4. Dentine matrix is entirely organic when first secreted by the odontoblasts, consisting of the fibrous protein *collagen* and a complex mixture of proteins and other molecules called *ground substance*. Tiny crystallites of the mineral *apatite* are seeded into this and grow until the tissue is 60% mineral by weight. Growing dentine therefore has a matrix-forming surface lying directly on top of the odontoblasts and, a little deeper (10–40 μm) to this, a mineralising surface. Enamel matrix is about one-third mineral, one-third protein and one-third water when first secreted by the ameloblasts. The ameloblasts of the internal enamel epithelium therefore also lie on top of a matrix-forming surface, although it already has tiny crystallites of apatite seeded into it. Each ameloblast has two parts to its career: matrix secretion and maturation. When it has secreted all the matrix it will ever make, it switches to maturation, during which it removes protein from the underlying matrix and increases the proportion of mineral, until eventually 99% of the tissue is mineral.

Enamel matrix secretion continues down the sides of the tooth crown until the last ameloblasts, just above the cervical loop, have completed their full thickness. Dentine formation has been continuing in parallel, inside the crown, with the layers dividing to make the pulp chamber. The cervical loop grows downwards in a tube that maps out the form of the root, which then starts to form inside it. This is called *Hertwig's epithelial root sheath*. The epithelial cells of the sheath trigger the underlying dental papilla to differentiate odontoblasts and continue dentine formation down the root. The root sheath therefore maps out the shape of the root. Multiple roots are formed by 'tongues' of root sheath growing towards one another and meeting at the junction of the roots.

Issues in studying dental development

Stages of development and their variation

There are four key chronological events in the development of each tooth:

1. *crown initiation* – the age at which dentine and enamel matrix secretion start;
2. *crown completion* – the age at which enamel matrix secretion is completed at the base of the crown;
3. *eruption* – the age at which the crown emerges into the mouth;
4. *apex closure* – the age at which dentine matrix secretion in the root is completed at its apex.

As will be seen in the following discussion, none of these events is entirely clearcut. There are fundamental differences between observation of dental development in dry museum skull specimens or fossils, direct observation of freshly dissected tooth germs, radiographs and examination of living patients, or histological preparations of teeth for study under the microscope. When first deposited, enamel and dentine matrix are not fully mineralised, so the most recently formed tissue in an actively growing tooth may not be apparent either in a radiograph or in a fossil tooth. Eruption is another difficult issue because, although it may be inferred in a fossil specimen from the initial appearance of wear on the cusp tips of a tooth, this is not how it is gauged in living primates. One of the main difficulties for archaeology and palaeontology is that by far the easiest dental development data for a dentist or zoologist to collect is gingival emergence – the first pinpoint penetration of the tooth through the gums. There are, however, no gums present in fossils and the challenge is to infer this point from the appearance of the underlying bone or development of the root.

As described in Chapter 2, the tempo of growth varies between individuals so that, for example, a slow developer will be small for any given age, will enter puberty later and will attain dental development stages at an older age than others in the same population. Similarly, a fast developer will attain development stages at a younger age. Variation between populations works in a similar way. Liversidge (2003) found that human populations whose development was, on average, faster than others erupted all their teeth at younger ages.

Other general patterns can be noted. There is no evidence of a systematic difference between left- and right-side dental development within each individual. One side is not consistently later than the other and the difference appears to be random, although the size of the difference becomes more variable in the later formed teeth in the sequence. For most studies, as can be seen from the following sections in this chapter and Chapter 5, teeth in the lower jaw achieve their development stages on average somewhat younger than teeth in the upper jaw, although the range of variation overlaps. Teeth formed earlier in the developmental sequence are less variable in the age at which they attain their development stages than teeth formed later. Similarly, in any one tooth type, crown initiation is less variable than crown completion, which in turn is less variable than apex closure (Figure 3.2) and all three are less variable than eruption. In general, however, age at attainment of dental development stages appears to be less variable than most other aspects of development, including the skeleton (Lewis and Garn, 1960).

Variation in dental development is well illustrated in the rhesus macaque dental eruption study of Hurme and van Wagenen (1961), in which the animals were checked every month. A boxplot (Figure 3.3) of their results for permanent teeth shows clearly that the first molars, earliest to form of the teeth, had by far the smallest variation of ages at eruption. Similarly, the third molars were the latest to form and had by far the greatest variation. The variation of other teeth was intermediate between these two extremes, but increased with median age of eruption. It is also apparent that the range of variation was greater above the median value than below it; that is, those individuals that erupted a particular tooth relatively

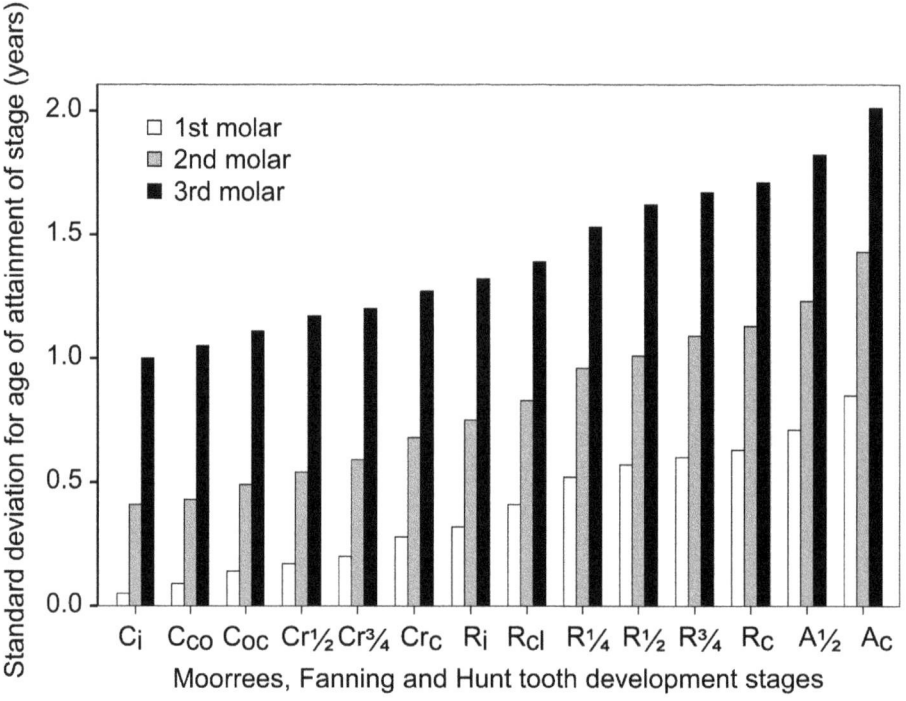

Figure 3.2 Variability of human permanent lower molar formation stages. Standard deviations for attainment of stages, which are in sequence from the earliest stage on the left of the horizontal axis to the latest on the right. Data from Moorrees *et al*. (1963a) and Harris and Buck (2002). See Figure 3.16 for explanation of stages.

late showed a greater variation in eruption age than those which erupted it early. It is also possible to follow individual animals in Hurme and van Wagenen's detailed data (Figure 3.4). Some monkeys were fast developers and some slow, but there are inconsistencies. For example, individual 10 shown in Figure 3.4 was a fast developer, always erupting its teeth at the younger end of the range. Individuals 12 and 13 were slower developers, at the older end, but they varied in their placement: mostly 13 was later, but sometimes it was 12. Individual 70 occupied an intermediate position for most teeth, but varied and was by far the latest to erupt its lower third molars. These comparisons show that, although the difference in eruption age between teeth was fairly consistent for one individual, some individuals did not plot out as expected. Overall, there was a consistent difference between teeth in median age of eruption, but there was overlap between teeth in their range of variation. This overlap was greatest between the canines, premolars and second molars, and between the first and second incisors. It is particularly important when discussing single specimens, as often in the fossil record, to consider how representative this individual might be of the population as a whole. It might simply be a fast or a slow developer, or even anomalous like individual 70.

Issues in studying dental development

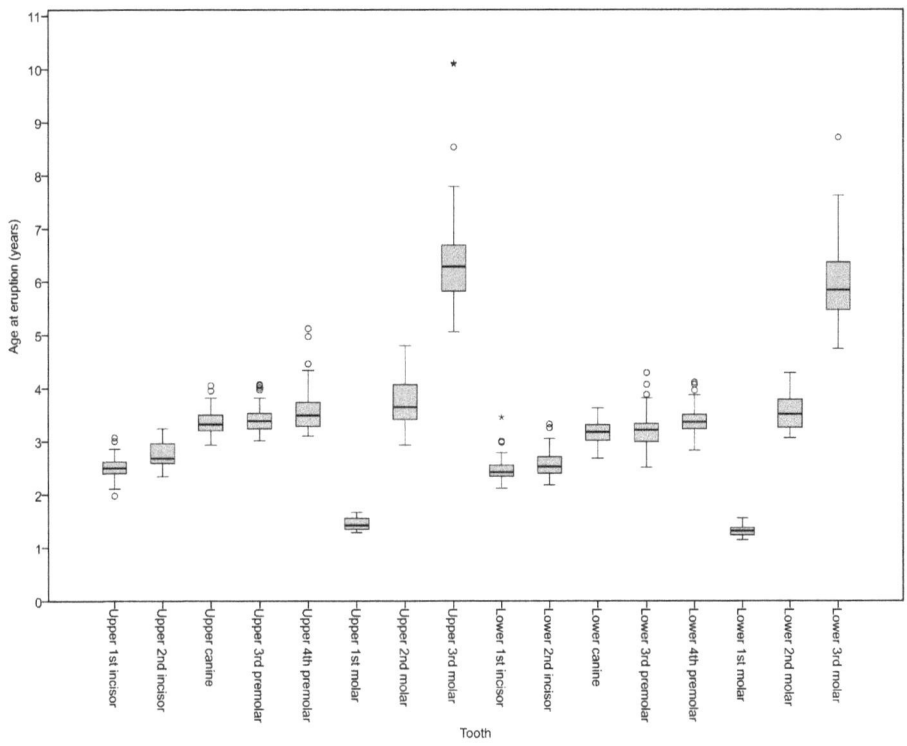

Figure 3.3 Boxplot of ages of permanent tooth eruption in female rhesus macaques (*Macaca mulatta*). Data plotted from Hurme and van Wagenen (1953). The height of the grey boxes represents the interquartile range and the line inside them represents the median. The whiskers above and below the boxes represent the main range of values, whereas outliers are plotted separately as circles and stars.

Statistical methods

The main problem in interpreting dental development data is that records are only made at intervals. In the rhesus macaque study discussed in the preceding section the interval was monthly, but in most studies it is once every six months or once a year. The examinations almost never catch a developmental stage as it is actually changing – for example, the point at which the tips of the cusps just break through the gums, or the last enamel matrix is deposited at the base of the crown. Nearly always, the individual has either not yet achieved the stage in question, or has achieved it sometime in the past. It is therefore necessary to interpolate.

The most common way to present such developmental data is as a cumulative percentage graph or ogive (Figure 3.5). A graph of this kind is prepared for each developmental stage of each tooth and every point plotted represents one individual. The ages at which the examinations took place run along the horizontal axis and the vertical axis represents the percentage of individuals in the study group as a whole, so it totals 100%. The leftmost point plotted represents the youngest

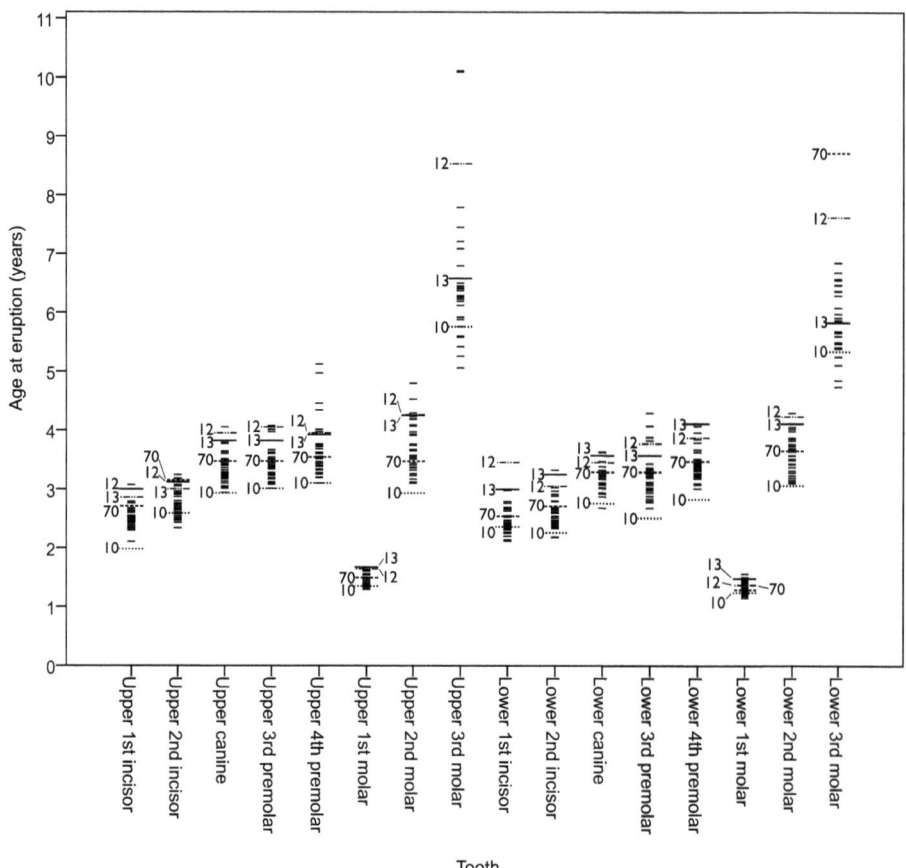

Figure 3.4 Plots of individual values from the data shown in Figure 3.3. The short horizontal lines each represent the age of eruption for a single individual. The longer horizontal lines represent four individuals which have been separately identified: individuals 10, 12, 13 and 70 (see text for discussion).

individual at the time of examination in which that stage was first recorded as completed and the next point, the next youngest, and so on. For each point, the height on the vertical axis represents the cumulative percentage of individuals in the study group that attained the stage at their plotted age, or younger (in other words, it includes all the individuals plotted on its left as well). Where enough individuals are included in the plot, this produces a characteristic 'lazy S' shaped curve, starting with a gentle gradient, then passing through a steep section and finally settling back towards a gentle slope. It is apparent from its shape that just a few individuals attained the stage at a young age, and just a few at an old age – these make the curving 'tails' of points at the top and bottom of the line. The central part, where the points make a relatively straight, more vertically arranged line, represents the majority of individuals who attained the stage within a relatively narrow range of ages and its slope depends on that range. In most graphs of dental development

Issues in studying dental development

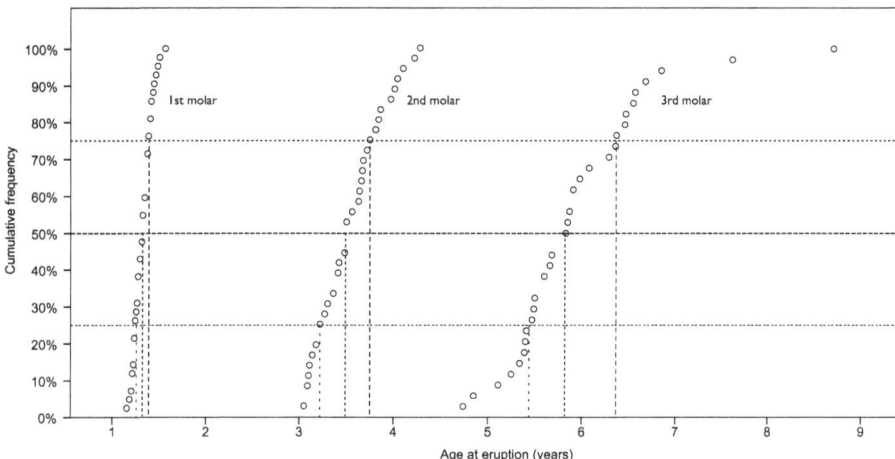

Figure 3.5 Cumulative frequency plots for permanent lower molars shown in Figures 3.3 and 3.4. Each point in the plot represents the eruption age for the given tooth, for a single individual, along the horizontal scale of age in years. The vertical scale is the cumulative frequency expressed as a percentage of the total number of individual values for each tooth. Horizontal dashed lines represent the median (50th percentile), lower quartile (25th percentile) and upper quartile (75th percentile). Vertical dashed lines show how the values for these percentiles can be read from the age scale. The slope of the central portion of the curves decreases with increasing interquartile range. It is also possible to see, particularly in the third molar, how the individuals whose teeth erupted in the older half of the range showed greater variability in their ages of eruption.

stages (see the third molar in Figure 3.5), the tail at the top of the plot is longer than the tail at the bottom, suggesting that those individuals who attained the stage relatively later than the majority did so over a wider range of ages than the few who attained it relatively young. This is known as positive skew. It is found in many developmental phenomena and is usually treated by a logarithmic transformation of the age axis which makes the tails of the plot equal in length to create a symmetrical curve.

One of the advantages of a cumulative percentage graph is that it automatically leads to the median and quartiles, which can be read off the graph by fitting a smoothed curve and then finding the point at which the 25, 50 and 75% lines intersect it (Figure 3.5). The median determined in this way is the age by which half of all individuals in the growth study attained the stage in question, which makes a good measure of what is required – that is, the age from which the majority exhibit that stage completed. Similarly, the quartiles provide a good measure of the variability of attainment as they define the range of ages within which half the group first showed evidence of completing a stage. Several early studies hand fitted curves to cumulative percentage plots to determine medians and Hurme (1948) felt that this approach coped better with the true nature of the data – missed observations and variability – than statistical modelling techniques.

Most studies, however, have transformed the cumulative percentages into *probits* (an abbreviation of 'probability units'). When plotted against the logarithm of age, the probit values for each individual lie along a relatively straight diagonal, rather than a sigmoid curve. It is then possible to fit a regression line to this scatter of points, so the median, quartiles and other percentiles, mean and standard deviation can be calculated from the regression equation (Dahlberg and Menegaz-Bock, 1958; Gates, 1966; Finney, 1971; Smith, 1991c). This procedure is relatively robust and is not greatly affected by irregularities in the spread of observations between age groups. The study group should, however, have a relatively even division between those in which the tooth in question is erupted and those in which it is unerupted and this is not always the case (Kuykendall *et al.*, 1992). There are several alternatives to the probit transformation for estimating the mean from cumulative distributions, mostly producing similar values. One of these alternatives which has been used in studies of dental development is the Spearman-Kärber method (Hamilton *et al.*, 1977; Finney, 1978), preferred by some because it makes no assumptions about the underlying distributions.

Means are affected by the 'leveraging' of outlying values and the longer tail of the graph at the older age end tends to make the *mean age at attainment* for a stage higher than the *median* age. In view of this, the median is probably the better estimator to use in growth studies (Tanner, 1989) but, if the mean was determined through a probit transformation which uses the logarithm of age, it should be the same as the median because the transformed distribution of ages would be symmetrical (as discussed earlier). For that reason, both are presented in the following discussion as measures of 'average' age at attainment of development stages. Many studies contain only a few individuals, so the observations may not be evenly spread across the age range for a given stage, thus giving rise to a step in the line of points rather than a smooth curve. Smith (1991c) has suggested that the best way to deal with this is by estimating the midpoint in age between the oldest individual in the study not having attained the development stage and the youngest individual having attained it. Kuykendall (1996) called this the *midpoint age of attainment*.

Where possible, only studies based on the methods discussed above have been used in this book, but there are several other possibilities. One is to calculate the mean for each tooth of the ages at which each stage was first noted as completed and then to adjust them for the length of the interval between successive radiographs. When there are long intervals between observations in a longitudinal study, or big gaps in the age of individuals in a cross-sectional study, the attainment of a stage in any one individual may have come months before it was recorded as 'attained'. The mean age at first noted completion will therefore always be older than the real mean age at attainment for the group and it is usually adjusted by subtracting half the interval between radiograph examinations (Anderson *et al.*, 1976). There are still difficulties, however. For the adjusted mean to be a good representative of the age at which a development stage was usually attained, the study group needs to include an even spread of ages, including children considerably younger than that age and older than that age. In most growth studies this is not so and the uneven spread of data points strongly affects the mean. It would affect the cumulative percentage and probit methods discussed earlier as well, but at least this would be visible in the plot. The same problem affects another approach (Gleiser and Hunt, 1955), which is to calculate the mean age of individuals that have attained the stage in question, but not the next stage. This is known as *age for stage*. It is strongly affected by the evenness of spread by age of individuals within the category from which the mean is calculated. In a cross-sectional study this is just a subset of the individuals in the project as a whole, whereas age at attainment as defined above is based on all the individuals in the project. Age for stage tends to be slightly older than age at attainment (Kuykendall, 1996). As Smith (1991c) observed, it is important to understand the nature of the group of children included and the methods used when comparing different studies.

Captive versus wild, living versus dead and rich versus poor

Most studies of development have been based on laboratory animals or colonies, rather than wild animals which are now rare and endangered. There must be considerable differences between captive and wild animals in nutrition, disease, parasite load and other factors of the environment within which their development takes place. Perhaps also captive animals derive from only a few wild individuals and are therefore not wholly representative of the genetic variation in the wild population. For these reasons it might be expected that there would be differences in the timing of developmental events, including dental eruption, between captive and wild groups.

There are, however, other potential complicating factors. One of these is the fact that most studies of captive groups are longitudinal and based on living subjects. By contrast, wild animals may be trapped alive and then tranquillised before examination, or be part of a museum collection of dead individuals. Such studies are necessarily cross-sectional. Specimens now found in many museums were deliberately shot, but in other collections (see following discussion of chimpanzees) they were known individuals found dead in their study area. All these different modes of collection are likely to have an effect on the estimated ages at attainment for dental development stages. For example, a young chimpanzee that died from disease in the wild is unlikely to be a good representative of development in the healthy living population.

Individuals shot by game hunters, preserved in a number of important museum collections, are likely to be those who were close to human habitation. There is some evidence that these individuals raided crops and food stores (Miles and Grigson, 1990) so they might well have had a different level of nutrition from more isolated groups of animals. Altmann and Alberts (2005) made a longitudinal study of three groups of baboons (*Papio cynocephalus*) at Amboseli in Kenya. Two were fully wild, whereas the range of the other group included a tourist lodge from where they collected waste human food. Ages were independently known because the juvenile animals had been observed from birth and they could be weighed periodically by attracting them to stand on scales which could be read remotely. The 'lodge' group had a greatly enhanced growth rate in comparison with the others. Phillips-Conroy and Jolly (1988) compared dental eruption in wild yellow baboons (*Papio hamadryas cynocephalus*), trapped in Ethiopia, with a captive group originally from southern Kenya, but born and raised at a research facility in San Antonio, Texas. The study in the wild baboons was necessarily cross-sectional, but the study of captive baboons was longitudinal. They found that mean age of eruption for all teeth in the captive animals was younger, by over a year in the case of the canines and premolars.

Recently, there has been extensive discussion of a group of wild chimpanzees from the Taï Forest in Côte d'Ivoire, Africa. They have been studied over many years (Boesch and Boesch-Achermann, 2000), with individual animals followed from birth and known by name. Members of the project found dead animals from time to time and, depending on the state of decomposition, identified them partly from their knowledge of which individuals had not been seen for some time and partly on estimated age, sex and other individual characteristics. The skeletons and their associated dentitions were collected and, at the time of writing, numbered 30 individuals. There are smaller collections of a similar kind from Gombe in Tanzania (Goodall, 1986) and Bossou in Guinea (Matsuzawa *et al.*, 1990). There is some discussion of identification but, in effect, these are known age-at-death assemblages of wild chimpanzees. Zihlman and co-workers (2004; 2007) estimated eruption timing in jaws from Gombe, Taï and Bossou. They assumed a four-month interval between the appearance of a tooth through the bone and the first penetration through the gums so they could compare their results with Conroy and Mahoney's (1991) longitudinal study of 58 captive chimpanzees born and raised in a laboratory at New York University. The deciduous teeth in the wild animals erupted at similar estimated ages to those of the captive chimpanzees, but the estimated ages at emergence for the earlier forming permanent teeth (first incisor and first molar) were at the older end of the laboratory animals' range and the later forming permanent teeth emerged outside this range.

Smith *et al.* (2010a) re-examined the dentitions from the Taï group and identified a number of inconsistencies. They made micro-CT scans and studied teeth from a number of individuals histologically (Chapter 5). When they compared tooth formation in the Taï chimpanzees with data from captive laboratory animals (principally from Kuykendall, 1996), they found that the ages for crown formation overlapped. There was some evidence that formation of the roots was accelerated in the laboratory animals, but their new estimates for dental eruption suggested that the wild chimpanzees were within the range of variation of the laboratory chimpanzees. Smith and Boesch (2010) confirmed that, although the wild animals on average erupted their teeth at an older age than the laboratory animals, the difference was not as great as Zihlman and co-workers (2004; 2007) had suggested. Their ranges of ages for wild chimpanzees fitted comfortably within the range for captive animals. In addition, they pointed out that, amongst the young animals which provided the key information on development in the wild group, the factors that caused their early death would also have been likely to cause delays in development.

Might a collection built from natural deaths such as the Taï Forest data actually be a better comparison for fossil primates, which are, after all, also from death assemblages rather than living populations? What needs to be remembered here, as Smith and Boesch (2010) concluded, is that there are different kinds of fossil assemblages. There are deliberate burials, carnivore accumulations, or fossils accumulated by the force of water in a sedimentary basin. All of these have different characteristics that will affect the infants and juveniles accumulated – for example, to what extent will they be selected from the slower rather than faster developing individuals in the living population? The modern human comparative sample is also a problem. All data, from radiographic studies of the living or direct observation of museum specimens, come from people who lived in houses of one kind or another, ate the products of agriculture and animal husbandry, and were to a greater or lesser extent subject to health care throughout their lives. In effect, we are all equivalent to captive or laboratory animals and have been so for thousands of years. If such animals are not a good comparison for fossil hominoids, then neither are modern humans. It is not clear whether any comparable collection of modern human teeth could be found. In any case, so little is known about the environment in which fossil hominoids grew up that it will never be entirely clear what an appropriate comparison would be.

Is dental development particularly affected by environment compared with other aspects of growth? In fact, the opposite seems to be true. Tonge and McCance (1973) made an elegant comparison in laboratory pigs between the bony development of the jaws and the development of the teeth that they contained. One group was fed a severely calorie-deficient diet, whilst the control group was fed normally. Dental development was certainly delayed in the calorie-deficient group, but not by nearly as much as growth of the jaw. Similarly, Garn and co-workers (Lewis and Garn, 1960; Garn et al., 1965) found that whereas tooth development seemed to be strongly related to inherited factors, it was little related to dietary variation in human children from the large Fels Growth Study (discussed later in this chapter), although they did point out that this might be more pronounced in a group of children in which there were larger contrasts in diet. They also found that dental eruption was more variable than stages such as completion of the crown. In fact, dental development was rather poorly correlated with skeletal development. It does therefore seem that comparison between fossils and living primates might most safely be carried out in terms of crown and root development, rather than eruption, which involves development in the skeleton and other tissues of the jaws.

One potential strength of teeth as a subject for developmental study is that even minor disruptions to growth are recorded as defects in the enamel of the crown (Chapter 6). Enwonwu (1973) compared two groups of Yoruba children from Nigeria. The 'optimal group' lived in the city of Ibadan and came from well-to-do families.

The 'Osegere village group' came from a much less well-off rural background. The optimal group were larger for their age than the Osegere children. In addition, their deciduous teeth erupted at younger ages. For example, at 22–24 months of age, the lower fourth premolars were erupted in almost all the optimal group children, but this was true of only 18% of the Osegere children. One important additional observation, however, was that the defects of enamel hypoplasia were present only in the Osegere group (page 194). This suggests that such defects could be a useful indicator of possible environmental effects on growth when comparing the rate and pattern of dental development in different groups.

Dental eruption

The age at attainment for eruption forms an important part of the discussion of primate dental development because it is much better known than other stages. It can be recorded by direct observation of the open mouth, but it is not without problems. One is the simple matter of persuading the child, chimpanzee or monkey to hold its mouth open so that the observations can be made without damaging the observer. This can be a serious business as the animals grow in size and strength so that examinations may become more infrequent as it is necessary to anaesthetise the subject (Nissen and Riesen, 1964).

The other difficulties lie in the nature of eruption. It is the most variable of all dental development stages. Partly, this is because it involves the development of the jaws and soft tissues in addition to the development of the teeth themselves. It is also, however, partly due to the fact that dental eruption is a continuum. As the teeth grow inside the alveolar process of the jaws, the bony crypts which contain them expand. In general, the growing edge of the root stays at more or less the same level within the maxilla or mandible, whilst the tips of the cusps progressively rise towards the crest of the alveolar process. This can be monitored on radiographs, but does not have any clearly defined stages. Bone remodels around the crypt as it expands and the overlying covering becomes gradually thinner. Eventually the top of the crypt opens out into an aperture which is covered only by soft tissues. This is known as *alveolar eruption* (Figure 3.6) and is the first stage that can be seen in dry bone museum specimens, archaeology or fossils. Continued remodelling of the bone widens the aperture and the tips of the cusps start to press into the soft tissues. Children are aware of hard bumps in their gums. The soft tissues too are remodelled around the cusps, which are eventually exposed as tiny white points of enamel. This is known as *gingival emergence*, or *clinical eruption*. Other cusps and parts of the crown emerge gradually. Most studies record emergence as the very first appearance of any part of the tooth, but there are other interpretations and Schultz (1935, p. 494), for example, recorded the point at which the whole top of the crown was first exposed.

As it develops, the new tooth rises higher and higher relative to the gums and other teeth in the arcade. Eventually, the cusps arrive in the occlusal plane and

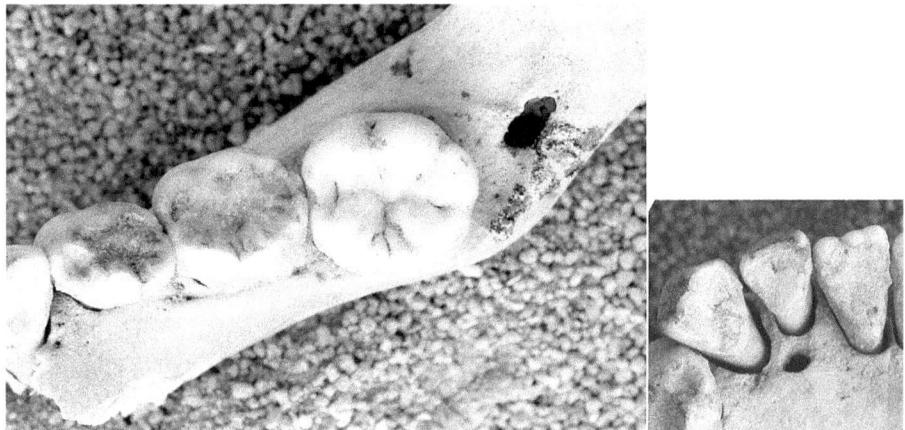

Figure 3.6 Alveolar eruption in a human mixed dentition at a development stage equivalent to 6.5 years of age in a modern child. Left image from left to right: well-worn deciduous lower third and fourth premolars; permanent first molar in full occlusal eruption with slight wear on its cusp tips; the crypt of the permanent second molar just breaching the bone on the alveolar crest to make a small opening. Right image: deciduous lower canine; second and third incisors; small opening for the crypt of the permanent second incisor. Neolithic specimen from Çatalhöyük, Turkey.

start to meet teeth from the opposing jaw. This is known as *occlusal eruption*. The exposed cusps start to wear, creating little occlusal attrition facets on their tips. This happens very slowly in modern children, although in archaeological and fossil dentitions it seems to have been rapid. In theory, the appearance of small occlusal wear facets can be used in fossils to mark the point of occlusal eruption, but it is difficult to estimate the time elapsed since the tooth first arrived in the tooth row. For non-human primate specimens in museums, it is sometimes possible to argue from staining that unworn cusps had emerged through the gums (page 146).

One particular issue is the eruption of the replacement teeth – the permanent incisors, canines and premolars – which emerge underneath the deciduous teeth. As the crypts grow, the roots of the deciduous teeth are resorbed by a similar mechanism to that which removes bone in the remodelling process. As the deciduous teeth lose the support of their roots, they become wobbly and eventually fall out. The permanent premolars emerge into the gap left by the deciduous premolars. The permanent incisors and canines, however, typically emerge on the tongue side of their deciduous predecessors. They often emerge before the deciduous teeth have been lost, complicating the definition of gingival emergence for the permanent teeth.

Studies of eruption in living primates are based on gingival emergence and need to be compared with studies of museum specimens and fossils which are based on alveolar and occlusal eruption. It is therefore crucial to understand the scale of the difference in timing between these stages. In a study of Finnish children, Haavikko (1970) showed that in most permanent teeth the gingival emergence stage occurred between 0.3 and 1.3 years later than the alveolar eruption stage. The difference was largest in the additional teeth, the permanent molars, but much smaller in the successional permanent teeth. Kelley and Smith (2003) described four individuals from a captive colony of olive baboons (*Papio anubis*), where gingival emergence was taking place at the time of examination. They observed alveolar emergence in radiographs around three months earlier and concluded that the observations of Zuckerman (1928) suggested an interval of 4–5 months for a chimpanzee. Garn *et al.* (1958) showed that occlusal eruption occurred up to 0.8 years later than alveolar eruption in human permanent lower third premolars, gradually increasing to 1.6 years in second molars. Liversidge and Molleson (2004) found that the difference between alveolar and occlusal eruption was between 0.3 and 0.7 years in the deciduous teeth of children from the crypt at Christ Church, Spitalfields in London, whose age-at-death was independently known.

Gingival emergence in living humans

Studies of gingival emergence are well summarised from many different parts of the world by Liversidge (2003). They vary widely in study group size, the range of ages of children included, the teeth recorded (depending largely on the age groups included), cross-sectional or longitudinal design and the statistical approach used. Some overall trends can, however, be pointed out. In any one study group, the range of variation overlaps with that of most other human groups so that the variation within groups tends to be larger than the difference between them. Even so, most groups (but not all) of black children from sub-Saharan Africa, or the large black communities elsewhere, on average erupt their teeth at younger ages than other children. The same can be said of the children of Australian aborigine communities and from Papua New Guinea. White children from Europe, or descended from Europeans, tend to erupt their teeth at older ages and Asian children tend to be somewhere in between.

There are irregularities between teeth in different studies but, in general, all permanent teeth show these differences in average emergence age. For any given population, girls are on average slightly in advance of boys for all but the later stages of gingival emergence, although the difference is relatively small in comparison with the strongly overlapping ranges of variation. Most studies collected by Liversidge did not include enough older adolescents and young adults to provide adequate coverage of the permanent third molars. It has therefore been difficult to select representatives for this book, but Hassanali and co-workers (Hassanali and Odhiambo, 1981; Hassanali, 1985) carried out studies in Kenya which included both African and Asian children from schools in Nairobi, with a group of students between 13 and 23 years of age. Within the one study, therefore, both a 'fast' and a 'slow' development group were included, as well as boys and girls (Figure 3.7). There are far fewer studies of the deciduous dentition.

Human deciduous teeth (Appendix A, Table 4) erupt in the following sequence: first incisor; second incisor; third premolar; canine; fourth premolar. First incisors

Dental eruption

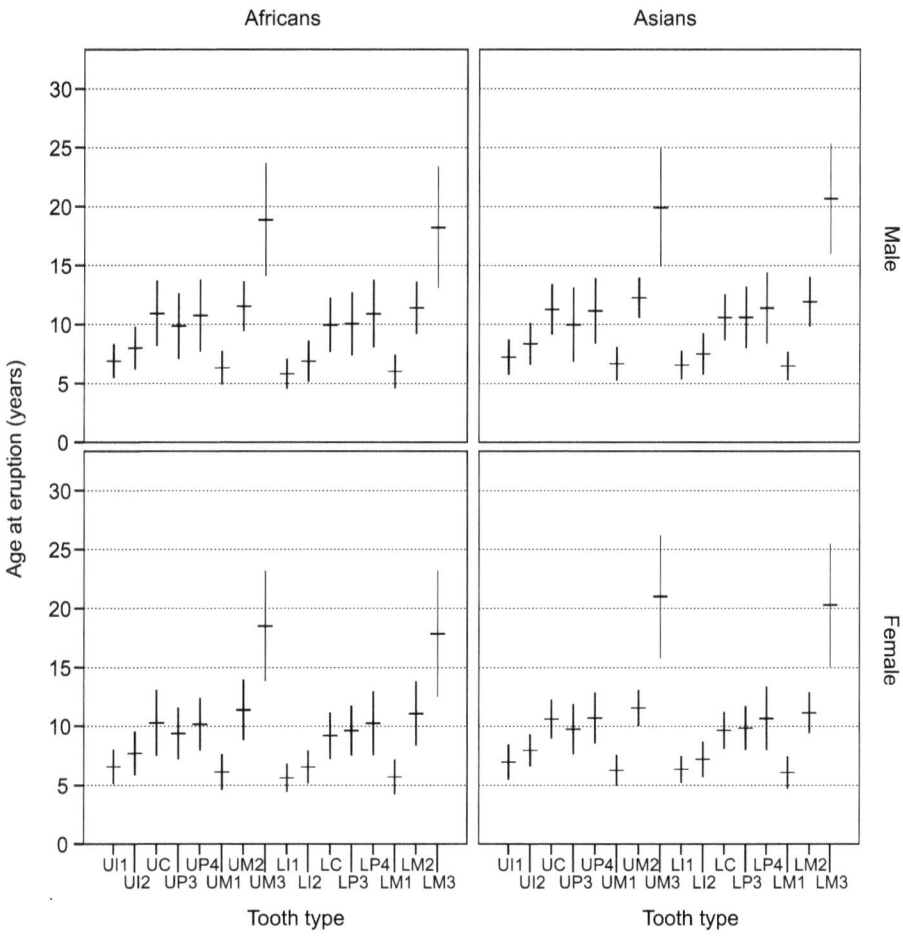

Figure 3.7 Mean age at eruption (gingival emergence) for recent African and Asian children from Kenya. For each permanent tooth, the horizontal bar represents the mean and the vertical bars represent 95% confidence intervals calculated from the mean and standard deviation. Data from Hassanali and Odhiambo (1981) and Hassanali (1985). See Appendix A, Table 5.

emerge gingivally between 0.6 and 0.9 years after birth on average. At around 1 year of age come the second incisors, then the third premolars at 1.3 to 1.4 years, followed by the canines at 1.5 to 1.7 years. The fourth premolars emerge on average at 2.1 to 2.4 years of age. There are no consistent differences in age at emergence between boys and girls, and only minor differences between populations.

In the *human permanent teeth* (Appendix A, Table 5), the first molars emerge around 6 years of age on average, the lower teeth being slightly earlier than the upper teeth. They erupt behind the deciduous fourth premolars, which remain in place; the first molars look very like an enlarged version of the deciduous premolars, extending the cheek tooth row. The deciduous incisors are lost (exfoliated) before 6 years of age, leaving a gap at the front of the dentition. This is the so-called 'ugly duckling' phase. Permanent incisors emerge into this gap; sometimes they appear to lingual of deciduous incisors which have not yet been exfoliated. Human lower first incisors emerge at practically the same time as the lower first molars, whereas upper first incisors may on average emerge a good six months later than the upper first molars. In both jaws, the second incisors emerge on average around one year after the first incisors. This is the first phase of eruption of the permanent teeth, which takes place with the deciduous canines and premolars still largely in place, during the period of mixed dentition (page 29).

The remaining deciduous teeth are exfoliated and then replaced by permanent canines and premolars, on average between 10 and 11 years of age. The permanent molars follow in sequence at the back of the jaws, with the second molar emerging around 12 years of age and the third molar at 18, 19 or 20 years. Once again, the lower teeth tend to be slightly in advance of the upper teeth. The first molars are the first permanent teeth to emerge and the third molars the last, so the whole process of emergence runs from 6 to around 19 years of age, lasting about 12 to 14 years. The 95% confidence intervals calculated from the standard deviation (Figure 3.2) suggest that the range of variation increases greatly, from 1.5 years either side of the average in the first molars, to a little more than 2 years in the second molars and around 5 years in the third molars. Everything about the third molars is highly variable, from their size and shape, to development timings and eruption – even whether or not the tooth is formed at all. They are by far the most commonly missing congenitally of any teeth. In a very few individuals, third molars may even emerge as young as 12 or as old as 24 years. In other words, some young adults who thought they had left teething behind them are surprised to find that they are teething in their early twenties.

Smith and Garn (1987) developed a useful formula for summarising the most common variations in eruption sequence in human permanent teeth which can be adapted here to:

Upper jaw – M1 I1 I2 (P3 C P4) M2 M3
Lower jaw – (M1 I1) I2 (C P3) (P4 M2) M3

Teeth are labelled as in Appendix A, Table 3. The teeth enclosed in brackets may erupt in the order shown, or in any other order so, for example, the lower first molar (M1) may erupt before the lower first incisor (I1), or may erupt later. It can be seen that the order of canines and premolars varies considerably.

Gingival emergence in chimpanzees, gorillas and orangutans

Chimpanzees are the most studied of non-human primates. Nissen and Riesen (1946; 1964) made a longitudinal study of 16 animals kept at the Yerkes Laboratories of Primate Biology in Florida, which were regularly checked at intervals of one week. Most individuals were apparently content to open their mouths, although a few objected and could only be examined occasionally when they were anaesthetised for

Dental eruption

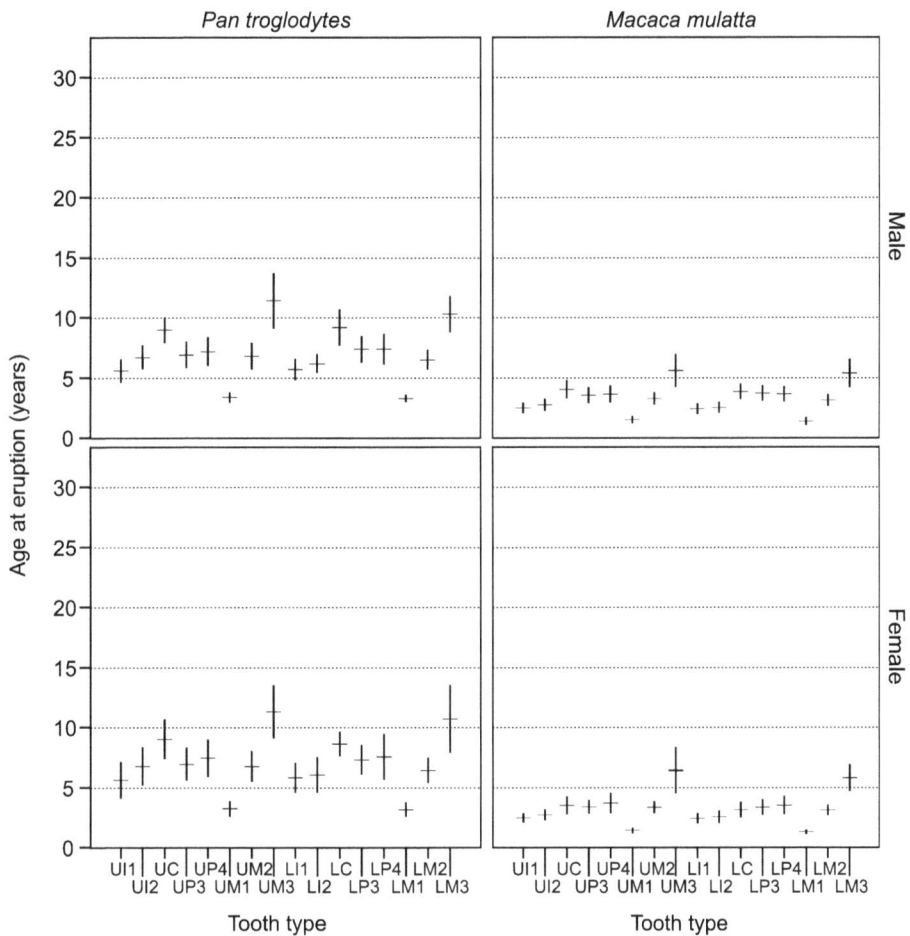

Figure 3.8 Mean age at eruption (gingival emergence) for chimpanzees and rhesus macaques. For each permanent tooth, the horizontal bar represents the mean and the vertical bars represent 95% confidence intervals calculated from the mean and standard deviation. Data from Nissen and Riesen (1964) and Hurme and van Wagenen, 1961. See Appendix A, Tables 7 and 9.

a general examination. Another longitudinal study (Kraemer et al., 1982) involved yearly examination of 17 animals in large outdoor enclosures over three years at the Stanford Outdoor Primate Facility. The largest longitudinal studies of chimpanzees were carried out during a ten-year observation of 58 animals in the colony at the Laboratory for Experimental Medicine and Surgery in Primates of New York University Medical Center (Conroy and Mahoney, 1991; Kuykendall et al., 1992). Smith and Boesch (2010) reviewed evidence for gingival emergence in wild chimpanzees from the Taï Forest in Côte d'Ivoire whose age was independently known (page 40).

In chimpanzees the deciduous first incisors, second incisors and third premolars emerge on average one after the other between 0.2 and 0.4 years after birth (Appendix A, Table 6). They are followed by the fourth premolars at 0.6 or 0.8 years and finally the deciduous canines just after 1 year after birth. The first molar is by some margin the first permanent tooth to emerge, at between 3 and 4 years after birth (Appendix A, Table 7 and Figure 3.8). The first incisors do not emerge until almost 6 years of age. In the lower jaw, the second incisors emerge on average just after 6 years and then the premolars and second molars at about 7 years of age. In the upper jaw, the second incisors, premolars and second molars all emerge at approximately 7 years of age. The canines emerge at about 9–10 years, overlapping in range with the third molars, which emerge on average between 10 and 11 years after birth. Thus the full sequence of permanent tooth emergence lasts from 3–4 to 10–11 years and so is around 7 years in duration. There seems to be little difference between males and females. Smith and Garn (1987) have suggested the following sequence of eruption for both chimpanzee and gorilla (using the same notation as for humans, but without separating upper and lower dentitions):

M1 I1 I2 M2 (P3 P4) C M3

Kuykendall et al. (1992) found the following sequence (without the third molar):

Upper M1 I1 I2 P4 P3 M2 C
Lower M1 I1 I2 M2 P4 P3 C

Much less is known about gorillas and orangutans. In the deciduous teeth, the sequence appears to be similar to that of chimpanzees. Orangutans show a longer gap in emergence between the deciduous first incisors and the second incisors. For the permanent teeth, both the sequence and timing of emergence in gorillas appear to be similar to those of chimpanzees (Appendix A, Table 8). Very limited data for orangutans suggest that, although the overall sequence and timing are also similar to chimpanzees, the second and third molar may erupt at a younger age. As a result, Winkler et al. (1996) suggested that the order of eruption might vary from that of chimpanzees and gorillas in the incisors and canines:

M1 (I1 I2 M2) (P3 P4) (C M3)

so, in some males, the canines may emerge after the third molars.

Gingival emergence in rhesus macaques and baboons

Dental eruption in laboratory rhesus macaques is particularly well known through the work of Hurme and van Wagenen (1953; 1956; 1961; Hurme, 1960) on a colony at Yale University Medical School. Over five generations of monkeys, the younger animals were observed weekly and the older individuals monthly. They published their full data tables so it is possible to see that, for the deciduous dentition (Appendix A, Table 9), the teeth emerged in the sequence

DI1 DI2 (DC DP3) DP4

because the average age of eruption of the deciduous canines and third premolars was very close. The deciduous first incisors erupted on average just 2 weeks after

Tooth formation

birth, with about 2 weeks variation either side. The canines/third premolars emerged on average at 10 weeks and the fourth premolars about 20 weeks after birth, with a variation of around 13 weeks either side. There was very little difference between male and female monkeys.

Once again, their data tables make it possible to see that the emergence order (Appendix A, Table 10 and Figure 3.8) for the permanent teeth was

M1 I1 I2 (M2 P3 C P4) M3

because the second molars, canines and premolars all emerged at around 3.1 to 3.6 years after birth. The permanent first molars emerged first, on average at 1.3 to 1.4 years, with a small range of variation. First and second incisors emerged at 2 to 3 years, one after the other. The third molars emerged on average around 6 years of age, so the whole process of eruption lasted about 4.5 years.

Baboons are less well known, but seem to follow a similar sequence of eruption, with an extended schedule. Olive baboons in the wild, for example, erupt their permanent first molars, first and second incisors at similar ages to rhesus monkeys, but the premolars, canines, second and third molars emerge at later ages (Kahumbu and Eley, 1991).

Differences in gingival emergence between humans and other primates

Dental eruption in humans has a number of very distinctive features. The whole sequence is much longer than the other primates described here, with the deciduous dentition not fully erupted until well over 2 years of age (compared with 1 year in chimpanzees) and the permanent dentition not until 19 years of age or even older (compared with the chimpanzee's 11 years). There is also a marked difference in the pattern of eruption. In humans, the permanent first molars and first incisors erupt at nearly the same age, but in all the other primates considered here the first molars are far in advance of any other permanent tooth. So, whereas permanent incisors erupt at a similar age (6–7 years) in both humans and chimpanzees, they overlap in humans with first molar eruption and, in chimpanzees, with permanent second molar and premolar eruption. For humans, permanent canine, premolar and second molar eruption together form a later phase, which is followed at an interval by third molar eruption. The much larger canines of chimpanzees erupt later than premolars and overlap eruption of the third molar. In orangutans, canines may even erupt after the third molars.

Tooth formation

Assessment of tooth formation

Stages of development may be observed directly in extracted teeth or those dissected from post-mortem specimens, in anatomy museum collections, or in dentitions from archaeological sites. Isolated teeth are among the most common fossils and some fossil jaws are broken to expose the state of development within. In living specimens, or in fossils where it is not possible to see the development state from the surface, it is necessary to use radiography. For most studies of dental development, this has been a conventional X-ray image (radiograph) on film, and there are substantial archives of dental films. Increasingly, film is being replaced by digital sensors and it is also straightforward to scan films into a digital format. Computed tomography (CT) techniques create serial slices or three-dimensional models of the internal structure (Appendix B). Increasingly, fossil specimens are being scanned with micro-CT machines, which are really more suitable for recording the tiny details of dental development.

A careful distinction needs to be made between development stages as seen by radiography and those seen by direct observation. Radiography shows structures because of contrasts in the absorption of X-rays passing through the specimen. This is partly due to variations in the density of mineralisation. Dentine is slightly more heavily mineralised than bone, but the difference is not large enough to create a large contrast. Enamel is considerably more mineralised than either bone or dentine and so shows as relatively *radio-opaque* – brighter in a traditional film because the photographic emulsion underneath is protected from the X-rays. The pulp chamber does not normally contain mineralised tissue, so it creates a *radiolucency*, which is darker on the film because more X-rays pass through.

The difficulty in comparing radiographs with direct observations and the histological methods outlined in Chapters 4 and 5 is that, to show as a contrast in radiography, the different tissue components have to be sufficiently mineralised. This is perhaps less of a problem with dentine, which mineralises relatively close to the matrix-forming surface (page 32). Enamel mineralisation (maturation) is a gradual process, so the early stages seen by histology can be missed in radiographs. There is therefore a concern that radiography may give a delayed picture of development. Winkler (1995) compared orangutan development as seen in dissected tooth germs with radiographs taken before dissection. She found that the radiographic appearance could be delayed by several months, particularly in the early stages of development, although this varied between teeth.

With conventional radiographs there is also the issue of projection. A developing tooth is essentially a hollow cone (Figures 3.9 and 3.10). The surface on which dentine matrix is secreted slopes strongly inside the tooth and the edge along which the crown or root is growing in height is wafer-thin. That edge is the crucial feature for assessing development, but the shape of the shadow it casts on the film will depend on its orientation relative to the axis of the X-ray beam. Ideally, it needs to be imaged fully in profile. Standard clinical radiographic projections can maximise the chances of this in living patients, but there is considerable variation. Micro-CT radiography allows the section plane to be positioned anywhere in the tooth and thus resolves the difficulty.

Tooth formation

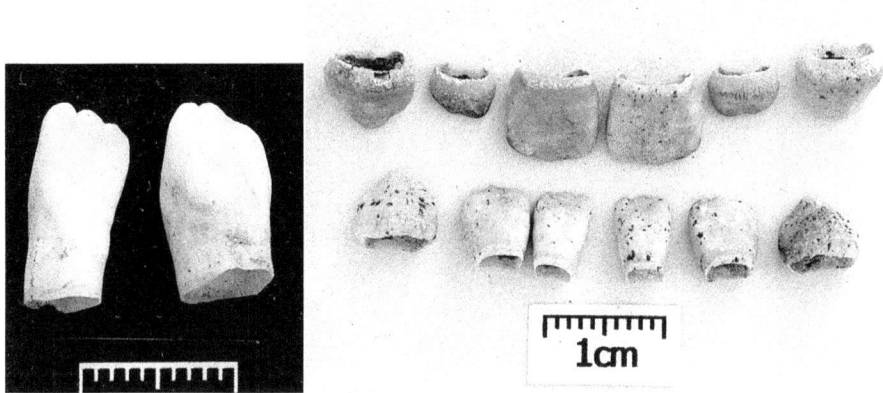

Figure 3.9   Developing human permanent teeth. Left image: lower first and second incisors with completed crowns and developing roots. They were clearly not erupted because the mamelons are still sharply defined. Right image: upper and lower right canines, second and first incisors; upper and lower left first and second incisors, and canines. Their crowns are still forming. Upper second incisors initiate crown development much later than lower incisors and this is apparent from the lesser extent of their development. Neolithic specimens from Çatalhöyük, Turkey.

Evolution of methods for studying dental development

Logan and Kronfeld, Schour and Massler, and dental atlases

William Logan was Dean of Loyola University Dental School in Chicago (Kremenak and Squier, 1997) and, during the 1930s and 1940s, attracted a number of researchers from Vienna Medical School, which at that time was a world leader in dental histology. One of these was the young Rudolf Kronfeld. Logan wanted to improve surgical techniques for the repair of cleft palates and, with Kronfeld, collected 30 specimens of the face from autopsies of children. Most were under 2 years old when they died, with a few at various ages up to 15 years. The specimens were demineralised in acid as complete blocks of tissue, impregnated with colloidin and sectioned in a microtome. The resulting serial sections, stained with haematoxylin and eosin, were some of the largest ever made and represented a technical triumph. Their report on jaw development (Logan and Kronfeld, 1933) included a table of ages for permanent tooth development. Kronfeld (1935) later published a schedule for the development stages of deciduous and permanent teeth. He referred to his work with Logan, but did not explain where his figures came from. The inclusion of deciduous teeth is puzzling because no foetal specimens are mentioned in their study (Smith, 1991c). Their key new observation was that the permanent upper second incisors initiated development much later than the other incisors (Logan and Kronfeld, 1933, p. 419).

Kronfeld died in 1940, but in 1941 the famous *Schour and Massler dental development chart* was published. Isaac Schour was also part of the Vienna group on the

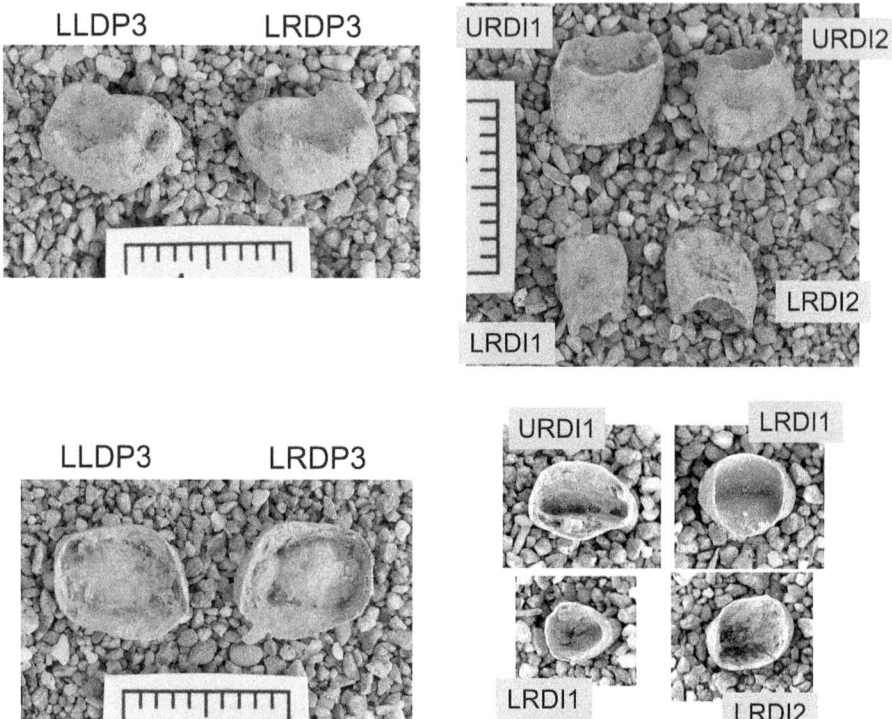

Figure 3.10 Developing deciduous tooth crowns from a human neonate. Left images: left and right deciduous third premolars. Upper image shows the cusps of the occlusal surfaces and the growing lingual crown side. Lower image shows the inside of the crown with the concave developing dentine surface exposed – the four depressions are the diverticles (pulp horns) underlying the cusps and the connecting grooves underlie the positions of the occlusal ridges. Right images: upper and lower right first and second incisors. Upper image shows the lingual surfaces with, clockwise from top left: upper first incisor and second incisor; lower second incisor and first incisor. Lower image shows the inside of the crown with, clockwise from top left: upper first incisor; lower first incisor; lower second incisor; upper second incisor. The developing edge is uppermost and the wedge-shaped concavity inside is the surface of the developing dentine. Specimens from the Kylindra cemetery site, island of Astypalaia, Greece.

faculty of the Loyola Dental School. He was a laboratory scientist as well as a dentist, with an interest in dental development. Maury Massler was a paediatric dentist at the University of Illinois College of Dentistry, also in Chicago. They published an extended account (Schour and Massler, 1940a; 1940b) of the development of the dentition, including a table giving a chronology for which they cited Logan and Kronfeld's work. It is similar to Kronfeld's (1935) table with some unexplained slight modifications. Schour and Massler also included an early draft of their development chart, which was later published with a short summary article (Schour and Massler, 1941) and as a coloured wallchart by the American Dental Association. It is the classic example of the 'atlas' approach, in which the state of tooth development and eruption are summarised as diagrams for each year from birth to 12 years

Tooth formation

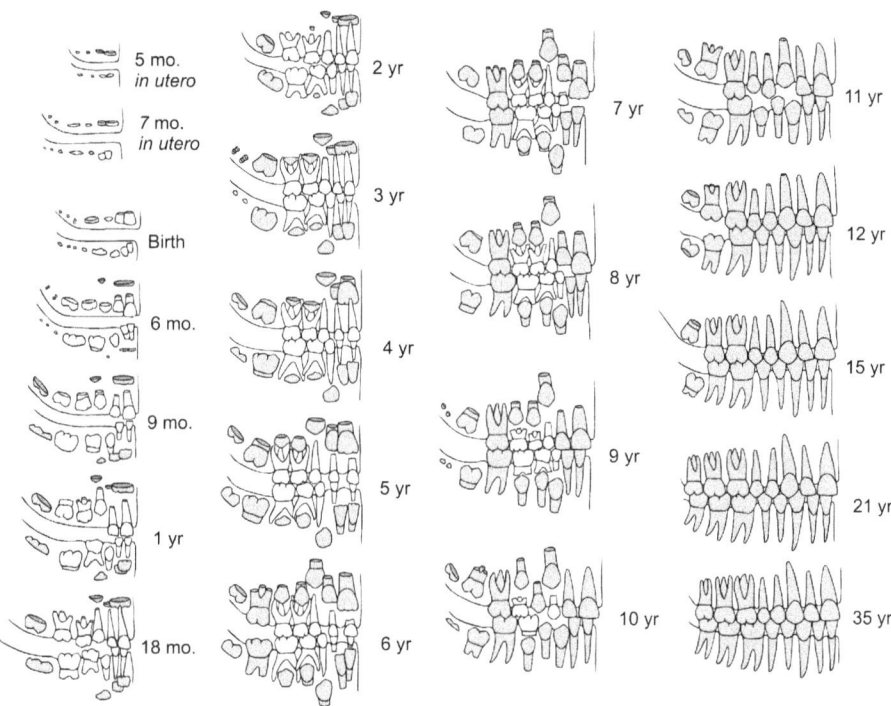

Figure 3.11 Schour and Massler dental development chart. Redrawn from their American Dental Association chart (Massler and Schoer, 1944), reproduced in Brescia (1961), Figure 25, pp. 76–77. Originally published in Hillson (2005), Figure 3.2, p. 224.

of age (Figure 3.11). Later versions (Massler and Schour, 1944) include the range of variation for the age of each stage as plus or minus figures. These were revised in the version of Ubelaker (1978), which is widely used in anthropology. Both the Schour/Massler and Ubelaker versions are the basis for age estimation in both forensic investigations and bioarchaeological work. They perform relatively well in comparison with other common methods (Liversidge, 1994), although anyone who has used them is well aware that few dentitions fit exactly into any one of the stages. In spite of the body of new data accumulated since 1941, the first entirely new chart of dental development, the London Atlas, was not published until 2010 (AlQahtani et al., 2010). It is divided into stages of six months each and is available in many languages for download as a pdf file: (www.dentistry.qmul.ac.uk/atlas%20 of%20tooth%20development%20and%20eruption/index.html).

There are issues with using any part of the developmental sequence to estimate age-at-death. Schour and Massler developed their original atlas as an aid for planning dental treatment: a simple guide to the state of development that could be expected in a child of a given age. It has grown far outside this original purpose and its use for age estimation makes a number of assumptions. Firstly, it assumes that the stages of dental development shown in the atlas are a good representative for any individual child. However, as anyone who has tried to find a stage to match an archaeological child's dentition can confirm, this is not so. It is rare to find a picture on the atlas that exactly matches – often two stages might suit equally, and different observers frequently make different choices. This is apparent when the results are compared with other measurements of tooth development discussed later in this chapter. This is partly because the atlas pictures are simply small cartoons and it is difficult to be sure what critical features are depicted. The London Atlas has gone some way to addressing this issue with tests of users. There is, however, a more fundamental problem that estimation naturally assumes that the ages given for stages are representative of the individual whose age is being estimated. Although it does appear that the average age at attainment of dental development stages (particularly the early ones) varies considerably less between populations than, say, equivalent skeletal development stages, individuals within those populations can depart in both timing and sequence from the average. This is what makes it so important to build detailed development sequences for individuals, as described in Chapter 5.

Histology and foetal tooth germs

Bertram Kraus was an anthropologist by training, but came to prominence in dental research and was interested in the ontogeny and morphology of human molar teeth (Krogman, 1970). He accumulated a large collection of specimens from human embryos and foetuses, sent to him from hospitals throughout the USA. The specimens were accompanied by crown-rump measurements from which Kraus and Jordan (1965) could estimate gestational ages (page 10) and they selected 787 dentitions between 10 and 36 weeks gestation. They removed the dental follicle and overlying layers of the enamel organ to leave the internal enamel epithelium lying on top of the enamel matrix which had been formed to that point. The calcium in the enamel and dentine was stained bright red with Alizarin Red S (page 114). A medical artist prepared drawings of the mineralised (calcified) tissue. The study remains the most detailed three-dimensional study of dental development ever carried out. They defined different morphological stages for the development of the deciduous premolars and tabulated them against estimated gestational age. The method can only work for the early development of teeth and would not, for example, be suitable for showing the completion of the crown because it could not distinguish between dentine and enamel matrix.

One crucial question is the order in which the cusps of the deciduous premolars and permanent molars start to form (Definition Box 2). Kraus and Jordan (1965) established this for humans and the cusps form in the same order in rhesus macaques (Swindler and McCoy, 1965), olive baboons (Swindler et al., 1968) and chimpanzees (Siebert and Swindler, 1991). It appears that the cusp development order is a primitive trait in the primates and it seems reasonable to assume that all hominins had a similar order. Kraus and Jordan gave little attention to the chronology of incisor and canine development, but were able to show that the initial formation of mineralised tissue started at the centre. Canines have a single central point, in effect a cusp, from which ridges spread to mesial and distal to make a cutting edge. The first increments of enamel and dentine are laid down under the

Tooth formation

central point. When they are first erupted, incisors have three or more low bumps along their cutting edge. These are called *mamelons* and they are rapidly worn away once the tooth is in the mouth. Once again, the first increments of dental tissue are laid down under the position of the central mamelon. Mineralisation then spreads to the mamelons to mesial and distal which develop like 'shoulders'.

Definition Box 2. Orientation of cusps and their order of development

A tooth has four sides. The *lingual side* faces the tongue (also called *palatal* in upper teeth), whereas the *buccal side* of molars and premolars faces the cheeks. The equivalent for incisors is the *labial side*, which faces the lips. Of the two sides which meet other teeth in the same jaw, one faces along the dental arcade towards its midline and is called the *mesial side*. The side facing the opposite way is called the *distal side*.

The four main cusps of *upper* molars and deciduous fourth premolars can be most simply labelled using these terms, but they also have other names, and they are initiated in the following order:

1. Mesiobuccal cusp, Cusp 2, the paracone
2. Mesiolingual cusp, Cusp 1, the protocone
3. Distobuccal cusp, Cusp 3, the metacone
4. Distolingual cusp, Cusp 4, the hypocone.

The five main cusps of *lower* molars and deciduous fourth premolars are similarly initiated in the following order:

1. Mesiobuccal cusp, Cusp 1, the protoconid
2. Mesiolingual cusp, Cusp 2, the metaconid
3. Centrobuccal cusp (between mesiobuccal and distobuccal), Cusp 3, the hypoconid
4. Distolingual cusp, Cusp 4, the entoconid
5. Distobuccal cusp, Cusp 5, the hypoconulid.

The permanent first molar is a key tooth for building development chronologies (Appendix A, Table 14). In human babies, Christensen and Kraus (1965) found that the mesiobuccal cusp was the first to show mineralised enamel and dentine after the 28th gestational week, but only in 30% of foetuses in their sample. This percentage gradually increased, so that by 36 weeks gestation, all showed dentine and enamel in the mesiobuccal cusp and, in some cases, the mesiolingual and distobuccal cusps. Using the same material, Butler (1967) illustrated a first molar in which four cusps had been initiated in a foetus of 42 weeks gestation. Christensen and Kraus (1965) found that the lower first molars were always in advance of the uppers. Thus the jaw of a neonate skeleton excavated from an archaeological site should yield tiny narrow conical caps of enamel and dentine for at least one cusp of the permanent first molars. They are, however, extremely difficult to find. In a well-preserved jaw of a neonate, the bony crypts for the permanent first molars are as well defined as those for the deciduous teeth. Even when a full set of the delicate part-formed deciduous teeth can be gently excavated from their crypts, experience over many years has shown that even the most careful emptying of the first molar crypt fails to reveal the developing cusps. They must be extremely small and delicate. As birth varies in relation to gestational age (see Definition Box 1) the state of development of the permanent first molar must be expected to vary between individual babies at birth. All babies with at least 36 weeks of gestational development at birth would be expected to show enamel and dentine under the mesiobuccal cusp. Some newborn babies will show these tissues under both mesiobuccal and mesiolingual cusps, although these cusps will not be large enough to have grown together. Still other newborn babies, particularly larger ones, will show enamel and dentine under three or even four cusps. This is also important information for understanding the position of the neonatal line seen in sections of permanent first molars, which forms an important marker (page 100).

Most newborn chimpanzees also show deposition of mineralised tissue under two or three cusps in their permanent first molars, suggesting that they generally start between 160 and 225 days (full term) gestational age (Siebert and Swindler, 1991). In orangutan too, three permanent first molar cusps show dentine and enamel formation in newborns (Winkler and Swindler, 1990; Winkler et al., 1991) and dissection has also revealed mineralised tissue formation in a permanent canine tooth germ (Winkler, 1995). This is a very early initiation for a canine, even taking into account the long formation time of this tall tooth crown in great apes. In the rhesus macaque the permanent first molars start deposition at 120 days gestation, where 168 days is the average gestational age at birth (Swindler and McCoy, 1965).

Kraus and Jordan further recognised that a variety of development states could be expected for deciduous premolars due to variation in gestational age at birth (Appendix A, Table 11). At a normal full-term birth (Figure 3.13), however, the third premolars would be expected to show a complete cap of mineralised tissue incorporating all the cusps and their connecting ridges, and extending some way down the crown sides. The cusps appear narrow and sharp, the occlusal surface is thin and there are no clear fissures – this is because more enamel matrix is to be added on top. Deciduous fourth premolars would be expected to show a ring of small, sharply pointed cusps connected by ridges, with an open area at the centre of the occlusal surface. Kraus and Jordan did not consider the deciduous canines and incisors, but evidence for their state of development at birth comes from the positioning of neonatal hypoplastic defects (page 192), as well as archaeological examples where all the teeth are preserved. At full-term development, it would be expected that the deciduous first incisor crown would be almost complete, with the second incisor crown perhaps two-thirds completed. Only the central point of the deciduous canine would be developed, with part of the mesial and distal ridges (as discussed earlier).

Tooth formation

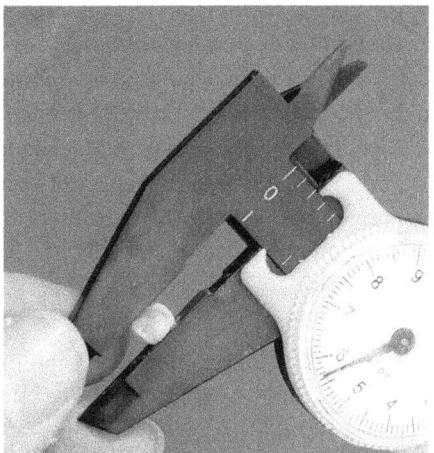

Figure 3.12 Measuring the development height of a human deciduous incisor. Simple plastic calipers are used because they do not scratch the delicate tooth. The developing edge of the tooth is placed flat on one jaw, which is then gently closed onto the highest point of the crown. The measurement is read from the dial to the nearest 0.1 mm.

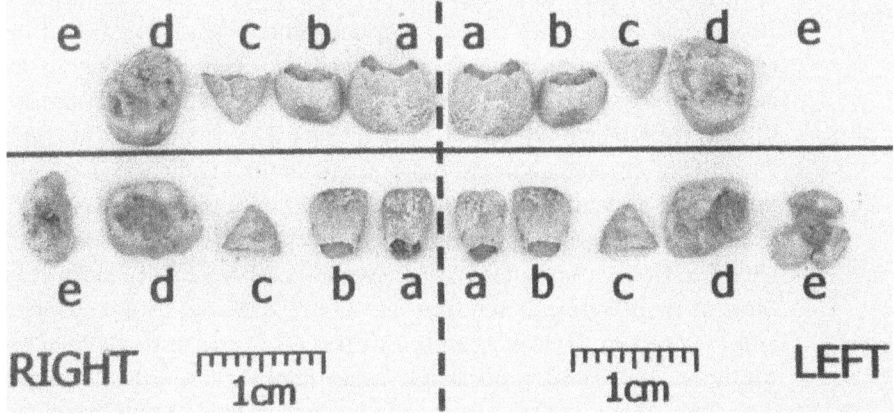

Figure 3.13 Deciduous dentition in a human neonate. Specimen from the Kylindra cemetery, island of Astypalaia, Greece. Upper crowns are above the line and lower below it. 'a' Denotes first incisors, 'b' second incisors, 'c' canines, 'd' third premolars and 'e' fourth premolars. Incisors and canines are showing their labial surfaces. Premolars are showing their occlusal surfaces.

The serial sectioning approach of Kronfeld discussed here was taken further at the University of Tokyo. Nomata (1964) investigated the schedule of deciduous crown formation in 140 human foetuses collected from Japan, using crown-rump length to estimate gestational age (Appendix A, Table 11). He confirmed the order of cusp initiation in deciduous premolars and permanent first molars, as well as the initiation of incisors under the central mamelon. Tadahiro Ooë (1981), Professor of Anatomy at Tokyo University, used serial sections of 172 embryos and foetuses, together with 16 infants, to create three-dimensional reconstructions of the epithelium in the dental lamina and tooth germs for both deciduous and permanent teeth. His procedure was laborious, stacking cardboard cut-outs of outlines drawn from the sections. This important study documented the initial development and organisation of tooth germs, and the developmental relationships between dentitions (page 30). It is one of the main pieces of evidence for the initiation of permanent incisor and canine formation during the months after birth, which is poorly known from radiographic studies (Appendix A, Table 16). Sunderland et al. (1987) also cut serial sections of maxillae and mandibles from 121 foetuses in the pathology collection at Sheffield Children's Hospital, UK. All specimens were documented with a pathologist's estimate of gestational age, ranging from 10 to 26 weeks. Their estimates for the initiation of deciduous tooth formation are a little later than those of Kraus and Jordan (Appendix A, Table 11).

Measurements of developing tooth height after birth

Deutsch et al. (1985) directly measured the height of developing deciduous incisors and canines in a collection of teeth from foetuses, neonates and infants from 5–10 months gestational age up to 1 year after birth. The measurements were strongly correlated with age and they derived regression equations from which gestational age could be estimated. This method is simple to apply to archaeological material (Definition Box 3), where there are often substantial numbers of partly formed teeth which fall out of their sockets. The clearance of a crypt underneath Christ Church in Spitalfields, London (Molleson et al., 1993; Cox, 1996) provided known age-at-death specimens against which development measures could be calibrated. The burials (AD 1729–1856) included 383 named individuals who could be identified from coffin plates, so their ages-at-death were independently known. Fifty-three of these were children from birth to 2 years and 28 were children between 2 and 19 years old. Liversidge and co-workers (Liversidge et al., 1993; Liversidge and Molleson, 1999) took tooth height measurements, not only of the deciduous incisors and canines, but also the deciduous premolars and all permanent teeth. Once again, they were highly correlated with age and it was possible to derive regression equations for estimating age from birth to 5.4 years (Appendix A, Table 13).

Developmental tooth heights are strongly correlated with one another (Figure 3.14), particularly between anterior teeth. It is therefore possible to represent dental development effectively by choosing one tooth. The deciduous upper first incisor is a good choice in archaeology because, as the crown is nearly complete at birth, it is more robust than the other deciduous teeth and is generally the best preserved. There is considerable variation (Figure 3.15) in measurements for teeth from dentitions assigned to different stages in the Schour and Massler chart (discussed earlier). It is therefore apparent that the measurements are a considerably sharper instrument for recording development.

Tooth formation

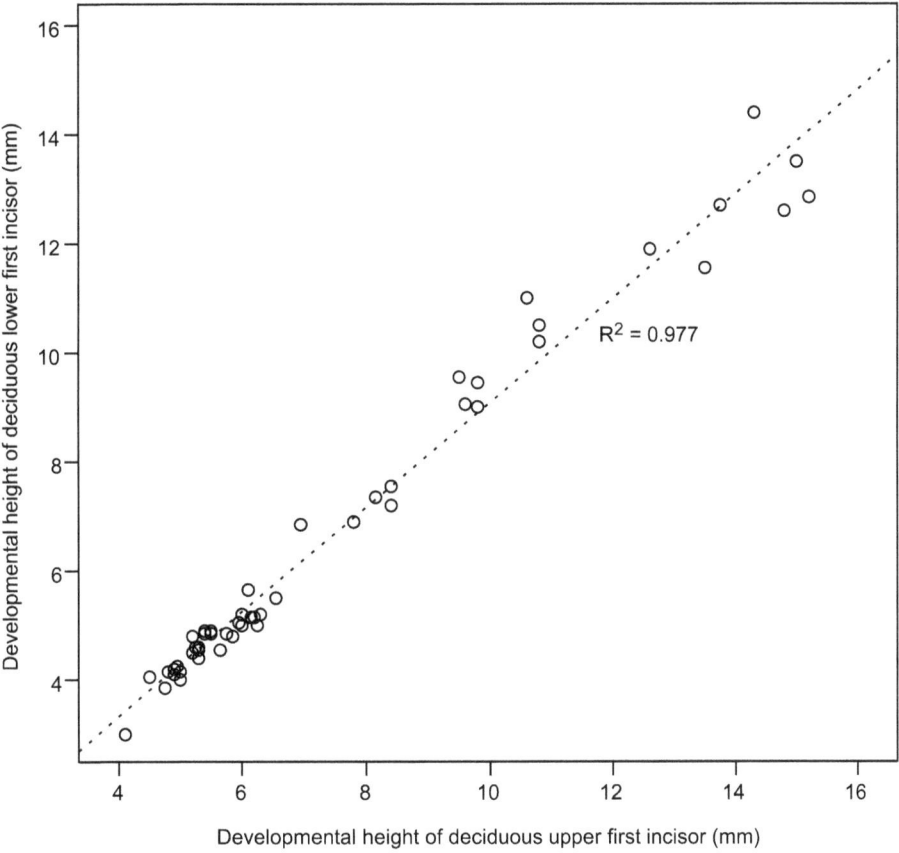

Figure 3.14 Developmental height of human deciduous upper first incisors plotted against deciduous lower first incisors. Data from the Neolithic site of Çatalhöyük, Turkey.

Definition Box 3. Developing tooth height measurements

Teeth can be measured at any development stage, when either the crown or the root are still forming. Deutsch et al. (1985, p. 203) measured tooth height 'perpendicular to the plane of the growing end (i.e., the cervical margin in teeth which contain no root, and root end in teeth where root formation had already begun) to the incisal tip, to an accuracy of ±0.1 mm'. In practice, the simplest method is to use plastic calipers, which are less likely to scratch than steel. Place the developing edge flat on the moving jaw of the calipers and close up gently on the highest cusp tip (Figure 3.12). This works well for all incisors and canines, permanent premolars and molars, where the developing edge sits flat on the caliper jaw. The exceptions are deciduous third premolars because their crowns have a mesiodistal bulge (the tubercle of Zuckerkandl), which creates a sinuous developing edge. If the developing crown is intact, then the measurement can be taken in the ordinary way, but these crowns often break at their narrowest point, which makes it impossible. Growing molar roots tend to splay, with their individual developing edges at different angles, in which case a simple maximum height from their lowest point to the cusp tips has to suffice.

Experience with teams in field laboratories suggests that dental development is recorded more consistently by such measurements than by scoring with the MFH (page 61) stages. The measurement should not be taken if too much of the delicate developing edge has chipped away. Where both sides of the dentition are preserved, it is possible to keep a running check because measurements of isomeres usually match exactly. In addition, the measurements are highly correlated between teeth and a simple scatter plot quickly highlights anomalous values.

In order to match with standard dental radiographs, Liversidge and Molleson (2004, pp. 172–3) proposed an approximal crown height 'defined as the distance from the cusp tip (or occlusal level) to a line drawn between the mesial and distal crown-root margins' taken with the sharp tips of the caliper jaws.

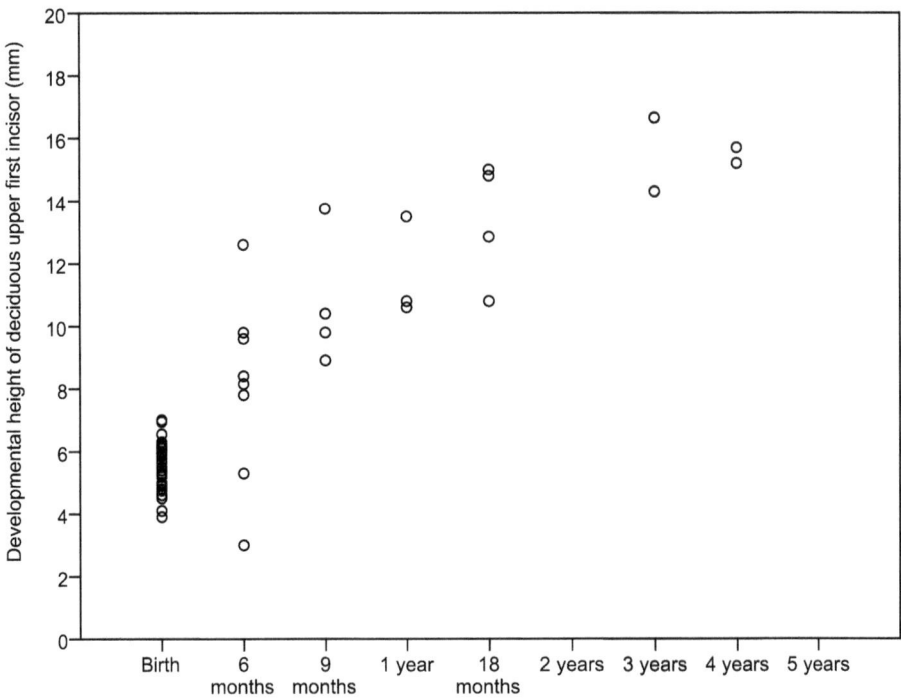

Figure 3.15 Schour and Massler stages plotted against development height of deciduous upper first incisors. Data from the Neolithic site of Çatalhöyük, Turkey. See Figure 3.11 for the definition of stages.

The Fels longitudinal growth study, Montreal, Helsinki and London: large-scale radiographic studies of permanent tooth development

A number of large longitudinal growth studies was started in the USA during the years of the Great Depression to examine its effects on the nation's children. One of the most comprehensive was established by the Fels Research Institute at Yellow Springs, Ohio, in 1929. It has moved several times, but the study is still running today. A similar longitudinal study was run by Harvard University, following children from the Boston area. In both cases, regular six-monthly examinations included dental radiographs, with closer intervals during the first 2 years after birth. Inevitably there were gaps, particularly during the Second World War, but, together, the two studies have produced a large collection of dental radiographs in which development can be studied. A scheme (Figure 3.16) was devised for scoring the stage reached by different teeth in each radiograph, initially by Gleiser and Hunt (1955), but later refined by Moorrees, Fanning and Hunt (1963a). This MFH system has 13 stages, defined by line drawings of the outlines seen in the dental radiographs, with an extra stage for molars. Moorrees and co-workers scored the radiographs from 134 children from the Harvard study and 246 from the Fels study, with the series running from birth until early adulthood. The data set is incomplete in places. The projection used for the dental radiographs superimposed the left and right upper jaws over one another, which made it impossible to score permanent upper canines, premolars and molars. Both permanent upper and lower incisors were recorded, but only for the development of the root (crown completion as well in the upper teeth). Deciduous lower canines and premolars were also scored (Moorrees et al., 1963b).

Moorrees and co-workers used cumulative frequencies and probit methods (discussed earlier) to derive the mean age at attainment for each development stage, with associated standard deviations. The results were plotted as a series of charts which remain the most well-supported evidence for the dental development schedule (Smith, 1991c), but the original figures were not reported. This has caused problems for those who wished to create tables of development. The only solution is to scale the values from the published charts. Fortunately, Moorrees and co-workers distributed large format versions of the graphs to interested parties and some of these have survived, making it possible to scale means and standard deviations with precision (Harris and Buck, 2002, see Appendix A, Tables 16–18).

The largest single investigation of dental development was that of Demirjian and co-workers; a cross-sectional study of over 5000 children from Montréal in Canada (Demirjian et al., 1973; Demirjian and Goldstein, 1976; Demirjian and Levesque, 1980). They developed a system of eight dental development stages (Figure 3.17), three of which can be matched with stages in the MFH system: start of tooth formation; completion of crown; completion of root apex (Smith, 1991c). The aim of the project was to provide an overall maturity scale of dental development, like the scale used for skeletal development. Each child was scored for all lower teeth except third molars and the scores were summed to calculate the dental maturity score. This maturity score has been widely used for other populations (Liversidge et al.,

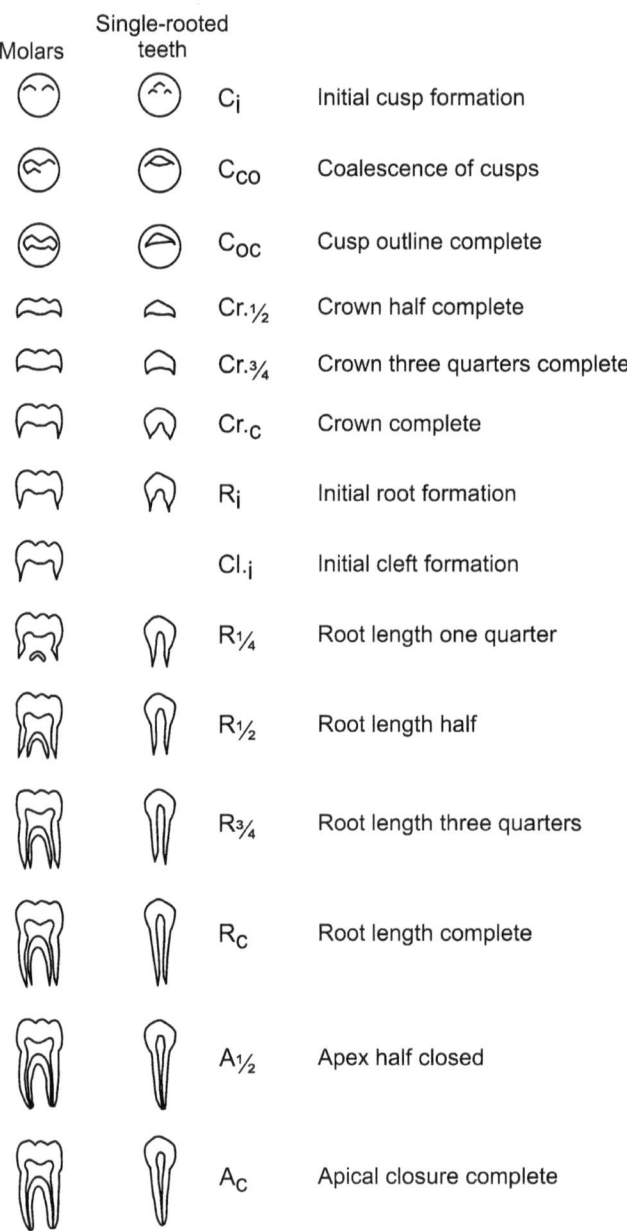

Figure 3.16 Moorrees, Fanning and Hunt (MFH) (1963a) scheme for dental development stages, redrawn from their Figure 1, p. 1492 and Figure 2, p. 1493, with descriptions from Table 1, p. 1492. There is a separate series of diagrams for single- and multiple-rooted teeth.

2006), but most appear to have been faster developers than the Montréal children. Liversidge (2010) instead used the individual Demirjian development stages for each tooth to score dental radiographs for children from Australia, Belgium, England, Finland, France, Korea, Quebec and Sweden. Only the later development stages of

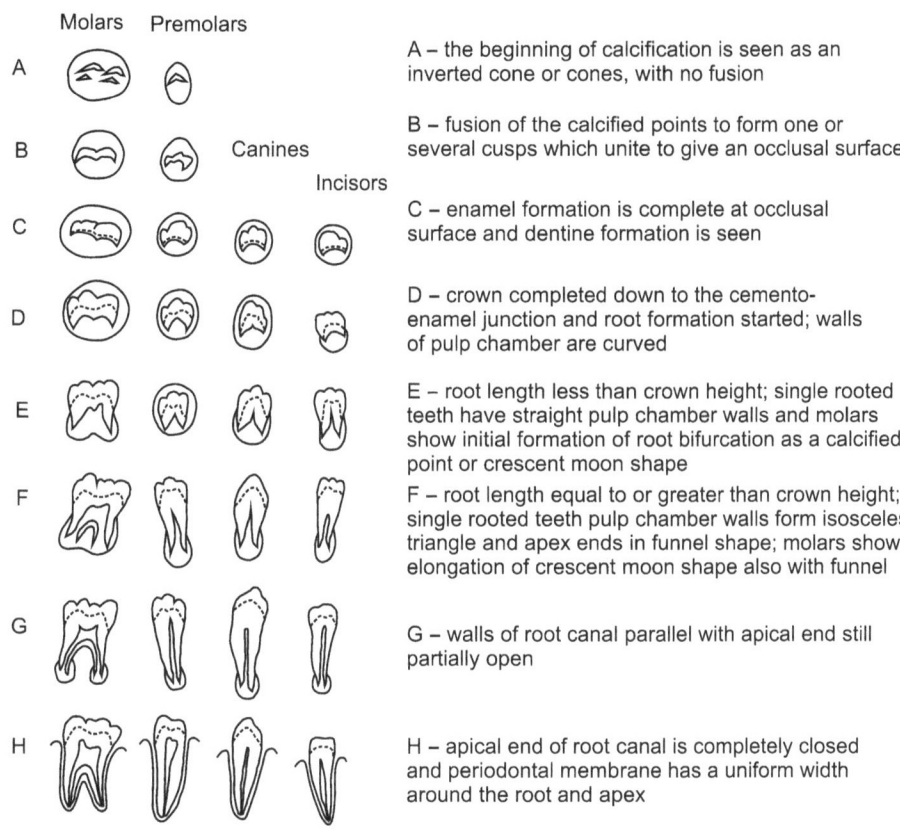

Figure 3.17 Demirjian *et al.* (1973) system for scoring development stages in dental radiographs, redrawn from Figure 1 p. 220, with descriptions from pp. 221–226.

lower teeth (except third molars) were scored, but they provide well-established statistics for age at attainment which are tabulated alongside those of Moorrees *et al.* (Appendix A, Tables 16–18).

The difficulty with the studies outlined here is that they include only lower permanent teeth. This issue can be addressed by the work of Haavikko (1970), who used panoramic tomography to capture the entire dentition in a single radiograph. This was a cross-sectional study of 1162 children and young adults, aged between 2 and 21 years, from Helsinki in Finland, with slightly more boys than girls in the group. Development stages not unlike MFH were used and median ages at attainment were estimated from cumulative frequencies.

Very few studies exist for post-natal development of deciduous teeth. Moorrees *et al.* (1963b) published graphs for deciduous lower canines and premolars from the longitudinal Fels growth study, based on six-monthly dental radiographs of 136 boys and 110 girls. Liversidge and Molleson (2004) made a composite cross-sectional radiographic study of 61 living children attending a London dental hospital, 121 children of known age-at-death buried in the crypt of Christ Church, Spitalfields in London (page 223), and a total of 133 Medieval children's jaws from two sites in Scotland. Data from boys and girls were combined, as it is practically impossible to distinguish them in archaeological material (Appendix A, Table 12).

Tooth development in humans

Deciduous teeth

All deciduous teeth are initiated *in utero* over a period of about 4 to 6 weeks. The earliest tooth to start forming mineralised tissue is the upper first incisor, for which the youngest estimate is 14 weeks gestational age (Appendix A, Table 11). This is soon followed by lower first and second incisors, upper second incisors, third premolars and finally canines and fourth premolars, which are initiated between 18 and 21 weeks gestation in most children. The amount of crown formed at birth varies with the length of gestation, but at normal full-term birth the incisor crowns are largely complete (Figure 3.13). In third premolars, the full occlusal surface is represented by a continuous sheet of mineralised tissue in which the cusps are not yet full height, so they appear narrow and pointed, without sharp fissures between them. Canines and fourth premolars are the least developed. The canines are represented by small caps which superficially look like the similarly small caps of the fourth premolar cusps, which are partially united into a ring by their connecting marginal ridges.

Both upper and lower first incisors, on average, complete their crowns 1–2 months after birth, followed by the second incisors at 3–4 months. There is variation, partly because the gestational age at birth varies, so in some individuals the first incisor crowns could be completed just before birth. The remaining teeth complete their crowns in the order deciduous third premolar, followed by canine, and then fourth premolar on average at 9–11 months after birth. In this way, the deciduous crowns are all completed during the first year after birth, although the range of variation in published figures suggests that some individuals complete their crowns early during the second year after birth. Completion of the roots follows a similar sequence, but the range of variation is greater and there is a marked difference between studies. It might, however, be expected that the root apices would be closed by the end of the fourth year after birth.

Additional permanent teeth: the molars

Few radiographs are taken during the first few years after birth and first molar crown development is not well known. Histological evidence (page 55) places its initiation at around 30 weeks gestational age, or 2.5 months before normal full-term birth. First molar crown completion (Appendix A, Table 17) varies, on average, between 2 and 3.5 years after birth. Initiation of the permanent second molar is reported on average between 3.5 and 4 years after birth, whilst crown completion age varies between 6 and 7.5 years. There are few consistent differences between upper and lower teeth, but girls tend on average to attain their development stages at younger ages than boys.

The human third molar represents a special problem because it is so variable in its development, perhaps more so than any tooth in any other primate. At around

19 years of age, it is also probably the tooth which erupts at the oldest age for any primate, and at the longest interval after the other teeth – about 6 years after the rest of the dentition is completed. Not only is the tooth interesting from the standpoint of human evolution, but it is also legally important. If 18 years is the age of majority then, as the third molar is the only tooth still showing developmental changes, it is important for estimating age in forensic investigations.

Two large forensic studies, Mincer *et al.* (1993) and Blankenship *et al.* (2007), did not use statistical methods compatible with the other studies described here. Harris (2007) did not include completion of the root apices. The largest comparative study using cumulative frequency methods is that of Liversidge (2008), which included both initiation and completion of crown and completion of roots, but concentrated only on lower third molars. In most instances, the initial deposition of dentine and enamel took place during the ninth to tenth year after birth, but the youngest child was 6 and the oldest was 14 years old. The crown was generally complete (and the root initiated) during the twelfth or thirteenth year and the apices of the roots were fully closed during the nineteenth or twentieth year after birth. Associated standard deviations varied from 0.8 to just over 1 year. Black children from South Africa achieved developmental stages in their third molars consistently younger than Cape Coloured children from Cape Town and white and Bangladeshi children from London. These studies suggested that boys attained their development stages in third molars earlier than girls, in contrast with all other teeth in the dentition, but Liversidge found that this varied between populations. In view of their later, larger growth spurt, it might make sense for boys to be in advance of girls, particularly in the root stages of third molar development, but it does not appear that this is consistently so.

Successional permanent teeth: incisors, canines and premolars

The development of permanent incisors and canines in living children is not very well known. Taking Ooë's (1981) figures (Appendix A, Table 15) it would be expected that both upper and lower first incisors, and lower second incisors and canines, would start to form dentine and enamel soon after 3 months of age. In the case of the lower canines this is a little earlier than the mean age of 0.5 years given by Moorrees *et al.* (1963a) (Appendix A, Table 16). Ooë's observations suggest that the upper second incisor and canine initiate later than the lowers, after 8–9 months of age. This is supported by Haavikko's study (1970), which suggested that both these teeth completed their crowns later than the lower second incisors and canine. Estimates for the completion of incisor crowns are, however, in short supply and rather variable, with averages ranging from 3.3 to 5 years for upper and lower first incisors. The incisor crowns are particularly important in the discussions that follow and, paradoxically, they may be better known from fossil material than they are in living children.

Permanent premolar crown initiation is known only from Moorrees *et al.*'s study of the lower dentition (Appendix A, Table 16). Their mean age for third premolar initiation was around 3.8 years, and 5 years for the fourth premolar. Crown completion for third premolars was on average 5 years or a little later and, for fourth premolars, between 6 and 7 years.

More data are available for the closure of the root apex of the permanent successional teeth, but they show considerable variation. Ages for the lower incisors are the youngest, at around 8 years for the lower first incisor and 9 years for the second. Haavikko's figures (1970) suggest that the median age of apex closure in upper incisors is between 10 and 11 years. The canines and premolars complete their apex, on average, at around 13 to 14 years, with the fourth premolars last.

Tooth development in non-human primates

Dean and Wood (1981) made the first comparative survey of the timing and sequence of tooth crown and root development in great apes. They radiographed 175 skulls of juvenile chimpanzees, gorillas and orangutans from UK museums. None of these specimens had an independently known age-at-death, but they ranged from individuals with a partial deciduous dentition, to individuals with all teeth fully formed except for the canine and third molar roots. From the coincidence of development stages in different teeth and the assumptions that the first molar would start to form at birth and that molar crowns would take 2.5 years to be completed, they constructed a comparative chart of dental development which has formed the basis for much discussion ever since. It was not until the 1990s that known-age animals were studied.

Tooth development in chimpanzees

Chimpanzees have by far the best known dental development of all non-human primates but, even for them, knowledge of crown and root formation is limited to the lower dentition. There are two radiographic studies of captive animals from laboratories in the USA (Appendix A, Tables 19–21; Figure 3.8). The results match reasonably well, even though their statistical basis was somewhat different.

Deciduous tooth development is not known for chimpanzees. It is, however, known that permanent first molars are initiated before birth (page 56) although, because they did not include foetuses, both radiographic studies suggested initiation 1 or more months after birth. The first molar crowns were completed around 1.5 years after birth and closure of the root apices between 6.3 and 7.3 years. The first evidence of dentine and enamel deposition under the cusps of the second molars was around one to 1.4 years after birth. Crown completion was between 3.7 and 4.2 years and, on the basis of only one individual, the root apex was complete before 10.7 years. The third molars were initiated about 3.5 years and crown completion at 6.9 years, but neither study had individuals old enough to show the apex completion stage.

For the permanent lower incisors and canines, the midpoint age for crown initiation was 0.4 years after birth. Both incisors also completed their crowns at the same age (3.1 years), well after first molar crown completion. Chimpanzee canines are much larger teeth than the incisors and they took longer for their crowns to complete, at around 5.4 years in females and 6.7 years in males. This sexual dimorphism in canine development is discussed further in Chapter 5. Midpoint ages of attainment for root apex closure in the first incisors, second incisors and canines were 8.6, 9.2 and 10 years, respectively. For both premolars, mineralised tissue was first visible at the cusps at 1.3 years and crown completion at 3.9 years in the third premolar and 4.6 in the fourth.

Tooth development in orangutans

As with chimpanzees, dissection of neonates has confirmed that dentine and enamel start to form before birth under the permanent first molar cusps (page 56). Incisor crown initiation at birth has not been observed, but one individual showed the initiation of a permanent canine. It may be that this is characteristic of the orangutan, but it may also arise from methodological differences, or individual variation. A radiographic study (Winkler et al., 1996) of 102 museum specimens without independent evidence of age-at-death has tabulated development in different permanent teeth against the stages reached in the permanent first molar (Appendix A, Table 22). This shows that, as in chimpanzees, the development of second incisors is delayed relative to the first incisors. It also suggests an overlap in formation time between the first and second molar crowns. In some individuals the third molar crown may be initiated at around the time the second molar crowns are completed, but in others they may overlap to a greater extent.

Tooth development in macaques and baboons

In living monkeys, permanent tooth development only has been studied in detail for two captive colonies at the Regional Primate Research Center Field Station at Medical Lake in Washington, USA. One followed pig-tailed macaques (*Macaca nemestrina*) and the other yellow baboons (*Papio cynocephalus*). The animals were X-rayed every three months for the first three years and then every six months for the rest of their development. The mean age for stage was calculated for initial calcification, crown completion and root apex closure (Appendix A, Tables 23 and 24).

In both species, the full permanent dentition formation sequence was completed at 6.6 years after birth with the closure of the third molar root apices. Permanent first molars were already initiated at birth and the mean age of crown completion was 0.8 years after birth. Second molar crowns started to form around 1 year of age and were completed by about 2 years. Third molar crowns were initiated at 3 years or a little younger, and completed by about 4 years. It is therefore apparent that molar crown formation timing does not overlap in these monkeys.

Incisors, canines and premolars were recorded only in the baboons, in which incisors started to develop around six months after birth and were completed about 2 years of age, considerably after the first molars. The tall canine crowns were initiated at about 11 months of age and completed in females around 2 years and males at 3 years. This strong dimorphism in timing is matched by the pronounced dimorphism in canine size in baboons.

Summary

Gingival emergence, which is visible by a simple examination of the living mouth, is by far the best known aspect of dental development. Other stages of eruption, the initiation of mineralised tissue formation and the completion of crown and root are known only from radiographic studies of living people and other primates, together with limited investigations of dissected tooth germs and sections of developing jaws. Altogether, dental development is much better known in humans than for any other living primate. Even so, there are deficiencies in knowledge; for example, in the timing of incisor development which is a key tooth in Chapter 5. One of the main lessons to be learned from large studies of living human populations concerns variations in the timing of developmental stages. For any one tooth, the earliest stages of its development are the least variable, and the latest stages the most variable. The earliest tooth to form in the dentition is also the least variable, and the latest tooth the most variable. As with other aspects of development, girls are in advance of boys for much of the sequence, but the difference is small for teeth. Similarly, even though there are population differences (for example, Africans on average tend to attain their development stages earlier than non-Africans), these differences are small and the variation within populations may match the variation between them. In general, dental development appears to be less variable than skeletal development, but it still needs consideration when judging studies including just a few individuals, or indeed single fossils. How representative can average ages at attainment be when any one individual might be at the extreme of variation? This is the rationale for using the histological methods discussed in Chapter 5, which focus on building detailed schedules for individuals, rather than statistics for populations.

To answer the question posed in the introduction to this chapter, the human dental development schedule is indeed prolonged compared with other primates. At around 20 years, the human eruption sequence is twice as long as that of the other large primates (chimpanzees, gorillas and orangutans), which in turn have longer sequences than any of the smaller species. This elongation of the sequence is mirrored in all stages of development; for example, in a greater interval between the initiation of molar crowns. In humans, first molars are initiated just before birth, second molars at 3.5 to 4 years and third molars at 9–10 years. In chimpanzees, the first molars are again initiated before birth, the second molars at about 1.3 years and third molars at 3.5 to 4 years.

The developing human dentition, however, is also distinctive in the order in which teeth develop. Human permanent incisors initiate and complete their crown development a little later than first molars, but they erupt at nearly the same age. In chimpanzees, the first molar crowns are completed a good year earlier than the permanent incisor crowns and they erupt about two years earlier. The result is that the permanent incisors erupt at around 6 years of age in both species. In a human child of that age, the deciduous incisors would normally be in the process of exfoliation and replacement by their permanent successors. For that reason, the state of wear on the permanent incisors and first molars should be similar in a slightly older

Summary

child. In a 6-year-old chimpanzee, the first molars would already be fully erupted and in wear, and should therefore be considerably more worn than the permanent incisors in an older animal.

In both species, the permanent premolars and second molars overlap extensively in their development and eruption sequences. This is the next main stage in development because the permanent premolars replace the deciduous teeth and the second molars extend the dental arcade. In chimpanzees they emerge about one year after the permanent incisors, whereas in humans they emerge a good four to five years later. Chimpanzees therefore show a considerably more rapid transition through the phase of mixed dentition.

Finally, the large permanent canines of chimpanzees develop over a slower schedule than in humans and they do not emerge until after the second molars have erupted, only a year or so before the eruption of the third molars. By contrast, human canines erupt at a similar age to second molars and permanent premolars. Less is known about gorillas and orangutans, but the pattern of eruption seems to be similar to that of chimpanzees.

4 Microscopic markers of growth in dental tissues

This chapter uses the example of human teeth because specimens are more readily obtainable, but the histological details are similar in chimpanzees, gorillas and orangutans.

The tooth surface

Take a permanent tooth from a human in your hand and examine it in good light. The crown is coated in white shiny enamel which overlies the yellow or brown of the root, often stained darker in fossils or archaeological specimens. Enamel, however, retains its glossy lustre in the great majority of ancient teeth. It is difficult to focus on the surface, which has a glass-like, translucent quality. Enamel is not actually a glass, but is very finely crystalline and contains microscopic discontinuities that scatter light back to the observer. Just below the surface, the light passes through without being much affected, but the deeper it penetrates, the more discontinuities it encounters and the more light is scattered, rendering the tissue bright white. To see features on the surface of the crown, it is necessary to tilt the tooth from side to side to make out the details amongst the bright reflections. In an unworn molar or premolar, the cusps form smooth mounds, separated in the occlusal surface by deep, narrow fissures. Incisors bear a row of small cusplets called *mamelons*, but these wear rapidly to make a smooth edge. The crown side bulges out below the cusps, then tapers gradually towards the root, which it meets at the *cervix* of the tooth. The boundary line with the root is the *cervical margin* of the crown. It is also the point at which the enamel of the crown meets the cement which coats the root surface, and so it may be called the *cement-enamel junction* (CEJ). In incisors and canines, the single root may bulge out a little below the crown before tapering down to the point of the apex. The multiple roots of molars branch out below the root trunk at the point known as a *furcation* and each root has its own apex.

Where development is complete, the apex of the root narrows down to a fine point in which there is a tiny *apical foramen* through which, in life, blood vessels and nerves pass to the pulp. If the root apex has a wider opening, with a wafer-thin, sharp edge, this shows that it was still developing at the time of death or when the tooth was extracted (Figures 3.1, 3.9, 3.10 and 4.1). The growing surface of dentine

The tooth surface

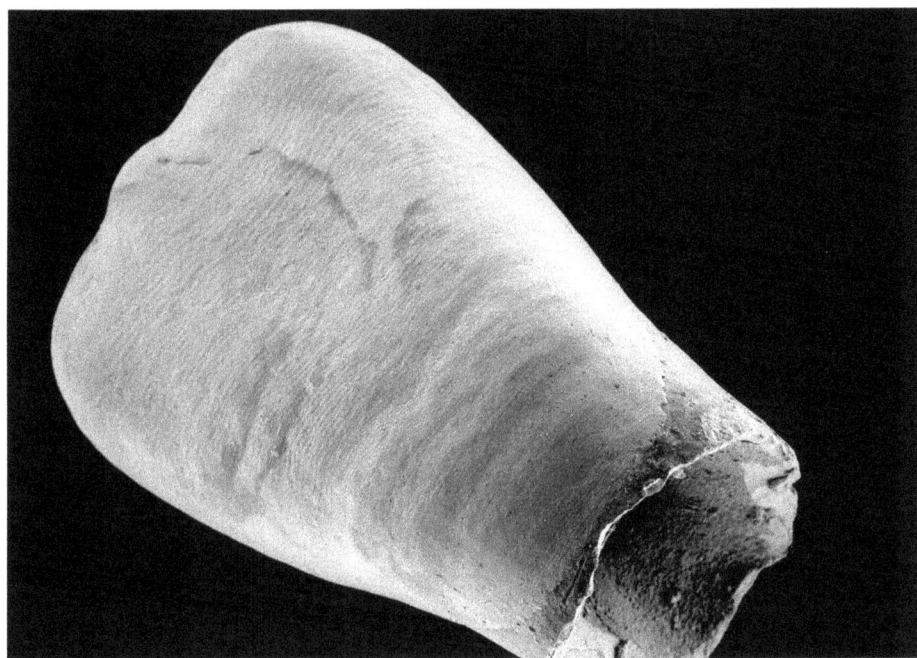

Figure 4.1   Unerupted, unworn human permanent lower first incisor with the incisal edge in the top left corner of the image and base of the crown lower right. Three low mounds, mamelons, are seen along the incisal edge. At the base of the recently completed crown, root formation has continued for a short while before death, with the forming surface of dentine left as a conical depression inside. The developing edge is chipped and abraded from contact with soil and no doubt excavation and cleaning. Perikymata cross the crown side and curve around to mesial and distal. The specimen was cleaned, but was not coated, and the image was produced in a scanning electron microscope under modest vacuum using an ESED (see Appendix B for details). Neolithic specimen from Çatalhöyük, Turkey. Field of view is 10.5 mm wide.

occupies a band lining the inside of this edge, angling sharply into the root. Where the root has not yet started to form, the developing crown has the appearance of a hollow cone. Enamel matrix being secreted at the time of death shows around the outer margin of this cone as a band of porous, irregular tissue with a fissured appearance. This represents the initial, poorly mineralised, enamel matrix before maturation (pages 32 and 87). A ring of dentine protrudes below this band because dentine development is always slightly in advance of the development of enamel.

Some living people show irregular white marks in their tooth enamel. This defect is called *hypocalcification* – the white marks are patches just under the surface that did not mineralise fully during maturation. They contain more light-scattering discontinuities than the surrounding tissue and thus appear whiter. Little is known about the cause but, although common in modern people, hypocalcification is rare in archaeological and fossil teeth. Perhaps it is lost through the process of fossilisation. There may also be dark stains on teeth, relating primarily to dietary, smoking or other habits. Most frequently, however, coloration of the surface comes from deposits of dental *calculus* or tartar. This is mineralised dental plaque, varying in colour from pale to dark brown or almost black. In life, calculus is strongly bound to the tooth surface but, in archaeological specimens, it more easily breaks away and fossils show calculus less commonly than teeth from living people, perhaps because it is removed when cleaning off matrix material.

Modern human teeth (and also teeth from chimpanzees and some monkeys) show lesions of dental caries. These are irregular cavities, commonly found at the contact points between crowns, in the fissures of molar occlusal surfaces, or around the cervix of the tooth. Often they are stained dark. Caries is much less common in prehistoric archaeological collections, and extremely rare in Pleistocene and Pliocene hominin fossils. Fractures of the crown are, however, frequently seen in these specimens, either post- or ante-mortem. The latter are distinguished by rounded, polished edges. Deciduous teeth often show signs of root resorption, a normal part of dental development in which the roots are removed to make way for the permanent teeth which are growing underneath. Resorption is recognised by its irregular edge and rough, acid-etched appearance, very different from the forming edge of a developing root. The process starts at the apex and ends near the base of the crown, to leave just a few remnants of root when the deciduous tooth is exfoliated and replaced by its permanent successor.

Teeth start to wear quickly once they erupt into the mouth. On the sides of the crown there is a general polishing, smoothing out the surface and reducing the prominence of surface features, known as *abrasion*. This is particularly apparent in modern teeth, where tooth brushing produces a lustrous gloss. The abrasion of archaeological or fossil crown sides is less strong, but they are often marked with coarse scratches, some large enough to make out with the naked eye. More general abrasion is due to small abrasive particles, such as grit from grinding stones, bones, soil or dust, or phytoliths from plant fibres. They are pulled across the enamel surface on food, the tongue or the cheeks, making microscopic marks which are so fine that the surface appears smooth to the naked eye.

Where the crown meets other teeth, either its neighbours in the same jaw or its opposite number in the upper or lower jaw, wear produces clearly defined facets (Figure 4.2). This tooth-on-tooth wear is called *attrition*. It creates facets at the contact points between neighbouring teeth (*approximal* or *interproximal facets*) and on occlusal surfaces where tooth meets tooth in the opposing jaw (*occlusal facets*). As wear progresses, the underlying dentine is exposed. This is pale yellow in living or recently extracted teeth, but in archaeological specimens it can vary from a darker yellow to brown or black, depending on the conditions of burial. Dentine is much softer and worn more rapidly than enamel, which is left standing as a raised rim. Archaeological and fossil teeth can be very worn, but the pulp chamber inside is not breached because it has a protective lining of secondary dentine. The living pulp retreats deeper inside the tooth and the advancing tooth wear eventually exposes secondary dentine at the top of what was once the pulp chamber. This can usually be recognised as a darker area within the dentine of the facet.

The tooth surface

Figure 4.2 Human permanent lower molars showing tooth wear. It is common for the molars to show a wear series from the last to first erupted teeth, with the third molar the least worn and the first molar the most. The polishing of wear facets where teeth meet is known as attrition. In the third molar the facet on the occlusal surface is confined to the mesial cusps, where small patches of the darker dentine are exposed, with only minor polishing on the distal cusp. In the second and first molars, large areas of dentine are exposed and the cusps have been removed. At the contact points between these molars are approximal or interproximal attrition facets where slight movements between teeth have polished flattened areas on their sides. Neolithic specimen from Çatalhöyük, Turkey.

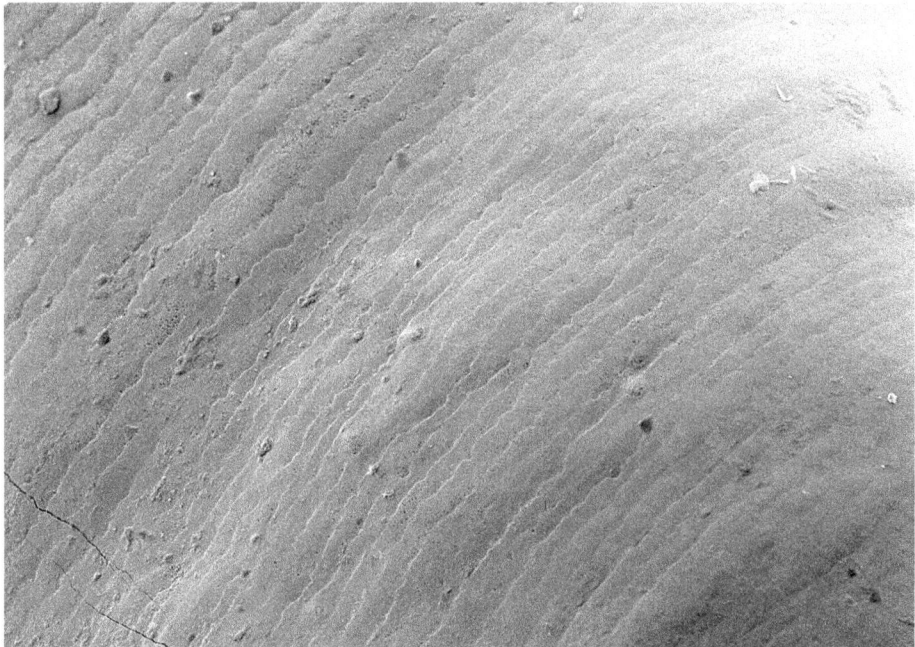

Figure 4.3 Closer view of the same specimen as in Figure 4.1 showing perikymata in the middle part of the crown. Variations in spacing are evident, both a general decrease down the crown side and local variations between neighbouring perikymata. Field of view is 1.7 mm wide.

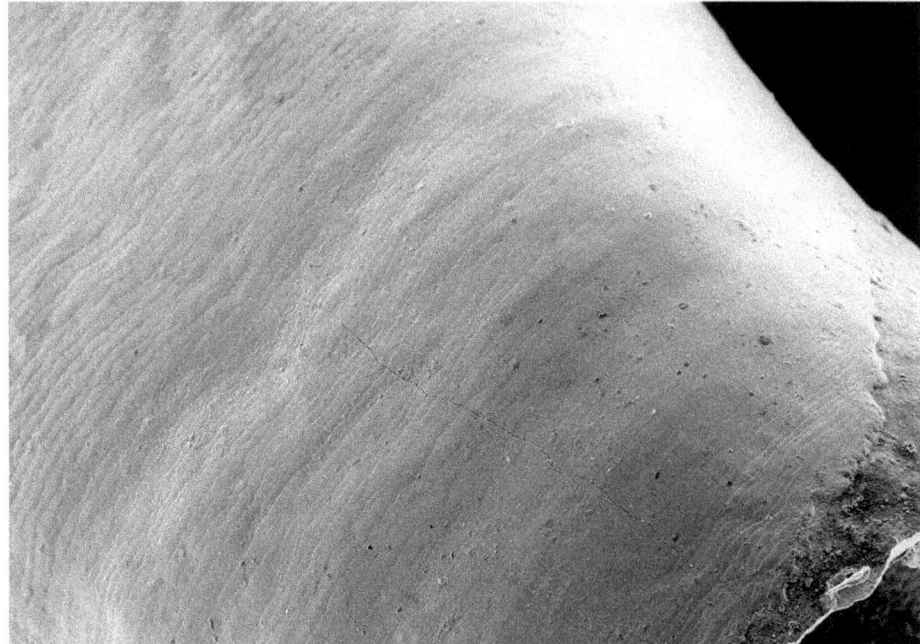

Figure 4.4 Closer view of the same specimen as in Figure 4.1 showing perikymata in the cervical part of the crown, with the CEJ at the base of the crown. Perikymata become much more closely spaced in this region and, in an incisor like this specimen, the curve of the CEJ up to distal and mesial cuts off the perikymata so they do not continue right the way around the circumference. There are two furrow-form defects of enamel hypoplasia about 1 and 2 mm up the crown from the CEJ. Field of view is 4.3 mm wide.

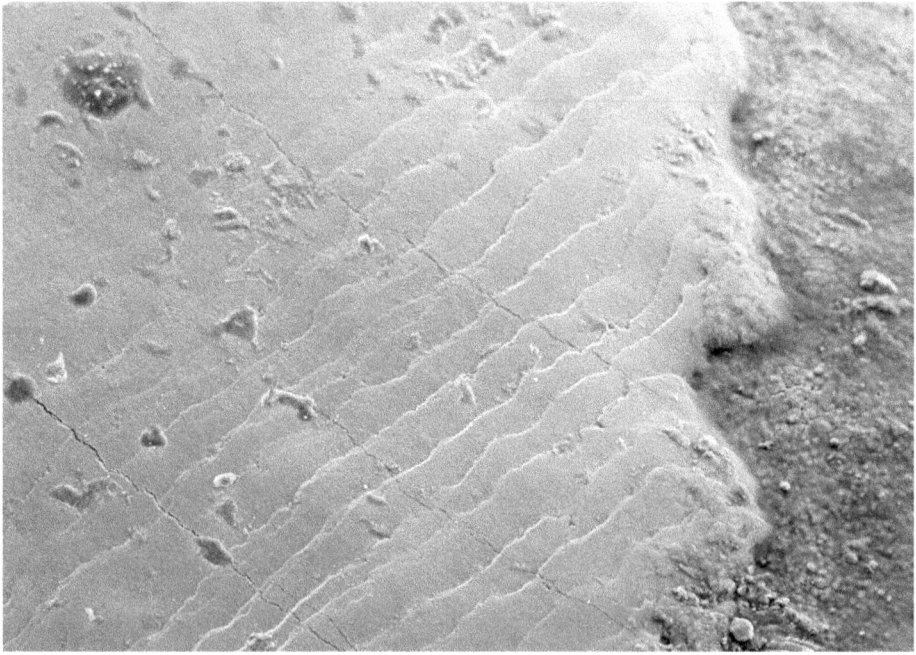

Figure 4.5 Closer view of the same specimen as in Figure 4.1 showing cervical perikymata at the CEJ. The scattered irregular clumps of material on the crown surface are dirt which it has not been possible to clean away. They cannot be dental calculus because the tooth was unerupted. Field of view is 0.75 mm wide.

Enamel growth layers at the surface

If an unworn human permanent tooth is tilted so that light strikes the crown surface obliquely, it is possible to make out small concentric ripples that spread around the sides of the cusps following the circumference (Figures 4.1, 4.3, 4.4 and 4.5). These alternating ridges and grooves are known as *perikymata* (see Definition Box 4). Deciduous teeth, in contrast, have a smooth enamel surface which rarely shows them. Most teeth extracted from modern adults are so highly abraded that perikymata are not preserved. Archaeological teeth show less abrasion of the crown sides, particularly to mesial and distal, but their heavy attrition obliterates whole portions of the crown. For these reasons perikymata are best observed in children's permanent teeth not long exposed to wear. The best specimens of all are unerupted teeth.

Definition Box 4. Perikymata and imbrication lines

The name *perikymata* comes from the graduate dissertation of Gustav Preiswerk (Preiswerk, 1895). Nowadays it is the most commonly used name for the incremental lines that run around the circumference of the tooth crown surface. It is not entirely clear that this is what Preiswerk originally meant but, like many anatomical names, this one clearly originates in Greek: κύμα (singular) or κύματα (plural), nowadays meaning wave or waves, although in ancient Greek also meaning any swelling, περι- meaning 'around'. Cuspal perikymata really do look a lot like ocean waves and it seems likely that Preiswerk invented the compound noun περικύματα (plural) to mean 'surrounding waves' or perhaps 'surrounding swellings'. It may be that he knew his Euripides, because there is a line in *The Trojan Women* (line 800) about the wave surrounded island of Salamis which uses a similar adjective περικύμων/ονος (*The Plays of Euripides*, translated by E. P. Coleridge (1891). Volume I. London: George Bell). Today, the word is pronounced in many different ways, but a modern Greek speaker would pronounce it 'perik<u>ee</u>mata' with short 'a' as in mat.

The phrase *imbrication lines* comes from H. Percy Pickerill (1912). Its origins lie in Latin: *imbrex*, meaning tile. *Imbricated* can be used to mean a number of things and often it implies that something is made up of parts that overlap like a tiled roof. This describes very well the circumferential incremental markings of the cervical part of the crown surface. Pickerill's method was to rub the enamel surface lightly with a little fine graphite powder, removing the excess with a piece of soft rubber. This left the furrows deep black and the ridges a strongly contrasting white, and a high level of detail was visible under the microscope. It is clear from his description that he used the term *imbrication lines* to mean the pattern of both ridges and furrows. Pickerill's accounts (1912, 1913) are remarkable in explaining for the first time the distribution of the lines down the surface of the crown, their relationship to brown striae of Retzius and their role in the defects of enamel hypoplasia. He also invented a simple tooth profiling machine, which he termed the 'odontograph'. With a few basic tools he had understood the nature of crown surface development in a way that few others have grasped since.

Perikymata are at the limit of what can be seen with the naked eye. Most people can make them out near the cusps if they get the lighting right. There, the ridges are about the width of a human hair apart (100 µm or one-tenth of a millimetre). Somebody with really good vision might make out individual pairs of perikymata half-way down the crown. At that point, the spacing has narrowed to perhaps 70 µm. Closer to the cervix, few can see more than an impression of fine ridging and for many people the surface just looks smooth. Only inspection with a microscope shows that the perikymata are still there.

Each ridge of the perikymata represents a layer of enamel matrix formation. These layers are deposited regularly to a fixed rhythm, approximately weekly, although the exact number of days varies between people. It is important to understand that the perikymata represent only a part of the layering seen when a tooth is sectioned. Many more layers lie buried underneath the cusps and do not emerge at the surface of the crown. For the most part, the regularity of the layering is apparent in the even progression of spacing between perikymata, but some teeth are marked by irregularities. Many are slight variations in the prominence of different ridges, but sometimes there is a furrow (Figure 4.4), perhaps 1 mm or more wide, or even a line of pits. Such defects are known as *enamel hypoplasia* (page 86) and record disruptions to development during growth in childhood. Chapter 7 shows how enamel histology can estimate the age of the child during the period of growth disruption. Similar methods yield charts of estimated age for formation of different parts of the crown surface (Figures 4.6 and 4.7). These provide an uncertain basis for establishing the age of the growth-disturbing episode, but they show where to look for matching defects in different teeth being formed at the time. For example, the mid-height part of an upper first incisor crown would be expected to form at a similar age to the base of the single main cusp of the upper canine from the same person. A defect found in the incisor should be matched by an equivalent defect in the canine because a growth disruption affects all the enamel matrix being formed during that episode. A similar chart has been constructed for chimpanzees (Figure 4.8).

Banding on the root surface

Some teeth show a regular pattern of ridges on the root, visible with the naked eye, known as *periradicular bands* (Figure 4.9). These run concentrically around the root, in a similar way to the perikymata on the crown. They are particularly prominent in some fossils, for example the lower third molar of the hominin OH 16, in which Dean (1995) was able to observe a regular spacing of 90 µm. It is not clear why the bands show up only on a small proportion of teeth, but the cement coating of the root varies in thickness and this might well mask all but the most strongly developed bands.

Microscopy of the crown surface

Actual surfaces, or epoxy replicas of them, can be examined by a variety of techniques outlined in Appendix B. The most effective is scanning electron microscopy

Microscopy of the crown surface

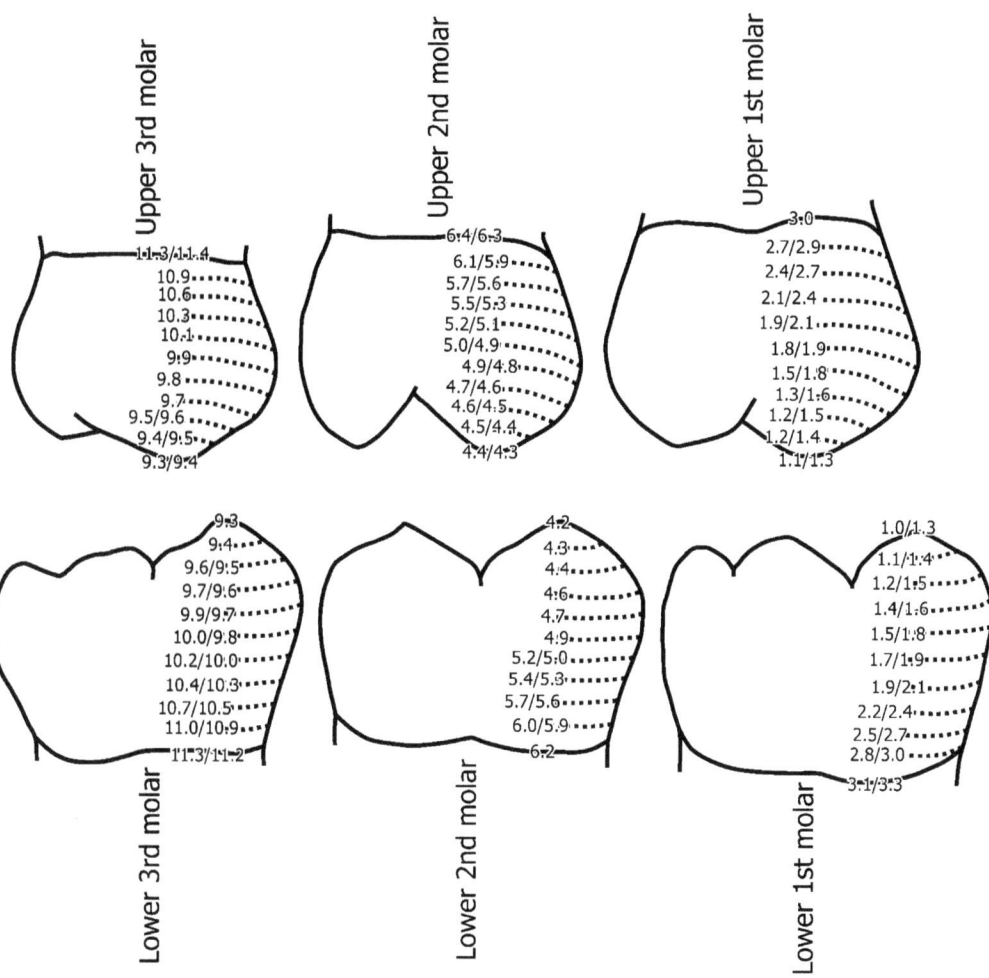

Figure 4.6 Human permanent molars at the protocone and protoconid cusps, with the crown height divided into deciles. The outlines represent a completely unworn tooth. The deciles are labelled with mean estimates in years for age after birth based on studies of groups of teeth from South Africans (to the left of the slash) and northern Europeans (to the right). Redrawn from Reid and Dean (2006), Figure 4, p. 344.

in SE mode. As this is not based on light, it is possible to see how rough the crown surface actually is, without the confusing bright reflections from the glossy enamel surface. Instead, it presents a smooth matte appearance, on which the perikymata stand out clearly, along with scratches from wear during life and in the ground after burial, and a range of other features.

Perikymata

Perikymata are best seen by scanning electron microscopy of unworn human permanent teeth, although experimentation is needed to find the best rotation and

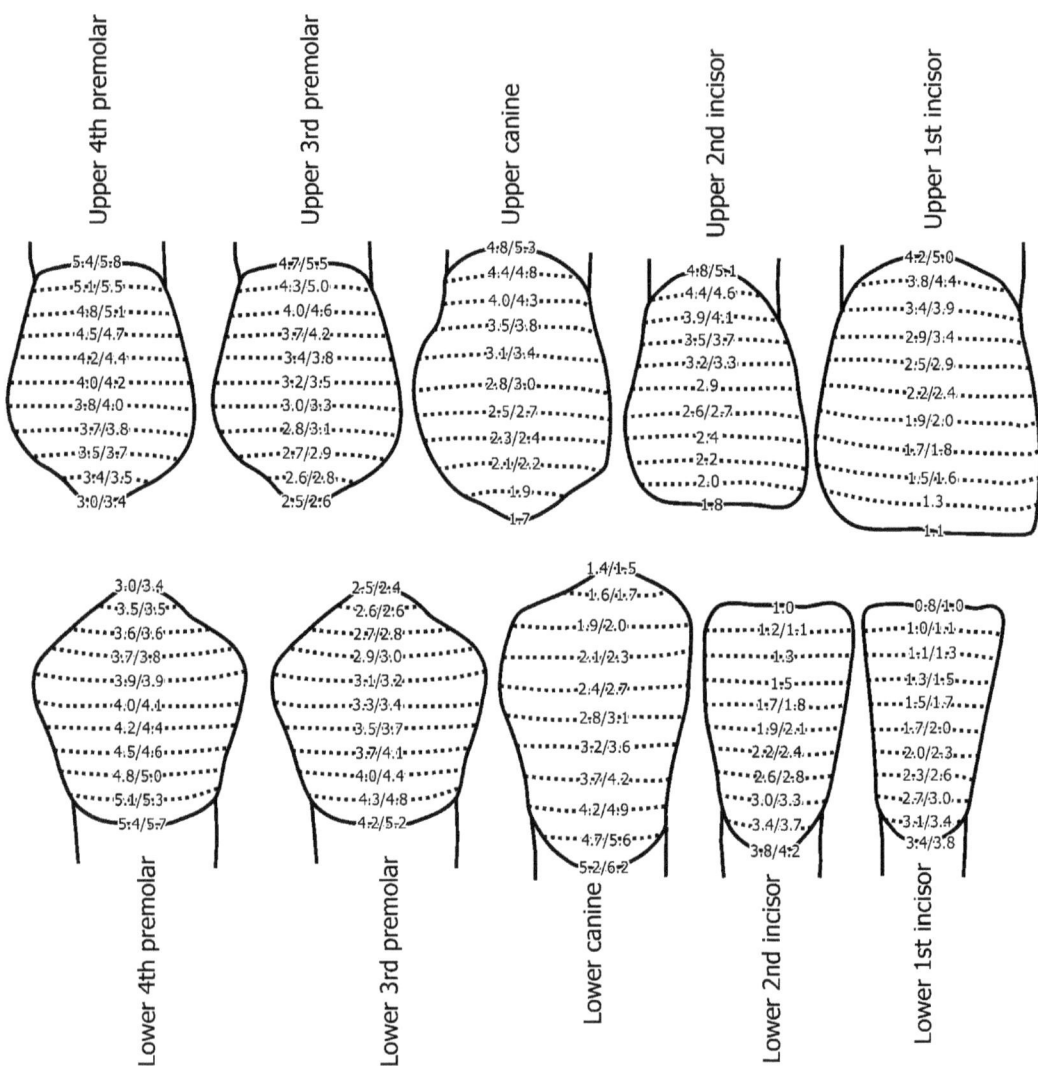

Figure 4.7 Human permanent premolar, canine and incisors seen from the buccal/labial side, with the unworn crown heights divided into deciles. See Figure 4.6 for explanation. Canines and incisors redrawn from Reid and Dean (2006), Figure 3, p. 343. Premolars redrawn from Holt *et al*. (2012), Figure 3, p. 6.

tilt of the tooth relative to the primary electron beam and detector. All the images presented in this book have the detector positioned at the top of the field of view. Perikymata imaged in this way can be categorised into three types: cuspal, mid-crown and cervical. *Mid-crown perikymata* (Figures 4.10, 4.11 and 4.12) can be seen as the more 'general' type and comprise four elements:

1. Increment margin. This is a line or step which marks the edge of each layer of enamel matrix. Its course is irregular, but the general direction runs around the

Microscopy of the crown surface

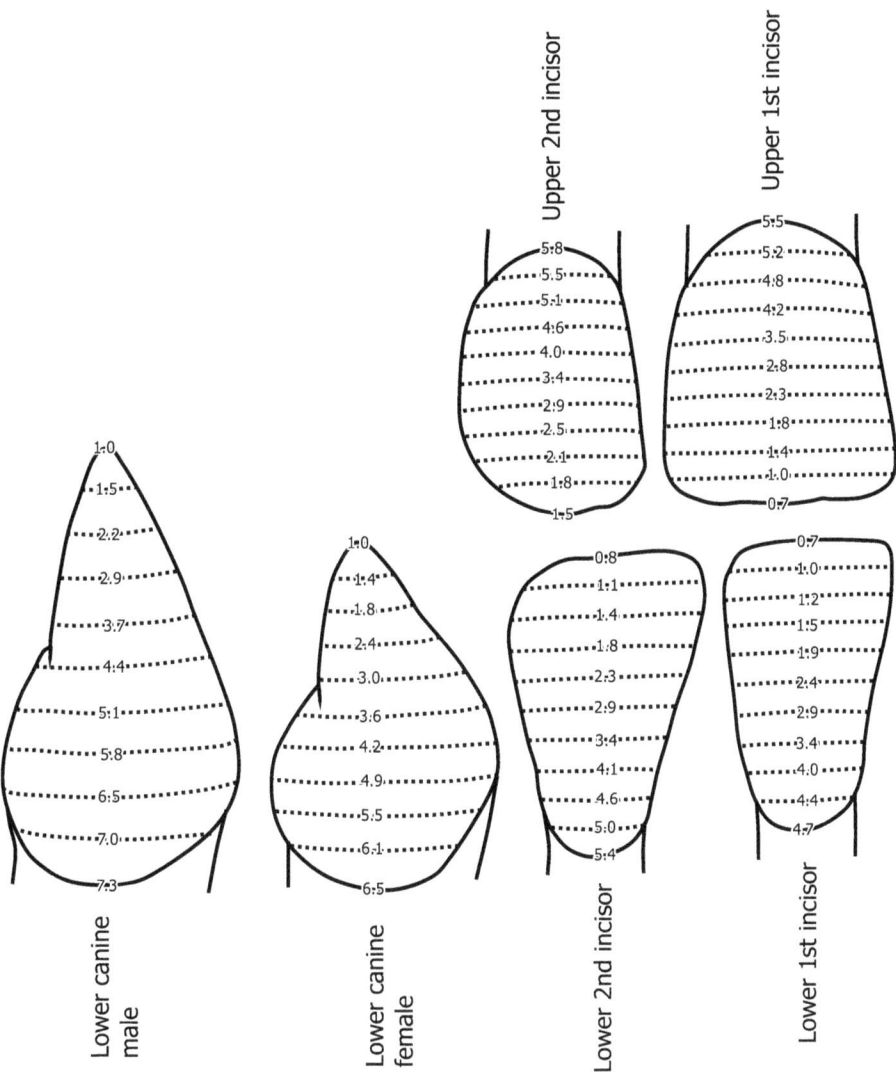

Figure 4.8 Chimpanzee permanent canine and incisors seen from the buccal/labial side, with crown heights divided into deciles. The outlines represent a completely unworn tooth. The deciles are labelled with mean estimates in years for age after birth based on a summary of published studies. Redrawn from Dean (2010), Figure 4, p. 3402.

circumference of the crown. It is bright in the image when it faces towards the detector; dark when facing away. There is considerable variation in prominence.
2. Perikyma ridge. Just below (to cervical) the increment margin, the enamel bulges slightly to stand up as a ridge with a relatively smooth surface.
3. Perikyma groove. Just above (to cuspal) the increment margin, the surface of the enamel is indented to make a groove.
4. Tomes' process pits. The surface of the enamel in the perikyma grooves is decorated by tiny shallow pits. Each is about 4 µm in diameter and represents the

Figure 4.9 Periradicular bands in a human lower third molar. The root was still developing and the tooth is impacted, i.e. developing in a position from which it would never have been able to erupt. Neolithic specimen from Çatalhöyük, Turkey.

depression made by the end of an enamel matrix forming cell, an ameloblast (page 32). The pits are generally arranged a little like fish scales, with the pits of each successive row fitting between the pits of the row below.

Typically, the increment margins of mid-crown perikymata are about 50 µm apart. They occupy a large proportion of incisor and canine crowns, but a relatively smaller proportion of premolar and molar crowns. Tomes' process pits on the completed crown surface vary considerably in depth, from 1.25 µm or more to a barely perceptible dimple just 0.5 µm deep. The fully developed pits are wedge-shaped depressions, with the deepest part on the cuspal side, their margins rounded and irregular rather than sharply defined (page 85). Within the perikyma groove there is a progression from the increment margin, against which the pits are deepest and most clearly defined, up the perikyma ridge to cuspal, where the pits become progressively shallower. Even the highest parts of the perikyma ridge are not completely smooth and the pits are represented by slight dimples. As explained in the following discussion, the depth of the pits marks variation in the loss of Tomes' process as the ameloblasts ceased matrix secretion. The form of the increment margin also marks variation in this cessation. It is irregular and meandering. Often there are little 'bays' in the margin (Figure 4.12) and their size of around 5 µm across shows that they represent the outline of those ameloblasts which ceased matrix secretion when their neighbours continued for a short while. This is borne

Microscopy of the crown surface

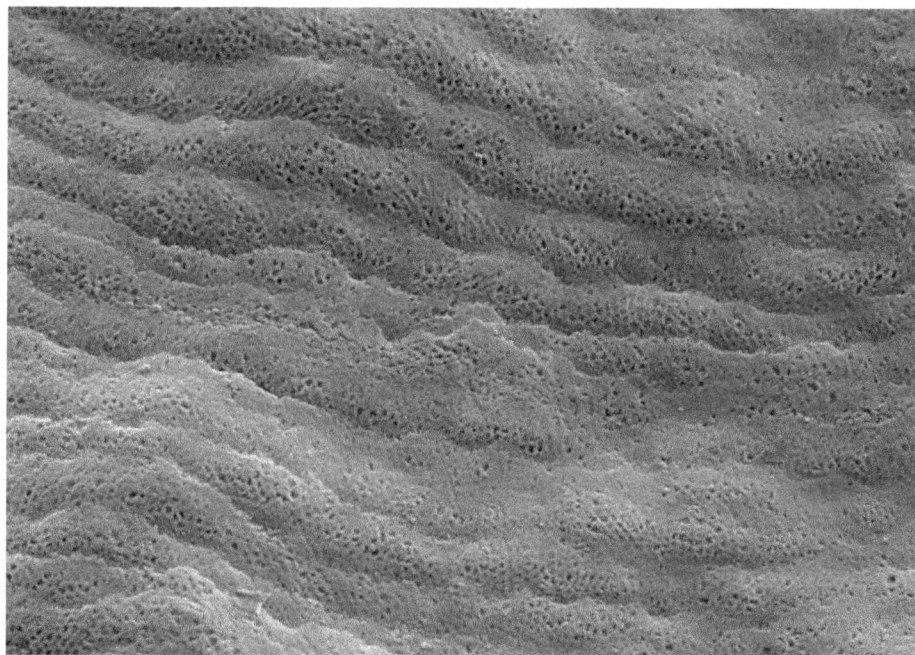

Figure 4.10 Mid-crown perikymata near the centre of the labial side of a human permanent lower canine crown. The occlusal tip of the tooth is above the top of the image and the base of the crown below the bottom. The increment margins are marked by irregular, sharp lines which are bright because they are like a series of steps angled towards the detector of the microscope. The image illustrates their irregularity and local variation in perikymata spacing. The tooth is from Medieval burials under the floor of the chapter house at Worcester Cathedral. It was cleaned gently with a cotton bud and acetone before being imaged in an environmental scanning electron microscope, without coating or any other pre-treatment. Operation of the microscope as for Figure 4.1 and outlined in Appendix B. Field of view is 706 μm wide.

out by the observation of Tomes' process pits inside many of these bays. As well as these smaller scale meanderings, neighbouring increment margins wander closer together and further apart, occasionally almost within touching distance, although the average spacing around the crown remains constant.

In *cervical-type perikymata* (Figure 4.13 and 4.14) the increment margins are more closely spaced, about 20 to 30 μm apart, with sharper and more step-like profiles. The perikyma ridges are not marked, so the whole arrangement is much more like a series of overlapping sheets (imbrication lines; see Definitions Box 4). This is emphasised because the Tomes' process pits are not well defined and the surfaces of these sheets are relatively smooth. The opposite is true of the *cuspal-type perikymata* (Figures 4.15, 4.16 and 4.17), in which the increment margins are wider spaced, sometimes more than 200 μm apart. The increment margins are also poorly defined and in some cases very difficult to discern, with a smooth transition from the broadly bulging perikyma ridges to the wide, shallow perikyma grooves.

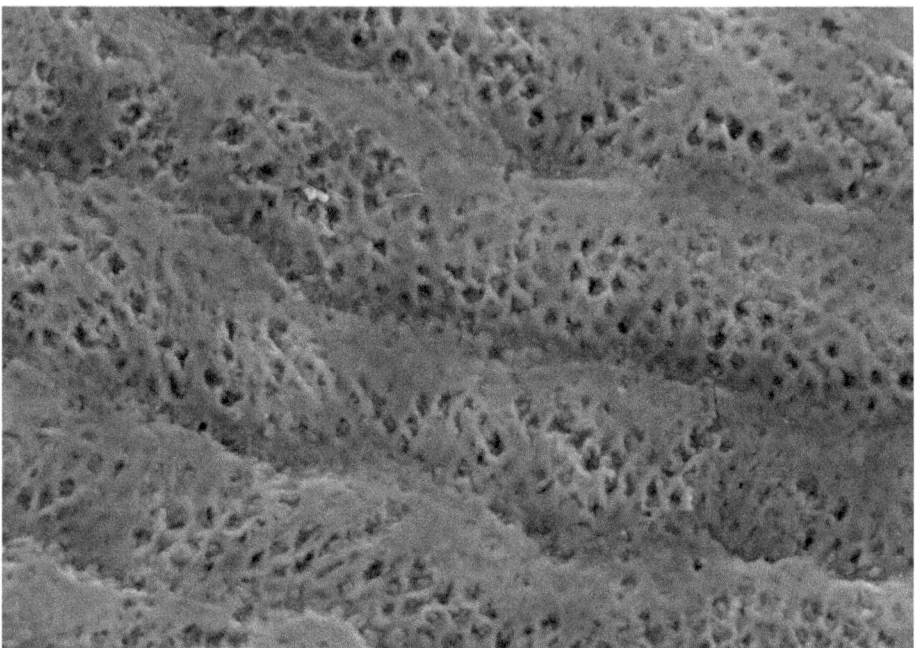

Figure 4.11 A closer view of mid-crown perikymata in the same specimen as Figure 4.10. Sharply defined Tomes' process pits cover the perikyma grooves, but they reduce in depth across the perikyma ridges until they are marked only by slight dimples. The pits are deepest on their lower sides, where the points of the Tomes' processes were located. The increment margins show tiny irregularities in their course as well as larger undulations. They also vary along their length in the sharpness with which they are defined. Field of view is 254 µm wide.

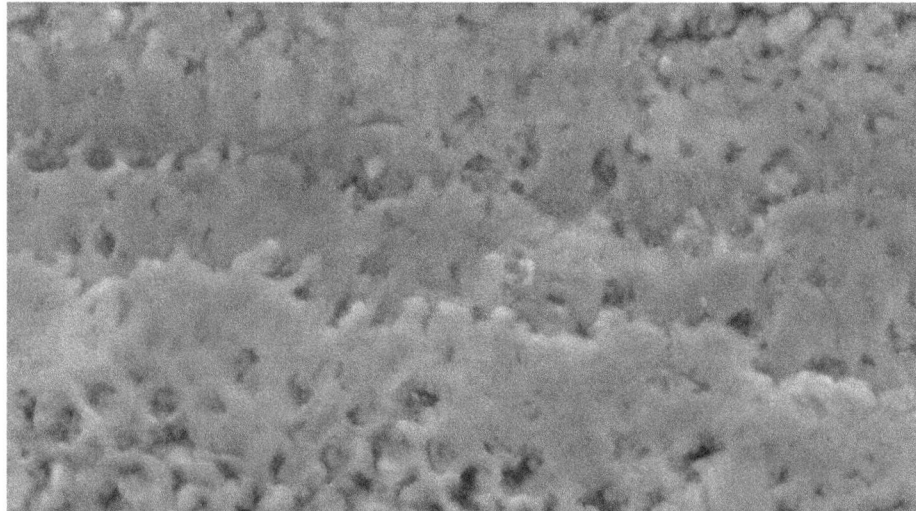

Figure 4.12 A closer view of increment margins in the same specimen as Figure 4.10. Their fine irregularities make little bays which enclose Tomes' process pits. This shows that, whereas one ameloblast ceased matrix secretion instantly, its neighbour continued for a short while afterwards. Field of view is 133 µm wide.

Microscopy of the crown surface

Figure 4.13 The cervical region in the same specimen as Figure 4.10. The cement-enamel junction is visible on the right-hand side, with cervical perikymata crossing the image diagonally. They resemble overlapping sheets of paper with irregular torn edges. Their increment margins are bright because they comprise sharp steps facing the detector of the microscope. They also show considerable variation in spacing and sharpness of definition. They are complicated by the low mounds of a field of brochs (see text) scattered along the margin of the crown. This leads to considerable irregularities of the increment margins as they pass amongst the brochs. Field of view is 1.4 mm wide.

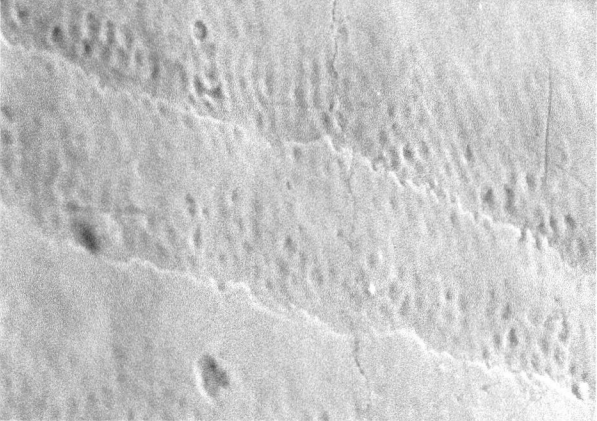

Figure 4.14 Detail of cervical perikymata from the specimen in Figure 4.13. The Tomes' process pits are only occasionally sharply defined, in the deepest parts of the perikyma grooves. Most are marked only by shallow dimples. Field of view is 180 μm wide.

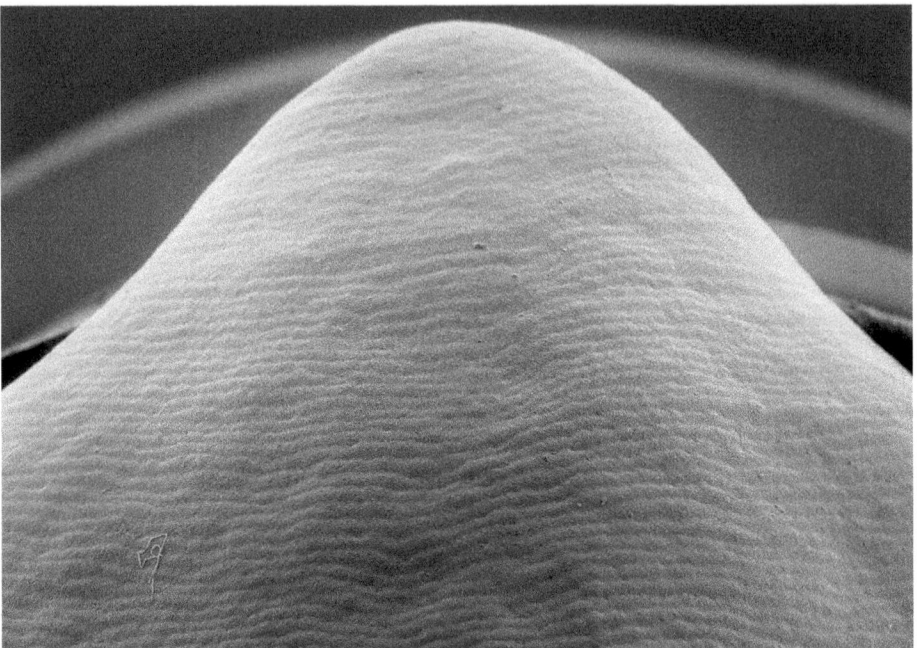

Figure 4.15 The occlusal region of the same specimen as in Figure 4.10. Widely spaced occlusal-type perikymata encircle the single cusp of this canine tooth. The first-formed perikymata can be seen, describing a small circle around the cusp tip. These are very soon worn away, so this tooth cannot have erupted into the mouth. Field of view is 3.9 mm wide.

Figure 4.16 Closer view of the crown surface seen in Figure 4.15. The increment margins of the perikymata are readily apparent, but marked only in places by a sharp step. Most of the crown surface is decorated by Tomes' process pits, with only a little difference between perikyma ridges and grooves. Field of view is 900 µm wide.

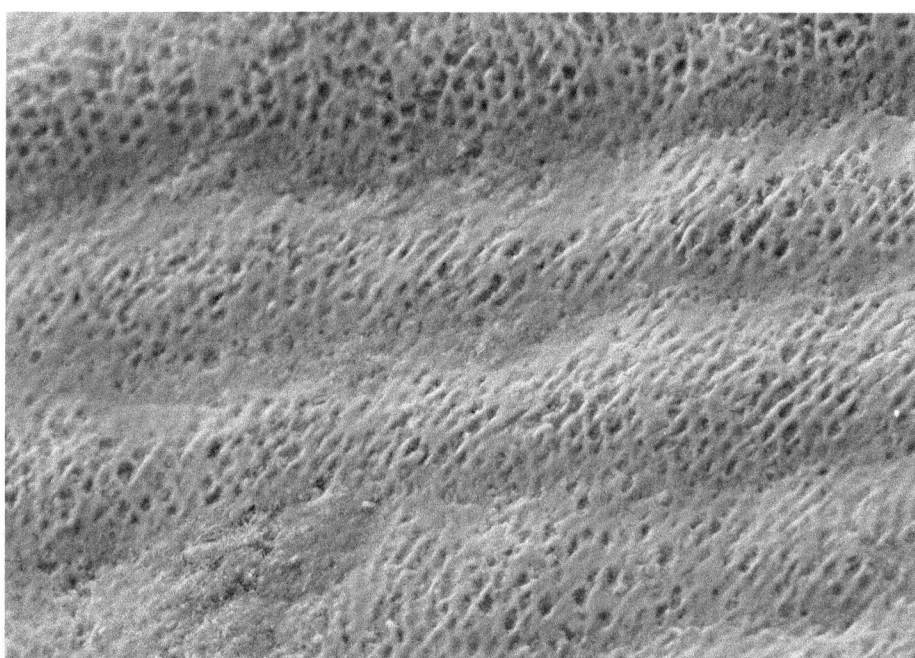

Figure 4.17 Closer view of occlusal perikymata seen in Figure 4.16. Where they are sharply defined, the increment margins show similar tiny irregularities to those seen in Figure 4.12. Tomes' process pits are marked throughout, except on the cervical side of the increment margins. Field of view is 282 µm wide.

The Tomes' process pits in cuspal-type perikymata are clearly defined, but spread over both ridges and grooves. In human permanent incisors and canines, relatively small proportions of crown height are occupied by cuspal-type and cervical-type perikymata. In total, the tall crown of an incisor might have between 100 and 250 perikymata down its side, whereas a canine might have 300 perikymata or more. In molars and premolars, almost half the crown height is occupied by cuspal-type perikymata and there is a rather rapid transition through a small zone of mid-crown-type perikymata to a large zone of cervical-type perikymata. The total count of perikymata for a human permanent molar might be 70 to 100.

Perikymata vary between individuals. In some, the perikyma ridges are very prominent and rough enough to be felt with a fingernail. In others, the increment margin of the cervical-type perikymata is very irregular, so that they look like overlapping tiles. These variations in 'style' characterise all crowns in one individual and can be highly distinctive. There are also other variations of the crown surface (Boyde, 1989), such as *brochs*, which are 30 to 50 µm diameter pimple-like bulges (Figure 4.13). They are found in the cervical region, particularly in premolars and molars. *Surface overlapping projections* are irregular, flattened blobs of matrix, up to 30 µm in diameter, overlying the normal surface. Sometimes they break away to expose *isolated deep pits*. Abrasion decreases the sharpness of definition to create a smooth surface marked by small scratches. Perikyma ridges are abraded first, so that it may be possible to make out the line of the perikymata from the scatters of deeper Tomes' process pits along the perikyma grooves. Attrition facets are flatter and smoother still and marked by a clear pattern of micro-wear, including microscopic scratches and pits.

Enamel hypoplasia

As seen under the microscope, there is a uniform progression of perikymata spacing down the crown. In addition, many teeth show irregularities, ranging from a slight variation in the prominence of individual perikymata, to large furrows involving many perikymata, to prominent steps or pits that fill the field of view (Figure 4.18). These are the defects of enamel hypoplasia (Chapter 7). They were classified into three main types in the detailed early description of Berten (1895): *fürchenformig* or furrow-form; *flächenformig* or plane-form; and *grübchenformig* or pit-form.

Furrow-form defects are smoothly defined grooves which parallel the perikymata, running around the crown (Figure 4.4). They vary in width and prominence, but the cuspal 'wall' of the defect usually slopes more sharply than the cervical wall. Under the microscope, it is apparent that the perikymata involved are not abnormal in form, but depart from normal spacing for that part of the crown. They are usually more widely spaced, although, in some defects, they are more closely spaced than usual. There is clearly a range of ways in which the defect is formed and its size and prominence are affected by its position on the crown because the perikymata are more widely spaced in the cuspal part of the crown than in the cervical part. The optimum part of the crown for showing furrow-form defects seems to be the middle part of its height, where the mid-crown perikymata are located.

Plane-form defects are much more prominent. One perikyma is very greatly widened in spacing to expose a whole surface of the enamel matrix that was developing at the time of the growth disruption. This makes a broad plane that dips down steeply into the body of the enamel. The perikymata that follow are often disturbed in a variety of ways – irregular, closely spaced and so on – but in some cases matrix secretion seems to have resumed normally. Microscopic examination of the exposed plane shows that the surface is very different to the usual appearance of a perikyma groove. It is much more like the form of the developing enamel matrix surface underneath fully active ameloblasts. This is seen particularly in the sharply defined Tomes' process pits (page 79). The plane itself often involves just one perikyma, although there may be a sequence of planes, like steps (Hillson and Bond, 1997).

Pit-form defects comprise a spread of isolated pits ranging from several hundred micrometres in diameter down to a few tens of micrometres, each exposing a plane of enamel matrix formation where the ameloblasts were abruptly interrupted. These pits are often arranged in a band around the crown, penetrating an otherwise normal surface in which perikymata are spaced as expected. In other cases, they may be part of a furrow-form defect, with a single row of pits running along the

Microscopy of the crown surface

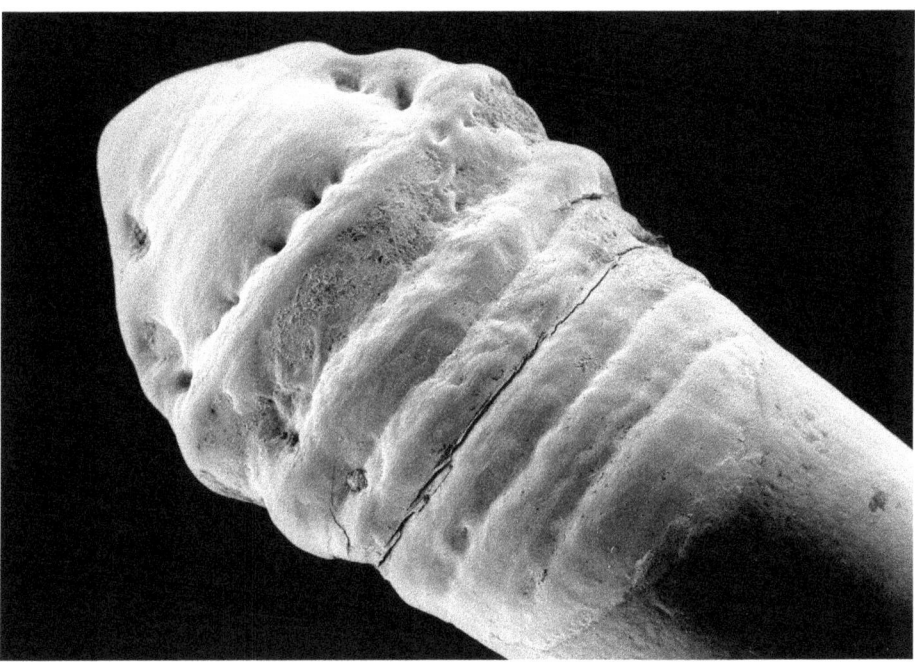

Figure 4.18 Unusually pronounced enamel hypoplasia in a human permanent lower canine from the Odontological Collection at the Royal College of Surgeons of England. Multiple defects represent disruptions at many points throughout the development of this tooth crown. Pit-form and plane-form defects are most prominent, but there are also furrow-form defects. This tooth was gently cleaned and then imaged uncoated, under low vacuum in a scanning electron microscope using an ESED (see Appendix B). Field of view is 11.5 mm wide.

deepest part. Closer inspection shows that these represent bay-like irregularities of a single increment margin. Sometimes a defect representing one growth disruption may include both pits and exposed planes. The distinguishing feature, however, is that these defects are spread intermittently around the circumference of the crown: they are not continuous like the plane-form defects.

Developing edge of tooth crowns

The characteristic features of developing enamel and dentine matrix can often be seen under the scanning electron microscope. For archaeological and fossil specimens, even if the tooth has been protected in a crypt, the wafer-thin developing cervical edge of the crown is usually damaged (Figures 4.19, 4.20 and 4.21). The most recently formed enamel matrix is only one-third mineral so, after death, this is lost down to the point at which the remaining matrix is sufficiently mineralised that it resists erosion. This surface forms a band around the outside of the cervical crown margin, relatively smooth, but marked with a characteristic pattern of fissures. Developing dentine matrix lines the conical opening inside the crown. Once more, it would have been covered in life with unmineralised dentine matrix, but in

Microscopic markers of growth in dental tissues

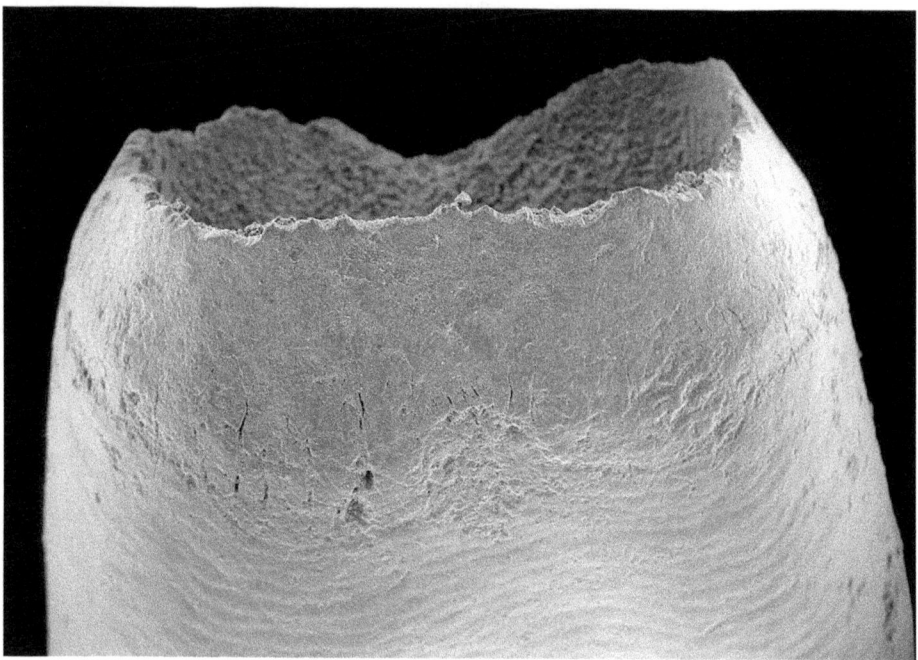

Figure 4.19 The cervical margin of a developing human permanent lower first incisor crown. The wafer-thin margin was turned uppermost in the microscope and shows taphonomic damage because this is an archaeological specimen. The surface of enamel matrix formation at the time of death occupies a smooth band around the top half of the image and below it are the perikymata preserved in more mineralised, previously formed matrix. The forming enamel matrix has been polished in the ground and by handling after excavation so it presents a smooth surface, without the marks of Tomes' process pits, which would be expected in matrix which was actively being secreted. It seems likely that the least heavily mineralised, most recently formed matrix has been lost, so that what remains is the matrix which had a high enough mineral content to survive. It still, however, shows the pattern of cracking which is a characteristic surface texture of the matrix-forming surface. Inside the hollow cone of the developing crown is the surface of dentine mineralisation (see Figures 4.20 and 4.21). The tooth is from Medieval burials under the floor of the chapter house at Worcester Cathedral. It was cleaned gently with a cotton bud and acetone before being imaged in a scanning electron microscope under low vacuum, without coating or any other pre-treatment, using an ESED. Operation of the microscope is outlined in Appendix B. Field of view is 5.2 mm wide.

archaeological specimens the mineralising front is preserved. The surface has a globular texture which represents the calcospheritic pattern of mineralisation (page 102).

Structures seen in sections of teeth

Techniques for cutting sections and microscopy are given in Appendix B. It is assumed here that a section has been cut along a plane running from lingual to

Structures seen in sections of teeth

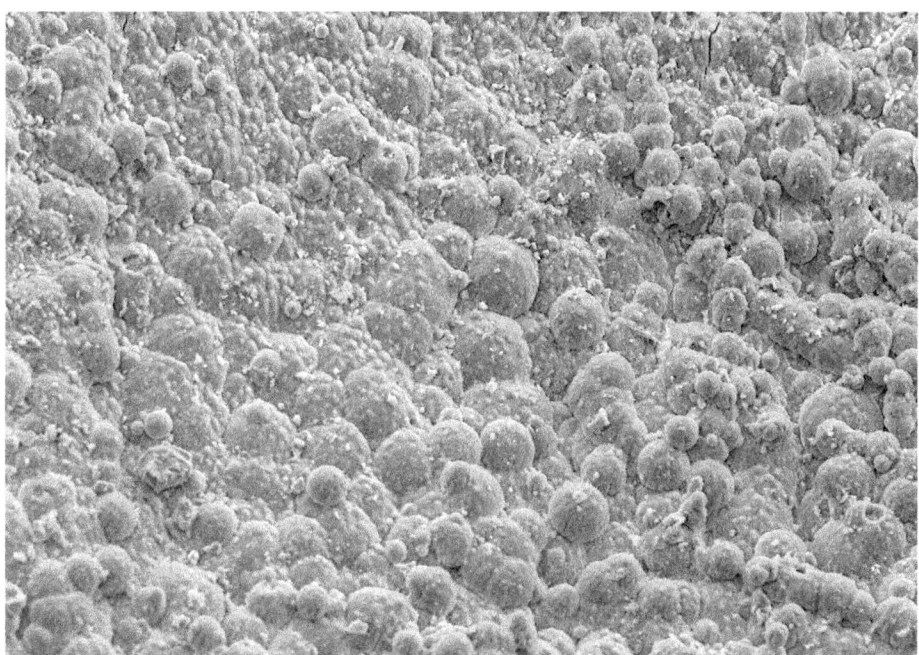

Figure 4.20 The mineralising surface of dentine from the specimen shown in Figure 4.19. The spherical bodies are the calcospherites: bodies of mineralised tissue which grow until they coalesce. The smallest calcospherites will be the most recently formed at the time of death in this individual. Field of view is 748 μm wide.

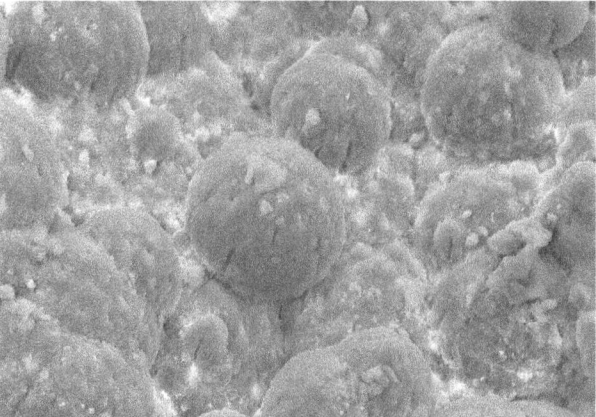

Figure 4.21 Closer view of calcospherites from the field of view shown in Figure 4.20. They are marked by the tiny holes and tunnels of the dentinal tubules. These contain the odontoblast processes, which continue to pass through the tissue as the calcospherites grow and coalesce. This is remarkable preservation of detail in an ancient specimen, imaged with very little preparation. Field of view is 180 μm wide.

buccal, from the tip of a cusp through the cervical region, down the root and through the apex.

Dental enamel

The *enamel-dentine junction* (EDJ, or amelodentinal junction, ADJ) is a marked line with a scalloped appearance. The discontinuity it represents scatters light strongly as it passes through the section, so it appears relatively dark to the observer. Enamel can be distinguished from dentine by its very different texture (Figure 4.22). The enamel layer is thickest over the cusps in an unworn tooth and tapers down towards the cervix of the crown. Three structures dominate the enamel section: *prism boundaries*, *prism cross striations* and *brown striae of Retzius*.

Prism boundaries

Prism boundaries are parallel lines running obliquely across the section (Figure 4.23), typically 4–6 μm apart. They represent discontinuities introduced into the enamel matrix by the Tomes' processes of the ameloblasts, separating bundles of the crystallites which are the main constituent of mature enamel. The name 'prism' suggests an isolated rod of crystallites but, throughout most primate enamel, prisms are linked along their length by complex interconnecting structures (Boyde, 1989). They run out from the EDJ across the enamel to the surface of the crown (Figure 4.24), following a long curve in the vertical plane of section, whilst at the same time also undulating regularly up and down relative to the plane of section. This is the phenomenon known as *decussation*. The undulations are so closely spaced and irregular under the cusps that they obscure the outlines of the prisms and brown striae. This is the zone of so-called *gnarled enamel* (Figure 4.22). Down the sides of the crown, the undulations are more regular and are less apparent when the tooth section is viewed under ordinary light microscopy. They become much more obvious under a polarising microscope (Appendix B), where they line up to form regular structures called *Hunter-Schreger bands*. This shows clearly that the prism undulations are more strongly developed in the inner two-thirds of enamel than they are near the surface or next to the EDJ.

Prism cross striations

In ordinary light microscopy, it is apparent that there are regularly spaced dark marks running across the lines of the prisms, rather like cross-hatching (Figures 4.23 and 4.25). The marks are known as *prism cross striations* (see Definitions Box 5). They are more strongly developed in some parts of the section than others, making it difficult to follow them along the whole length of every prism. The spacing of the marks ranges from 2 to 5 μm; that is, they are similar in size to the spacing of the prism boundaries and it is possible they might sometimes be an artefact of obliquely sectioned prisms. Other types of microscopy have, however, made it clear that cross striations really do exist (Appendix B). They are visible in confocal

Structures seen in sections of teeth

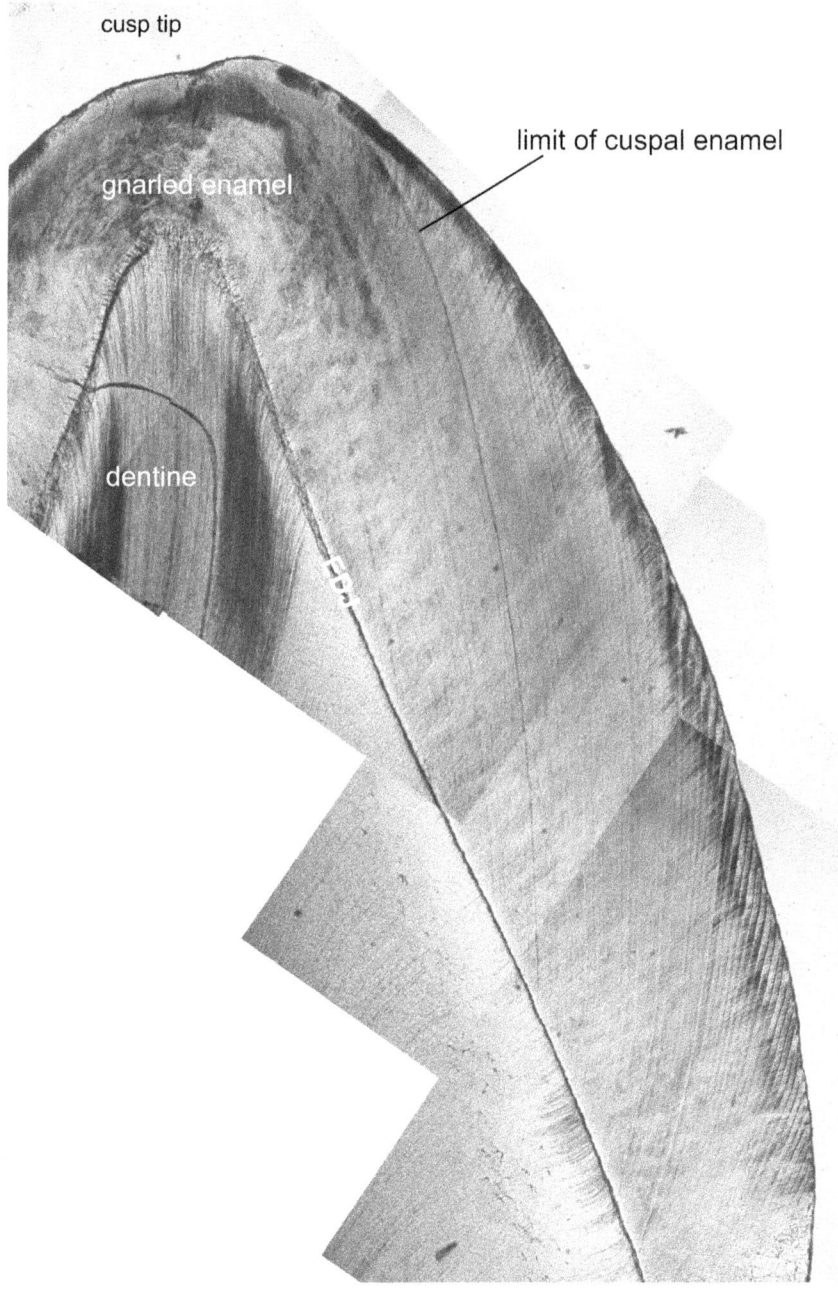

Figure 4.22 General view of a crown section made from a human lower third premolar, showing the buccal cusp tip and side of the crown. The brown striae of Retzius are seen as dark lines. Fortuitously in this tooth, a particularly prominent stria is placed at approximately the outer limit of the cuspal enamel. Beneath it, dome-like increments wrap around the dentine horn, but the brown striae are never clearly defined in this zone of gnarled enamel. Outside it is the lateral enamel, where striae angle up to the crown surface and meet perikymata. Polished section, imaged in a conventional light microscope as described in Appendix B. The image is a composite stitched together from smaller pictures.

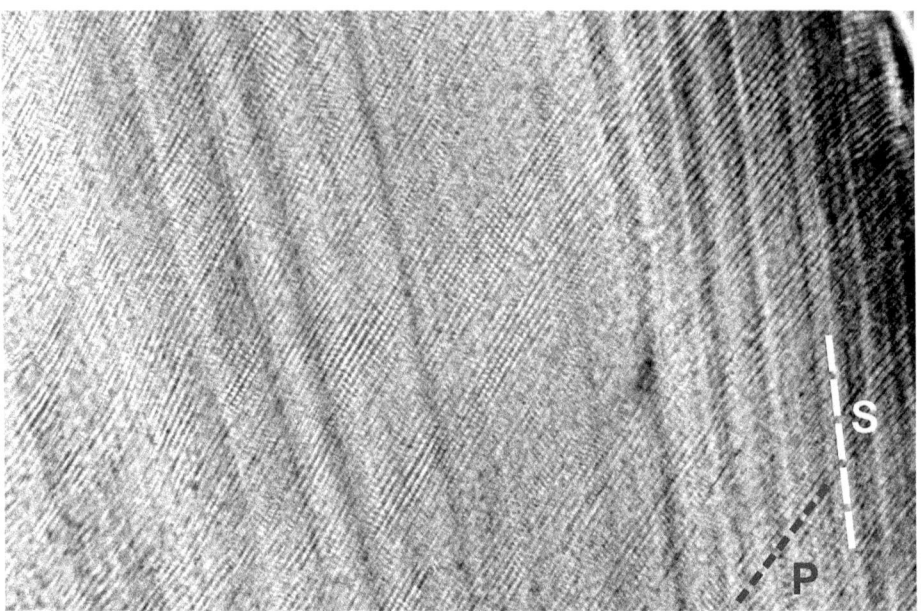

Figure 4.23 Detail of prism boundaries and cross striations. 'S' marks the line of the brown striae of Retzius and 'P' the line of the prism boundaries. The cross striations are clear in the centre of the image, crossing the prisms to divide them into segments. The brown striae vary in prominence and in regularity. The section and microscope are the same as those in Figure 4.22. Field of view is 600 μm wide.

light microscopy, for example, which focusses on a plane just 1 μm thick within an enamel section. This is considerably thinner than the prism boundary intervals. When fractured, enamel breaks up along the prism boundaries and, when examined in the scanning electron microscope, these are marked with a series of varicosities and constrictions identical in spacing to the cross striations. They may represent a regular variation in the rate of matrix secretion during the 24-hour period (Boyde, 1989). The varicosities and constrictions would create alternating angled and horizontal interfaces which would appear dark and light to the observer because of reflection at the boundary, so making the dark marks in light microscopy. Cross striations are also visible in micro-radiography and in synchrotron radiation micro-CT, so they must also represent variation in mineral composition (Boyde, 1989).

Definition Box 5. Brown striae of Retzius and prism cross striations

The brown striae take their name from the great Swedish anatomist Anders Retzius (1796–1860). He described and illustrated them (Retzius, 1837, p. 538, Fig. 7) as *bräunlicher Parallel-Striche*, or 'brownish parallel lines'. It often happens that a particular author's name – not necessarily the first to describe it – is attached to a particular anatomical feature. The name may be shortened to

Structures seen in sections of teeth

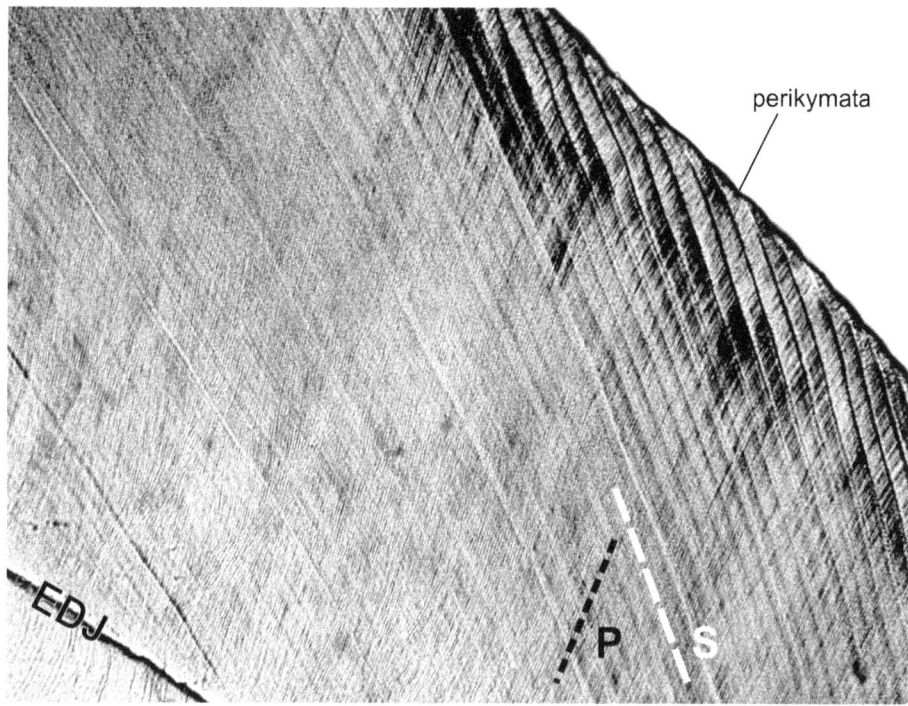

Figure 4.24 Brown striae relative to prism boundaries, the crown surface and perikymata. 'S' marks the line of the brown striae of Retzius and 'P' the line of the prism boundaries. Just below the surface, the striae are sharply defined and regularly spaced. Each is associated with a perikyma groove at the surface, so the perikymata are seen in section as a scalloped edge. As these brown striae run deeper into the enamel, they become less sharply defined. They are not really clear enough to count in this region, but it is possible to see a wide range of variation in prominence. The prism boundaries run a slightly curving course through the enamel. In the deeper parts, they weave to and fro, quite clearly showing the effect of decussation. Field of view is 850 μm wide.

'Retzius striae' or just 'brown striae'. The use of the word 'brown' has sometimes caused researchers to think that they represent lines of pigmentation. This is not true, as can be checked by shining light onto the surface of the slide under the microscope (rather than through it), when the lines show clearly as white rather than brown. This is because brown striae of Retzius are discontinuities in the enamel which scatter light. When the surface is illuminated they reflect more light back than the intervening enamel, so they appear white. When the section is viewed under the microscope in the conventional way, illuminated from underneath, the striae scatter light to the side, so they appear dark relative to the intervening enamel which scatters less light. The blue end of the spectrum is scattered more than the red end, so they appear as a dark reddish-brown mix of colours. Pickerill (1912, 1913) preferred to call them 'incremental lines' to avoid the connotation of staining.

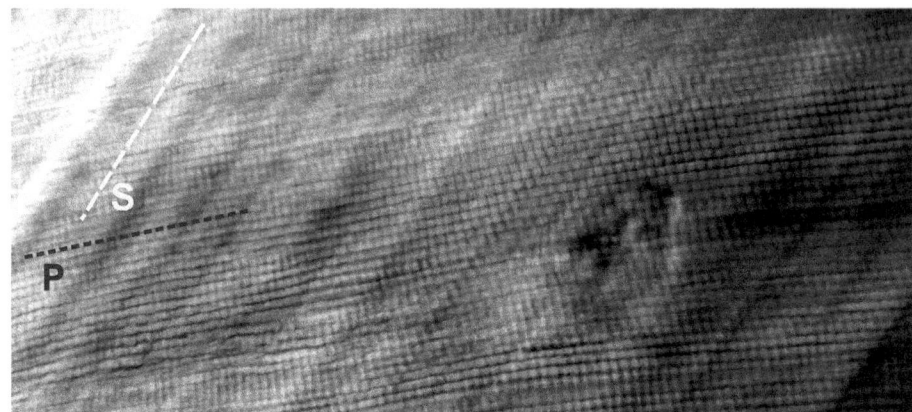

Figure 4.25 Prism cross striations in human enamel. Polished section, imaged in a polarising microscope. The prism boundaries are marked by 'P' and 'S' marks the line of the brown striae of Retzius. The cross striations can be seen as sharply defined marks which divide the prisms into segments. Where the prisms cross long-period lines, these are marked by a 'staircase' arrangement of neighbouring cross striations (see text). This is a very thin section in which the cross striations are especially well marked, but regular brown striae are difficult to make out. Specimen from the crypt of Christ Church, Spitalfields in London, kindly loaned by Daniel Antoine, The British Museum. Field of view 380 μm wide.

> Retzius (1837, p. 535) also gave one of the earliest descriptions and illustrations of *prism cross striations*. He called them *Querstrichen*, which means a dash or a stroke, and described how closely spaced they were and how they crossed the prisms (which he labelled *Fasern* (fibres) or *Nadeln* (needles)).

The cross striations are believed to represent an approximately 24-hourly (circadian) rhythm to the secretion of enamel matrix. Evidence for this is reviewed in Chapter 5. Occasionally, in ordinary light microscopy, additional faint marks appear in between, known as *intradian lines*. These may represent a less than circadian rhythm, but they might also be an artefact of section orientation and plane of focus. Cross striations themselves are readily distinguishable by their spacing, although this is closer near the EDJ and wider towards the crown surface. In addition, those in the cervical region have a closer spacing than those higher up the crown.

Brown striae of Retzius

The prism boundaries are cut across diagonally by another set of lines, angling up from the EDJ to the crown surface (Figures 4.23 and 4.24). In ordinary light microscopy they appear dark relative to the surrounding enamel, often with a slightly brownish tinge. They are formally known as *brown striae of Retzius*, or more simply as *brown striae* (see Definition Box 5). They are rarely continuous right through the enamel. Just under the crown surface they can be sharply defined and are clearly regular structures. If one stria is followed into the enamel, however, it usually becomes wider, less distinct and discontinuous. Only in some cases, when the stria is particularly prominent, can it be followed down to the EDJ. The striae are particularly indistinct under the cusps, but their trend shows how they are arranged concentrically one above another under the tip of the cusp. Down the sides of the cusp, middle and cervical parts of the crown side, they parallel one another running from the EDJ to the surface. Where well-defined brown striae meet the surface, each is matched with a very shallow depression in the edge of the section that matches a perikyma groove (Risnes, 1985a; 1985b). Like the perikymata, the spacing of brown striae decreases down the crown side, from around 22 μm apart near the cusps to 18 μm near the cervix.

Examination of developing tooth crowns (page 87) confirms that the brown striae represent successive positions of the surface at which enamel matrix was secreted during development (Figure 4.22). Under the cusp, they show that the crown grows initially as a series of dome-like layers, sitting one on top of another and gradually increasing in width as well as height, right up to the tip of the cusp. From then onwards, successive layers are sleeve-like, overlapping one another down the sides of the crown. The height of each sleeve-like layer decreases progressively into the cervical region, until matrix secretion ceases at the base of the crown.

In light microscopes, brown striae are seen best at low magnifications in thick sections (around 100 μm) and they are not clearly distinguished in techniques such as confocal microscopy that focus on a very thin focal plane. This is because they represent planes of discontinuity that run through the enamel, scattering light as it passes through so that a thicker section includes a greater expanse of each plane, which thus scatters more light and appears darker. If the discontinuity plane is perpendicular to the plane of section, the striae will appear narrower and darker and, if angled, wider and less dark. When examined in reflected light – simply by switching off the light under the microscope stage and shining a bright light onto the surface of the section – the striae show as white because the light is scattered back to the observer. Little is known about the discontinuity itself. At least near the crown surface, the striae can be seen in synchrotron radiation micro-CT, which images a very thin plane within the enamel (Appendix B), so they must be associated with a variation in mineralisation. In scanning electron microscopy of fractured specimens of enamel, the tissue generally breaks along the prism boundaries but, just under the surface of the crown, it also fractures along the brown stria planes which join the increment margin of the perikymata. The fracture-exposed prism boundaries show the varicosities and constrictions of cross striations, but often a particularly prominent constriction can be identified and followed from prism to prism along the line of the brown striae (Boyde, 1989). This suggests that the brown striae represent cross striations in which there was a more marked slowdown of enamel matrix secretion than usual. Whatever the nature of the discontinuity, it seems that the circadian rhythm of the cross striations continues unabated, because their spacing does not change as prisms pass through the planes of brown striae.

Brown striae of Retzius vary considerably in prominence through a section. In particular, they vary in their darkness and apparent width. The prominent striae make a clear pattern in one tooth, which can be matched in other teeth from the same individual in which matrix was being secreted at the same time. This shows that the variation in the light microscope appearance of brown striae is caused by a systemic factor which affects the body as a whole. Hypoplastic enamel defects are usually underlain by at least some prominent striae, which suggests that similar factors to those which are known to cause defects may be at work (see page 173). The matching patterns of brown striae are important as recognisable 'registration lines' (FitzGerald and Rose, 2000) that allow development sequences to be matched between teeth from one individual (page 120).

Relationship between brown striae and cross striations

At first sight, it is difficult to link the progression of cross striations with the line of the brown striae of Retzius. Cross striations cross the prisms perpendicularly, whereas brown striae cross at an angle. The answer can be seen in some sharply defined brown striae which are marked by a series of steps: 'staircase striae' (Risnes, 1990). The steps are cross striations of neighbouring prisms so, for a cross striation in one particular prism, the neighbouring prism towards the cusp is one cross striation more advanced in development, and the prism towards the cervix is one cross striation less advanced. If the cross striations represent a circadian growth rhythm, this stepped arrangement implies that the sheet of active matrix-secreting ameloblasts is increased by one additional ring of cells per day at its cervical edge along the EDJ.

Bearing in mind the stepped relationship, it is possible to count cross striations between brown striae. This is difficult where they are less sharply defined, but, by counting cross striations over several striae, it is possible to arrive at a consistent figure. It has always been found that the count is constant throughout the enamel in all the teeth of one individual, ranging in modern humans from 6 to 12 with an average value of 8 or 9 (page 118). It therefore represents another rhythm to enamel matrix secretion: a long-period rhythm rather than the short, circadian rhythm of the prism cross striations.

Relationship between brown striae and perikymata

Normally, each brown stria that emerges at the surface from the tip of the cusp to the cervical margin is marked by the increment margin of a perikyma. The change in the spacing and form of perikymata which is seen down the crown (discussed earlier) is related to:

1. the angle of the brown striae against the crown surface;
2. the spacing of the brown striae within the enamel.

In cuspal-type perikymata, both the wide spacing of brown striae and the shallow angle with which they approach the surface produce wide spacings of perikyma

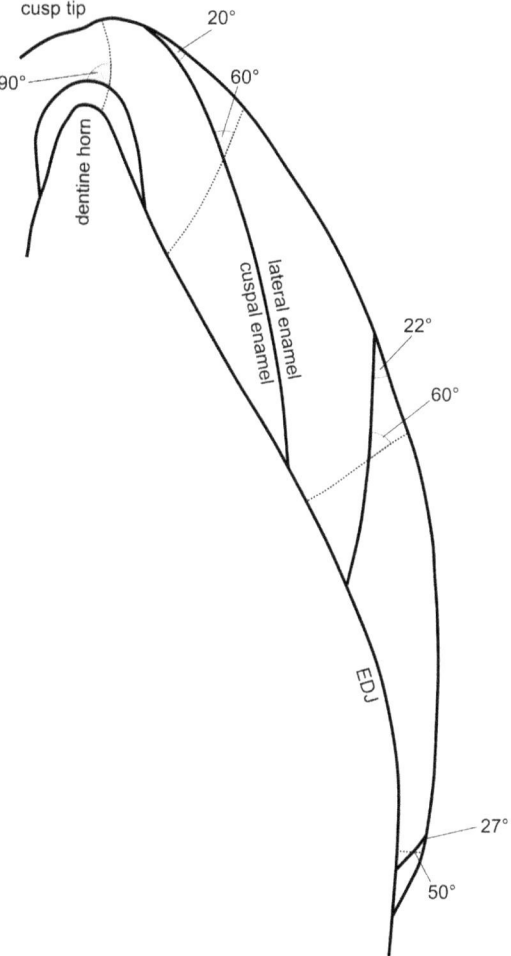

Figure 4.26 Angular relationships between prism boundaries, EDJ, brown striae and crown surface, traced from the specimen shown in Figure 4.22. Prism lines are dotted and brown striae solid. The angle that the brown striae make with the prism boundaries decreases steadily down the crown. The angle that the brown striae make against the crown surface is very shallow near the cusp tip, but increases down the crown side.

grooves (Figure 4.26). The shallow angle creates broad, low perikyma ridges. In a microscope section, it is possible to count about 30 prisms meeting the surface between each pair of increment margins. This arrangement contrasts greatly with cervical-type perikymata, in which the narrow spacing of the brown striae and the steeper angle with which they approach the surface produce the closer spacing of perikyma grooves and sharper perikyma ridges. Eight or so prisms meet the surface between each pair of increment margins. There is a steady transition down the crown, so that the mid-crown-type perikymata have intermediate counts.

As each prism boundary approaches the surface, it marks the point at which an ameloblast ceased enamel matrix secretion. In some areas, the ameloblast's Tomes' process was lost just before the end, so the surface is unmarked by a pit. The Tomes' process is also responsible for making prism boundaries so these are missing too, creating a thin layer *prism-free surface zone enamel* (Boyde, 1989). In other regions, the ameloblast ceased matrix secretion before its Tomes' process was lost and prism boundaries continue to the surface. For these, Tomes' process pits remain in the matrix surface and the depth of the pits reflects the extent to which the process had begun to withdraw. Even in the deepest pits, there is usually some overgrowth or rounding of the edges, indicating a little irregular growth at the finish. More sharply delineated Tomes' process pits, underlain by prismatic enamel, are found in mid-crown perikyma grooves, whereas no pits, or only slight dimples, underlain by prism-free enamel, are found in the perikyma ridges. The sharpest Tomes' process pits are found in plane-form or pit-form hypoplastic defects (see earlier discussion), where there seems to have been a very sudden end to matrix secretion. The pits have steep sides, with a crack running around their floors (Boyde, 1970).

Suppose a developing crown is just arriving at the point when there will be a beat of the long-period rhythm that generates planes of discontinuity in the enamel matrix which causes brown striae (Figure 4.27). Enamel matrix is being formed by a sheet of ameloblasts arranged in a band around the crown side. The highest (most cuspal) ring of ameloblasts, along the top edge of the band, have been secreting matrix the longest and are just about to cease secretion and start their maturation phase (page 32). The ameloblasts in the ring below that have been secreting the next longest, the ring below the next longest, and so on, down the band to its bottom edge, where a ring of ameloblasts have just started secreting matrix. When the rhythm beats, between eight and 30 rings (depending on the location on the crown side) of top edge ameloblasts cease matrix secretion altogether, whatever their current state, and pass into maturation. The very highest rings had already started a transition by withdrawing their Tomes' processes, producing prism-free matrix to form the perikyma ridge. Ameloblasts lower down are interrupted before this transition had started and so retain their Tomes' processes. This is what causes the increasing depth and sharpness of Tomes' process pits into the perikyma groove. The increment margin marks the point at which matrix secretion continued following the beat of the long-period rhythm. To cuspal of the margin, ameloblasts are now starting their maturation phase whereas, for the moment, the ameloblasts to cervical continue matrix secretion. In this way, the long-period rhythm of the brown striae causes the shaping of the crown surface by making successive rings of ameloblasts cease matrix secretion and pass into their maturation phase. In the cuspal region, each beat of the rhythm causes many rings to cease secretion, but the number of rings affected decreases progressively down the crown, resulting in closer increment margins. The long-period development rhythm is therefore an important mechanism in the modelling of the crown surface contour.

How to build a tooth crown

Putting all this together, how is a crown formed? Imagine a simple canine with a nine-day interval between brown striae. The internal enamel epithelium (page 31) marks out the shape for the crown inside a hollow cone. A very small group of cells

Structures seen in sections of teeth

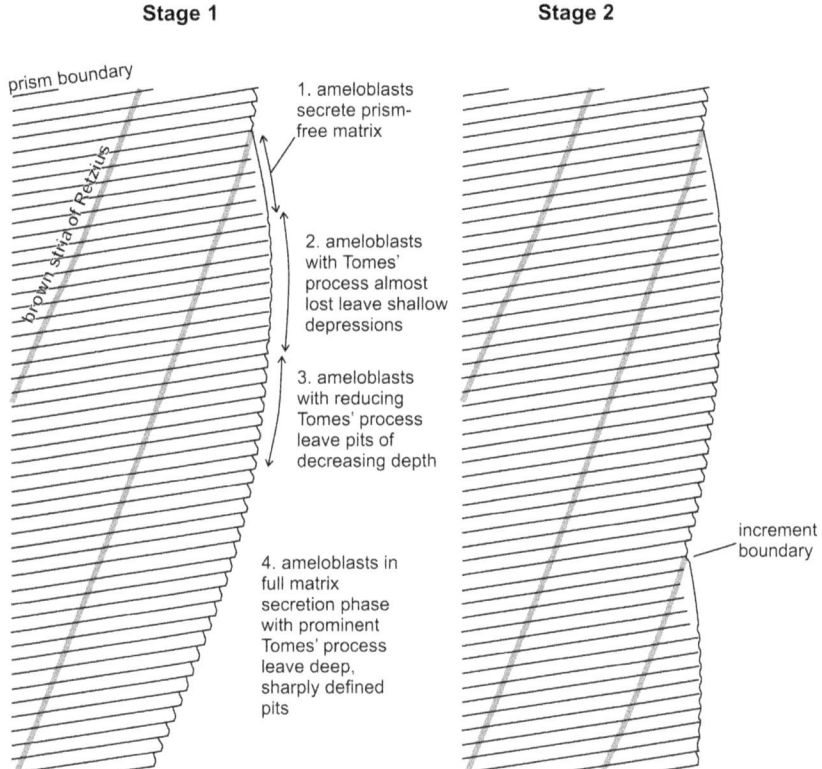

Figure 4.27 Model for the formation of perikymata. These diagrams represent two stages in the formation of the crown side. Stage 1 is the point at which the long-period rhythm that forms brown striae of Retzius is just about to beat. Ameloblasts overlying zones 1, 2 and 3 have already started to transition from matrix secretion to maturation by losing their Tomes' processes. In zone 3, there is just a slight reduction in the processes and the prism boundaries reach to the surface of the developing crown, with relatively sharply defined pits. In zone 2, the processes have almost been lost and the pits are only shallow depressions. In zone 1, the processes are completely lost and the ameloblasts are secreting prism-free true surface zone enamel. Stage 2 shows the brown stria that marks the beat of the long-period rhythm, with its associated increment boundary. Below that, the same pattern of Tomes' process loss is being repeated.

in the epithelium at the point of the cone differentiates into ameloblasts and starts to secrete matrix directly underneath. The next day, a single ring of cells around the original group differentiates and starts to secrete in turn. The following day, another ring of cells is added and so it goes on for the whole of crown formation. The first group of cells to differentiate continues to secrete matrix for perhaps 315 days, during which time 35 brown striae are formed (bearing in mind that they are difficult to see in the cuspal region). At the 35th brown stria, the first group of cells stops matrix secretion – let us say, a disc 30 rings of ameloblasts in diameter. At its cervical edge is the first increment margin on the crown surface. Nine days later at the 36th brown stria, the next band of cells ceases secretion. Once again, let us say it is 30 rings of cells wide and this creates the first true perikyma of the crown surface, with a broad ridge and shallow groove. Brown stria succeeds brown stria and the width of the band of ameloblasts slowly decreases down the crown side – 25, 20, 15 – so the perikymata become closer together. The loss of rows of actively matrix-secreting ameloblasts at the crown surface outstrips the recruitment of new matrix-forming cells at the EDJ and the angle of the brown stria planes becomes steeper. Not only are there changes in the angle of prisms and brown striae, coupled with decreasing spacing of daily increments, but each newly added ring of ameloblasts secretes matrix for a shorter time before being terminated in a perikyma. At the cervix, each cell secretes matrix for only a few days. Eventually, the last brown stria/perikyma event stops all the few remaining actively matrix-secreting cells, forming the cervical margin of the crown. By that stage some 155 brown striae might have formed in the canine. A total of 120 will have risen to the surface of the crown side and so made perikymata.

This model is very simplified, but shows how the physiological rhythm of the brown striae of Retzius and the perikymata acts as the mechanism by which the crown is shaped. It is a trajectory which starts with the first few ameloblasts and then follows some rules:

1. one ring of actively matrix-secreting ameloblasts is added per day;
2. a brown stria of Retzius/perikyma is formed every nine days (whatever the individual's personal rhythm);
3. the first group of ameloblasts to differentiate secretes enamel for a given period (315 days in this canine, but longer in molars);
4. for the first perikyma, a band of 30 rings of ameloblasts stops matrix secretion;
5. with successive perikymata, the width of the band decreases steadily down to the cervical margin.

If these rules are varied slightly, it is possible to produce different contours of tooth crown, from the broadly bulging crowns of molars to the taller and flatter-sided crowns of canines and incisors.

Deciduous teeth and the neonatal line

In general, brown striae of Retzius are not prominent in deciduous teeth. This is also true of perikymata, partly because the brown striae approach the surface at a shallow angle and partly because there is a thick layer of prism-free enamel on the surface. Brown striae can, however, be made out in places and it is also possible to count prism cross striations. One particular feature seen in all deciduous teeth is the *neonatal line* (Figure 4.28). This is a strongly marked brown stria, which corresponds with the expected position of development at birth. All deciduous teeth start to form *in utero* (see page 64) and so a considerable amount of prenatal enamel matrix has been secreted by the time of birth. The permanent first molar also starts to form around 1 month before birth, but it has a much thinner layer of prenatal enamel and the neonatal line can be difficult to identify unless the section is precisely centred over the dentine horn. Brown striae are less prominent in prenatal enamel matrix than they are in the post-natal matrix, but cross striations can be counted in places. The neonatal line itself is a dark, wide band in thick sections

Figure 4.28 Neonatal line in the cusp of a human deciduous fourth premolar. The dentine is heavily altered in this archaeological specimen from the Kylindra cemetery, island of Astypalaia, Greece. Under the neonatal line, the prenatal enamel is relatively featureless, with just a suggestion of incremental lines. Outside it, the post-natal enamel is quite sharply marked by lines, although they are not clearly regular. Polished section viewed with conventional transmitted light microscopy as described in Appendix B.

under the light microscope. If the section thickness is greatly reduced, then the line can be seen as a sharply defined mark in between cross striations, giving a 'staircase' appearance (Weber and Eisenmann, 1971).

Structures in dentine

In section, the appearance of dentine varies greatly with the degree of preservation and fossilisation. Seen with the naked eye, fresh dentine is a pale yellow colour and slightly translucent, but archaeological dentine is often opaque and chalky in appearance. This is usually due to invasion by microorganisms which create a series of 'diagenetic foci' through the tissue, forming a zone of altered dentine, particularly under the surfaces of the root, root canal and pulp chamber. In spite of the additional processes of diagenetic change seen in fossils, many preserve dentine structure well. The reason for this is not clear. Perhaps it is only the best-preserved specimens that survived to be fossilised; however, in some, the process of fossilisation has actually enhanced the contrast of the microscopic features.

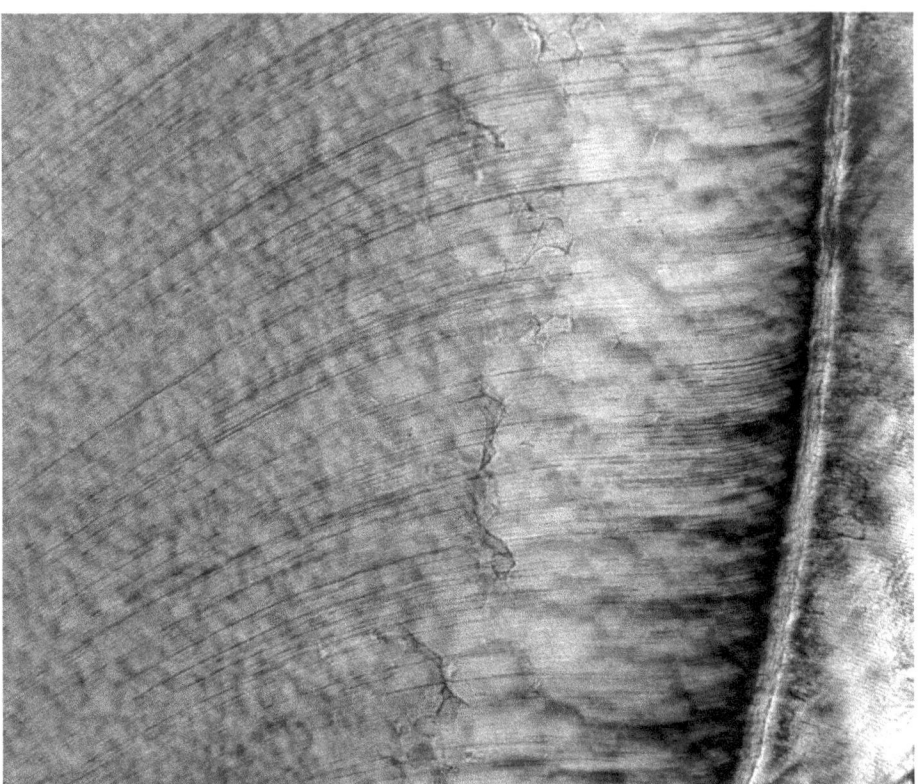

Figure 4.29 Dentine in the human lower premolar shown in Figure 4.22. The EDJ is at the right side of the image. The sinuous lines crossing the dentine are the dentinal tubules and the cloud-like masses are the calcospherites. The bright and dark patches outline their radial pattern of crystallites. In the centre are patches of interglobular dentine, where the calcospherites have failed to coalesce. Polished section imaged in a polarising microscope (see Appendix B). Field of view is 950 μm wide.

Under ordinary light microscopy, the main feature of fresh dentine is a series of closely spaced lines radiating out from the wall of the pulp chamber to the EDJ (Figure 4.29). They represent the walls of the *dentinal tubules*, which in life carry long, fine processes from the cells of dentine, the odontoblasts. Near the pulp chamber these tubules are 2 to 3 μm in diameter, but they narrow along their length so that near the EDJ they are less than 1 μm in diameter. Each tubule is separated from its neighbours by 4 to 8 μm of *intertubular dentine* (Figure 4.30). In polarised light microscopy, intertubular dentine has a texture of irregularly rounded masses representing a globular, or *calcospheritic*, pattern of mineralisation. As initially secreted, dentine matrix (predentine) is wholly organic. About 15 to 30 μm deep to the matrix-forming front is a mineralising front, where apatite crystals are seeded into the matrix at separate centres which grow radially to form spherical bodies of mineralised intertubular dentine matrix (*calcospherites* – see Figure 4.21). These grow until they meet one another to create a fully mineralised

Structures seen in sections of teeth

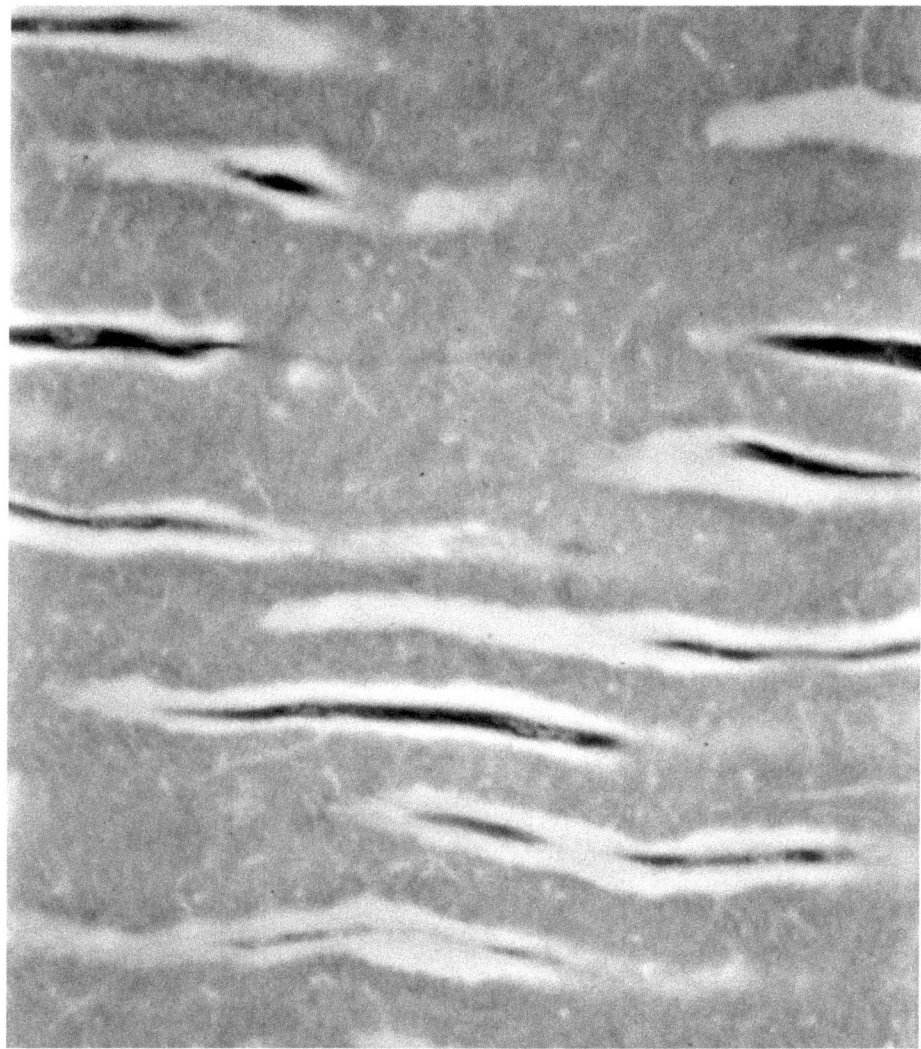

Figure 4.30 Cut and polished surface of human dentine, imaged in a scanning electron microscope operated in back-scattered electron mode. Details are given in Appendix B, but the specimen has been impregnated with acrylic resin and is polished as flat as possible. The microscope has been operated in a way that emphasises compositional contrast, here meaning variations in the density of mineralisation. Brighter features are more heavily mineralised and darker features are less heavily mineralised. The image shows dentinal tubules, sectioned longitudinally. The black stripes are the voids inside the tubules, which are occupied in this prepared specimen by supporting resin. Lining the tubules is a layer of heavily mineralised peritubular dentine matrix with tiny threadlike extensions marking branches of the tubules. Between the tubules is more lightly mineralised intertubular dentine matrix. The outlines of the calcospherites cannot be seen with this technique, but the tiny bright dots in the intertubular matrix represent their more heavily mineralised cores. Field of view is 43 μm wide.

intertubular matrix. In some places, however, the calcospherites never grow large enough to meet and this leaves irregular patches of unmineralised intertubular dentine matrix between them (Figures 4.31 and 4.32), known as *interglobular spaces* (Definition Box 6). In the dentine immediately underlying the cement-dentine junction (CDJ) in the root, such interglobular spaces are a normal finding. This is the *granular layer of Tomes* and its distinctive texture comes from small calcospherites which have failed to coalesce (Figure 4.33). Other patches of interglobular spaces may be found anywhere in the dentine, aligned parallel with the developmental layering discussed in the following. Defects of enamel hypoplasia (page 86) may be matched by corresponding layers of interglobular space in the dentine underlying the crown. At least in some cases, therefore, the incomplete mineralisation of interglobular dentine represents a systemic developmental disturbance, but not all patches of interglobular spaces under the crown can be matched with enamel hypoplasia.

> **Definition Box 6.** Contour lines of Owen, Andresen's lines, lines of von Ebner, interglobular spaces
>
> The contour lines were first described in the *Odontography* of Sir Richard Owen (1804–1892; more famous as the originator of the word 'dinosaur') (Owen, 1845). This is a great work of comparative anatomy, although in the introduction he described a matching undulation of neighbouring dentinal tubules which together made a line running parallel to the contour of the tooth. These lines are completely different phenomena from those described here.
>
> The dentine long-period lines were probably first shown by Viggo Andresen (1870–1950), Professor of Orthodontics in Oslo (Andresen, 1898). They usually take his name. He seems in addition to have described the shorter period lines, although they were also described by Victor von Ebner (1842–1925), who was Professor of Histology in Vienna (von Ebner, 1902). von Ebner was really describing the organisation of collagen fibres in dentine rather than the layered structure, but nevertheless the short-period lines usually take his name.
>
> Interglobular spaces were illustrated and named by Johann Czermák (1828–1873), who was a Czech physiologist (Czermák, 1850). His name for them was *Interglobularräume* and he described them as cavities between the globules of dentine. Nowadays it is understood that these gaps are occupied by unmineralised dentine matrix.

Incremental lines in dentine

In patches through a dentine section, it is often possible to make out a series of shadowy dark lines crossing the tubules (Figures 4.34 and 4.35). These represent incremental layering, but there has been discussion about their spacing and the

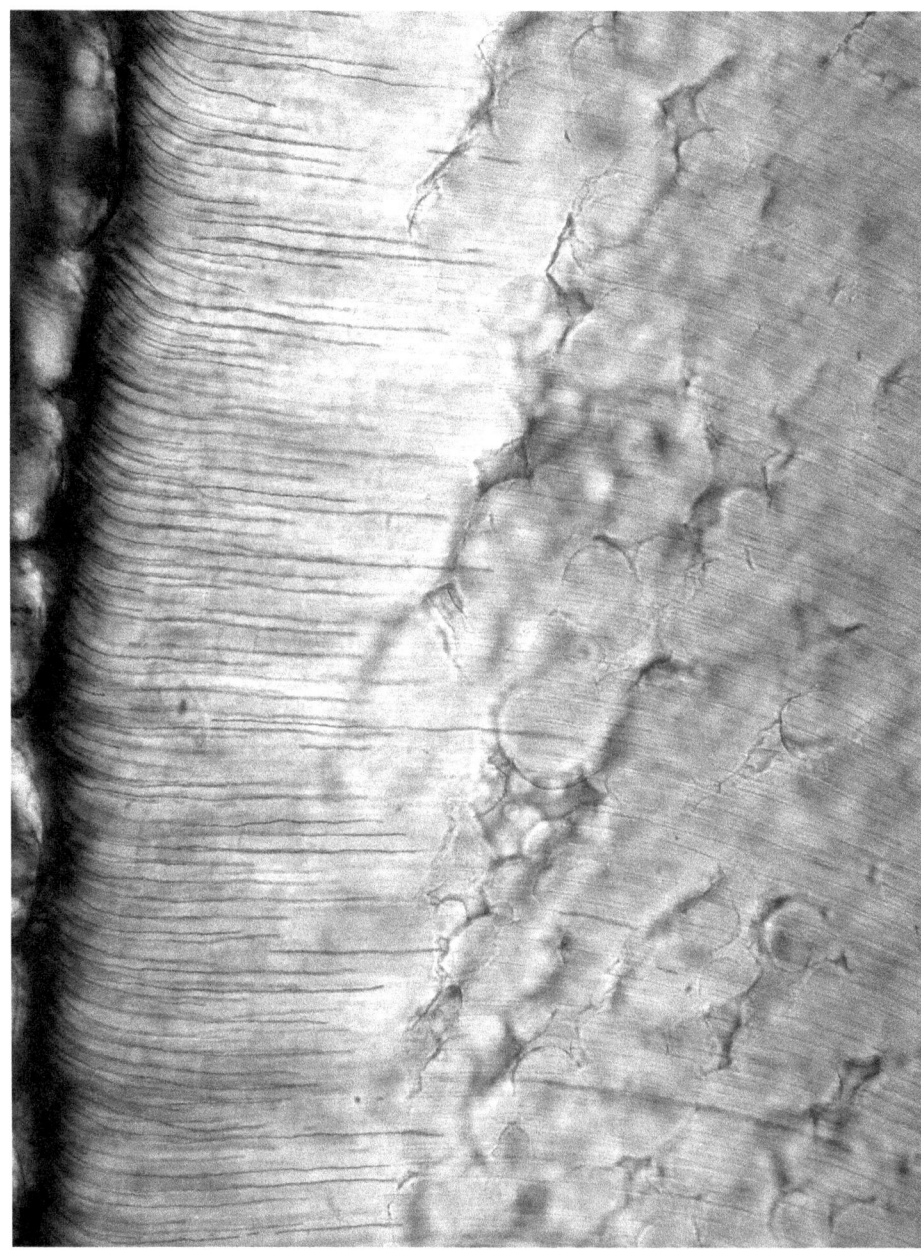

Figure 4.31 Interglobular spaces in human dentine. The EDJ is at the left edge and the dentinal tubules curve across the image. Individual calcospherites are visible amongst the mass of interglobular spaces on the right. Polished section of permanent lower second incisor imaged in polarising microscope. Field of view is 475 μm wide.

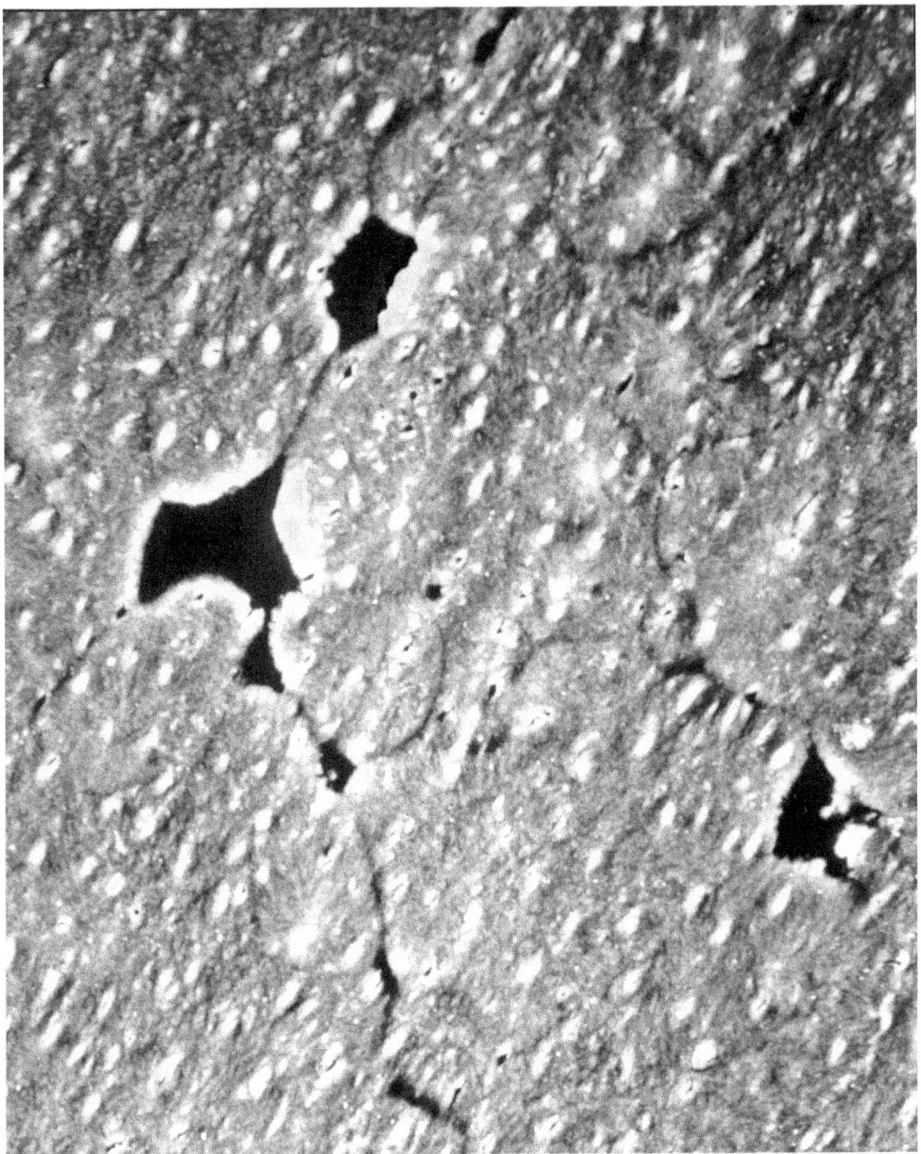

Figure 4.32 Cut and polished surface of human dentine, imaged in a scanning electron microscope operated in back-scattered electron mode (see Figure 4.30). The black areas are voids in the specimen which are occupied by supporting resin. These voids outline several calcospherites which have failed to coalesce and the larger voids are interglobular spaces. The scattered bright marks are the dentinal tubules which have been sectioned transversely. Field of view is 114 μm wide.

Structures seen in sections of teeth

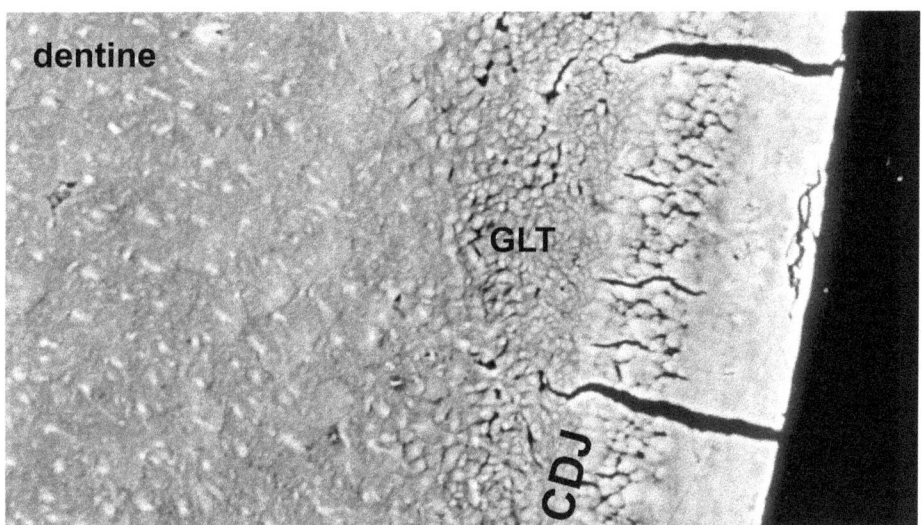

Figure 4.33 Dentine and cement in the same specimen as Figure 4.30. The brighter layer on the right, with cracks running through it, is the cement. Cracking during specimen preparation is normal for this tissue. CDJ marks the cement-dentine junction, although this is never well defined. In the dentine, the grey of the intertubular matrix is scattered with white dots marking the peritubular dentine of tubules sectioned transversely. GLT marks the granular layer of Tomes, where small calcospherites have failed to meet, leaving unmineralised (and therefore dark) gaps between them. Field of view is 900 µm wide.

rhythm of development they represent. Three names are commonly used to describe them: the contour lines of Owen; Andresen's lines; and the lines of von Ebner (see Definition Box 6). One of the difficulties is that the original workers were not always clear about what they were describing and later authors have applied the terms inconsistently. Dean (1995; 1998b) has carefully revisited the nomenclature and distinguished two types:

- more sharply defined *long-period lines* with a wider spacing
- more weakly defined *short-period lines* with a closer spacing.

In humans, the short-period lines are less than 2 µm apart near the root surface, increasing to 3 to 4 µm deeper into the dentine (Dean, 1998b). The longer period lines are typically 15 to 30 µm apart, so there may be between five and ten short-period lines between them. It seems reasonable to suggest that the dentine lines represent the same development rhythms as the brown striae of Retzius and prism cross striations of the enamel, although the patchy appearance of the dentine lines makes this difficult to test. Fluorochrome marker experiments (page 115) do, however, suggest that it is true.

It is not entirely clear what structural components the dentine incremental lines represent. They are seen with stronger contrast as a series of bright and dark bands under a polarising microscope, suggesting that they might be formed by cyclic changes in the orientation of collagen fibres. For this to happen, the odontoblasts

108 **Microscopic markers of growth in dental tissues**

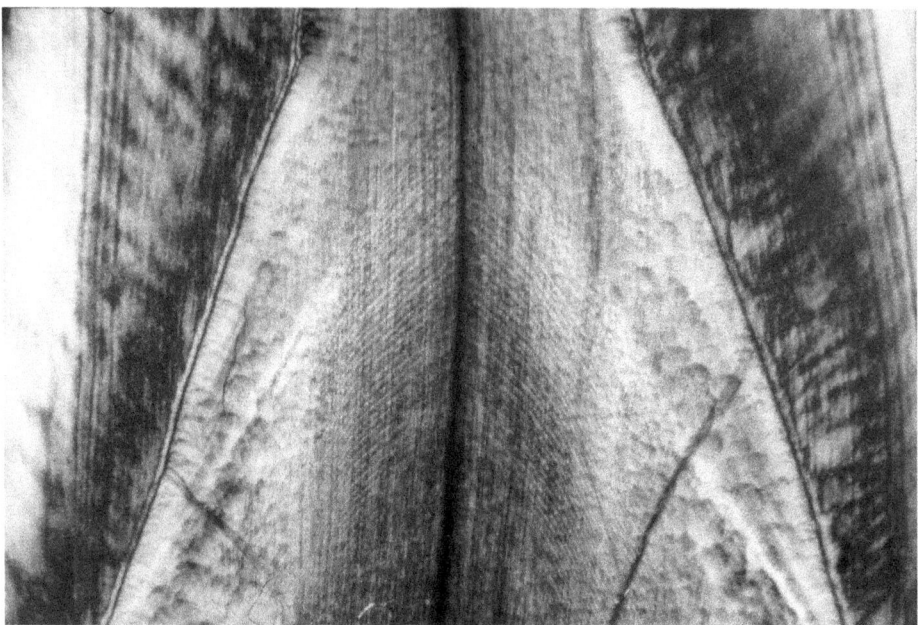

Figure 4.34 Incremental lines in human dentine. The image is located just under the dentine horn. On either side is the enamel, showing brown striae of Retzius, and the EDJ. The incremental lines of the dentine describe arches around the midline of the tooth. Their spacing suggests they are long-period lines. It is also possible to see the mineralisation pattern of the calcospherites and there are variations in mineralisation associated with long-period lines. Polished section of lower incisor, imaged in a polarising microscope as described in Appendix B. Field of view 2 mm wide.

would need to move relative to one another to change the weave of collagen during predentine matrix secretion. They cannot do this, however, as they are held firmly in place by their processes within the dentinal tubules (Jones and Boyde, 1984). The incremental lines remain visible in demineralised sections; that is sections which have been treated chemically to remove the mineral component and leave the organic component behind (Schmidt and Keil, 1971, Figure 84; Dean, 1999). They can be seen in these sections with polarising microscopy and by staining with haematoxylin (Rosenberg and Simmons, 1980). The latter suggests that there is a variation in the proportions of organic components of the matrix, perhaps between collagen and ground substance (Dean, 1995). However, they also remain visible in anorganic sections in which only mineral remains (Schmidt and Keil, 1971, Figure 88). How can this be? Some dentine mineralisation must follow the collagen fibres. This is a difficult concept, because the bulk of the mineralisation still takes place in the spheritic way described earlier, but the two patterns appear to occur simultaneously. To complicate things further, short-period layering following a wholly spheritic pattern can be seen in the granular layer of Tomes when the dentine is demineralised and then stained with silver nitrate using the Bielschowsky

Structures seen in sections of teeth 109

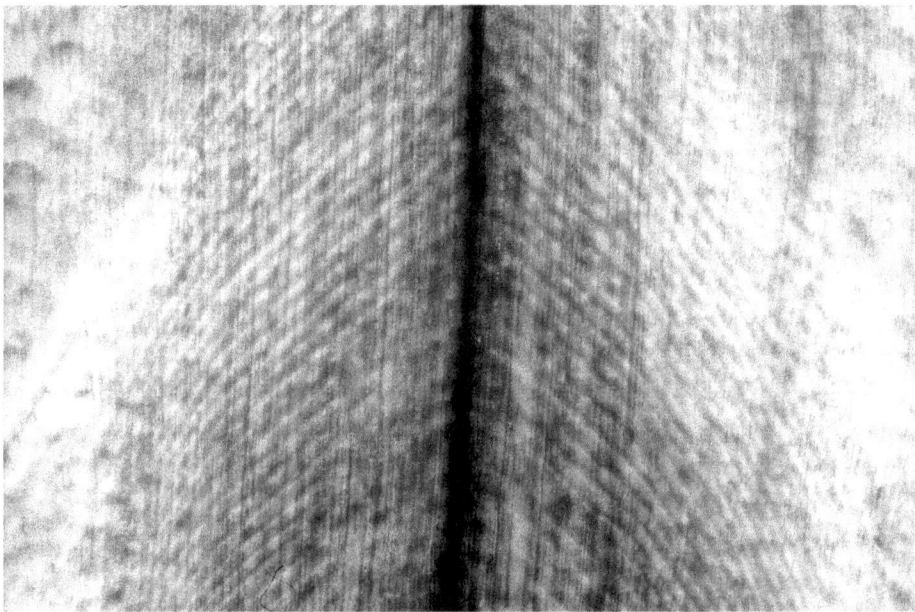

Figure 4.35 Incremental lines in human dentine. Closer view of the specimen shown in Figure 4.34. Field of view 836 µm wide.

technique (Kawasaki *et al.*, 1979; Dean, 1995), which in bone shows the degree of mineralisation (Tappen and Simmons, 1975). A series of concentric lines on average 1.8 µm apart marks each calcospherite. As the granular layer of Tomes merges with the bulk of the dentine, these lines coalesce to form longer irregular lines 2.8 µm apart (Dean, 1998b). These lines are difficult to explain, as in demineralised preparations the calcospherite structure must surely have been lost.

From the preceding discussion, it is clear that dentine developmental layering is very complicated. As Dean (1999) has observed, regular layering may be produced in the secretion of the organic predentine matrix, during its subsequent modification and also during mineralisation. In a cut and polished section, these might all be superimposed over one another. For the bulk of the dentine, however, where laminations can be seen, they show long-period lines running a fairly straight course, marking out a pattern of layering for the development of the crown and root, whose form matches well with that observed from whole tooth specimens at varying stages of development. Short-period lines may be seen in-between, but they are rarely so clearly marked. Dean (1999) defined eight rules for identifying them unambiguously, including such measures as checking that the count of short-period lines between long-period lines matches the count of cross striations between brown striae of Retzius in the enamel and checking that they follow the expected contours of root growth and that the spacing is close to the expected rate of dentine formation. The study of dentine incremental structures therefore remains a challenge, but it is important because the root continues to grow for a substantial time after the crown is completed and it thus greatly increases the range of ages over which development can be studied.

The long-period lines vary in prominence and, like the brown striae of Retzius in enamel, they make a signature which can be matched between teeth which were forming at the same age in one individual (page 121). This implies that their prominence is similarly under systemic control. Prominent long-period lines can often be matched with marked brown striae of Retzius. They do not meet exactly at the EDJ because dentine matrix formation is always slightly in advance of enamel matrix formation, so there is a slight mismatch.

Periradicular bands and dentine layering

Periradicular bands (page 76) are such slight elevations of the surface that it is difficult to recognise them in the edge of a tooth section under the microscope. Smith and Reid (2009) selected a tooth with pronounced bands on the root and marked a reference point with a blade. This was visible in the section and, by comparing measurements taken under the microscope with measurements between the marker and prominent bands on an impression of the root surface, it was possible to identify prominent long-period lines with those bands. Under the crown, it is also possible to see deviations of the EDJ associated with prominent dentine lines. In a study of the Scladina Neanderthal fossil, Smith *et al.* (2007d) suggested that minor defects of enamel hypoplasia could be matched with accentuated periradicular bands. On the basis of this evidence, several studies of dental development in fossils have assumed equivalence between perikymata and periradicular bands (page 142).

Summary

Dental enamel and dentine contain microscopic incremental structures, formed during development and preserved in the tissues well into adult life – until tooth wear removes them. In both tissues there are regular short-period and long-period incremental structures. The short-period structures are formed approximately every 24 hours and, as will be seen in Chapter 5, can be counted to determine how many days it took to form a particular part of the tooth. The long periodicity varies between individuals, in humans between 6 and 12 days, but remains constant in all teeth throughout one individual dentition. The striations produced by the long-period rhythm vary in prominence, in the same way in all teeth forming at the same age in one individual. This shows that they vary due to some underlying systemic factor which affects the whole dentition simultaneously. It also makes it possible to match the developmental sequence of one tooth with other teeth from the same individual. The highly mineralised structure of enamel preserves incremental structures with exquisite detail: the short-period structures are called prism cross striations and the long-period structures are called the brown striae of Retzius. The brown striae mark out successive positions of the enamel matrix-forming front at different stages in the development of the crown. They show that the first few months of crown development are buried under the tips of the cusps. Only later do the brown striae rise up to meet the surface, where they are marked with regular ridges and grooves called perikymata. Permanent teeth show enamel incremental structures much more clearly than deciduous teeth. Preservation of dentine is more variable and the incremental structures of dentine are much less well-defined than those of enamel. They do, however, appear to mark out the same growth rhythms and are important because of the longer period over which the root is formed after the completion of the crown.

5 Building dental development sequences

The first question outlined in Chapter 1 concerns the evolution of the slow tempo of growth that characterises living humans. Here in Chapter 5, the incremental structures of enamel and dentine which are described in Chapter 4 are used to address this by building sequences from which the rate of tissue formation and the timing of different developmental stages can be estimated.

Underlying principles

Five main conditions are required for building accurate dental development sequences:

1. Unambiguous counts of prism cross striations can be made;
2. prism cross striations and short-period dentine lines represent an approximately 24-hour development rhythm;
3. brown striae of Retzius represent a regular rhythm for which the periodicity can be established;
4. sequences of brown striae of Retzius and long-period dentine lines can be matched between teeth from one individual;
5. childbirth leaves a mark on the enamel and dentine sequences as a neonatal line.

Counting prism cross striations repeatably

First of all, it is necessary to be clear about what is being counted. Seen under the light microscope, cross striations represent the cumulative effect of reflections at all the prism boundaries in the enamel section through which the light passes. Within the depth of field, or thickness of tissue within which structures appear in focus, the striations seen are a combination of several prisms. As the microscope focus is adjusted up and down, the position and prominence of the striations changes because different prism boundaries are included. In a section cut along a buccal-lingual plane to pass through the cusp tip or central mamelon of the crown, prisms in general run parallel to the section surface. For counting cross striations they would ideally be perfectly parallel to it, but this is rarely so. Not only is it difficult to cut a section with such precision, but there is also the problem of decussation (page 90). This undulation (Risnes, 1986, figure 2 shows it well) causes the prisms to curve alternately up towards the section plane and down again a number of times as they pass through the enamel. The curves are relatively gentle in primate teeth, but their amplitude is greatest in the middle two-thirds of the enamel, gradually decreasing as the prisms run straighter towards the surface of the crown. Thus an individual prism's boundaries are at some points of this undulation relatively parallel to the section plane and at others slightly angled against it.

The discontinuities of the brown striae of Retzius shift with the prisms. Where prism boundaries are parallel to the section plane, the brown striae planes are perpendicular to it, but where the prism boundaries are not parallel, then neither are the brown striae perpendicular. This is one reason why brown striae are more sharply defined at the surface of the crown where decussation is least pronounced. There, the discontinuities which cause light scattering are superimposed one over another so their effect is reinforced. Deeper in the enamel, the discontinuities are slightly differently positioned at successively deeper levels through the section thickness so their overall light scattering effect is more diffuse, producing a fuzzy line in the microscope image. Decussation is also a likely reason why cross striations are only well defined in patches across the section, as it might be expected that they will be most strongly apparent when the prism boundaries are relatively parallel to the plane of section because, once again, the optical effects that cause them will be superimposed directly over one another. When the prism boundaries are inclined to the plane, then several overlying cross striations may be in focus at the same time, which might contribute to the appearance of intradian lines (page 94).

It is seldom practical to make cross striation counts continuously through the entire thickness of enamel. Instead, cross striation counts are made over several brown striae in many different places on the section, where both are sharply defined. Similarly, the average daily enamel matrix secretion rate (page 124) should be checked by cross striation counts and brown stria spacing measurements in many places. In the studies described here, it has been found that repeatable counts and rates can be measured in spite of the complex relationships of structures in the enamel.

Circadian rhythms and secretion rates

The regular *circadian* (approximately 24-hour) growth rhythm of prism cross striations and dentine short-period lines is crucial to everything that follows, but what is the evidence for it? One possibility is to inject a marker into subjects at known intervals, creating a recognisable line in the dentine or enamel between which tissue formation rate can be measured and short-period markings can be counted. Lead acetate is one of the oldest markers. Lead ions introduced into the bloodstream are 'bone seeking' and tend to accumulate in bones and teeth by substituting for calcium ions in the crystal lattice of the apatites, the calcium phosphate minerals that make up the inorganic component of bone and dental tissues (Hu *et al.*, 2007).

Demineralised preparations for microscopy can be arranged so that the lead ions react to form insoluble, black, opaque lead sulphide marker lines in tissue forming at the times of lead acetate injections (Schneider, 1968). It is combined with haematoxylin staining, which shows the incremental lines of dentine clearly. Okada and Mimura (1938; 1940; 1941; 1942) and Okada et al. (1940) found that short-period lines in the dentine of rabbit teeth corresponded to the number of days between lead acetate injections. Ohtsuka and Shinoda (1995) found circadian dentine lines in laboratory rats.

The central problem with lead as a marker, however, is that it is toxic, with effects on bone (Pounds et al., 1991) and dentine development. Furthermore, it is implausible that a single lead acetate injection could generate a visible lead sulphide line in either bone or dentine. Appleton (1991) investigated the effect of large lead acetate doses in laboratory rats using scanning electron microscopy in back-scattered electron mode (Appendix B). This technique clearly showed that the lead acetate injections were marked by sharply defined zones of poorly mineralised dentine. Blood tests demonstrated that disturbances to calcium metabolism accompanied each lead acetate injection. Using energy-dispersive X-ray spectroscopy in the scanning electron microscope, Appleton also found no evidence that lead was in fact concentrated in the region of the supposed 'lead lines'. It is therefore the toxic effects of lead that create the lines and, even though the count of lines between markers matches the number of days elapsed, it must be acknowledged that lead is a poor marker for normal development.

Another early marker was sodium fluoride. During the 1930s, naturally occurring high levels of fluoride ions in drinking water had been established as a cause of hypomineralised enamel defects (page 189). These were reproduced experimentally by adding sodium fluoride to the diet of laboratory animals and it was found that injections of sodium fluoride produced sharp lines of defective mineralisation in enamel and dentine. In an experiment which would not nowadays be contemplated, Schour and Poncher (1937) administered 25 doses of sodium fluoride to a terminally ill child. When the child died at the age of 9 months, the teeth were sectioned and the doses of fluoride were visible as lines in both enamel and dentine. Measurements between these lines were taken along the dentinal tubules and enamel prisms, and the rate of tissue deposition was calculated from the known intervals between injections of sodium fluoride (Appendix A, Table 25). The rate may well have been affected by the fluoride and, whilst the dentine was otherwise normal, the enamel did show associated sharp bends in the course of the prisms.

A further study by Schour and Hoffman (1939) tested fluoride markers against an alternative, Alizarin Red S. Alizarin was originally extracted from the root of the madder plant but, since 1869, has been synthesised artificially (Schour et al., 1941). Madder colours bone red and was used experimentally in the eighteenth century as a marker for bone development (Payton, 1932). Alizarin acts by reacting with calcium to form an insoluble pigment or 'lake' (Schour et al., 1941) in the mineral component of dental tissues and bone and, if injected, it creates a sharp red line representing the time of injection. Schour and Hoffman used this technique to find the rate of dentine formation in rats and rhesus macaques. They used sodium fluoride to do the same. The rates were identical, adding support to their use of fluoride injections, but Alizarin itself might also disrupt development, as suggested by Yen et al. (1971).

Most recent studies have used fluorochromes as markers. These are dyes which fluoresce in the microscope under ultraviolet illumination. When introduced into the bloodstream, they are carried to sites of new mineralised tissue formation where they are incorporated into the apatite crystals being formed at the time, binding to calcium ions to form complexes (Erben, 2003). Microscope sections are made by cutting and polishing, rather than by demineralisation and microtome, because the marker is bound to the mineral rather than the organic component. Under fluorescence microscopy, the glowing band made by the fluorochrome marker is superimposed over the other structures of the tissue. Alizarin acts as a fluorochrome, fluorescing red, although the form used nowadays is Alizarin Complexone rather than Alizarin Red S. For studies of bone development in humans, the most commonly used fluorochromes are the tetracyclines, a group of antibiotics for which the calcium binding is an unwanted side effect as it stains teeth if administered to children, although in the past tetracyclines were regularly prescribed for them. In animal studies, the most frequent fluorochromes are Calcein (which fluoresces green under ultraviolet light) and Calcein Blue (fluoresces blue).

One confusing factor is that enamel and dentine naturally fluoresce a faint yellow-green under ultraviolet light (Harcourt et al., 1962), creating a diffuse glow or instead following the outlines of some internal structures. This is known as autofluorescence. The artificially produced fluorochrome bands are much brighter, but they are clearly defined only in dentine. Enamel mineralisation changes greatly during maturation (page 32), which is a gradual process that 'smears' the fluorochrome effect through a broad area. A similar difficulty with dentine is that mineralisation lags behind the formation of organic matrix (page 87). Where short-period dentine lines are features of the matrix, fluorochrome markers must be positioned slightly to occlusal of each associated line. The lines are more directly marked where they are features of the mineral component. This issue may be resolved in some cases when the marker is associated with a pronounced incremental line which gives a clear starting point from which counts can be made. The implication of such a line is that the fluorochrome has some toxic effect that interferes with the sequence of development.

Many marker-based studies of dentine and enamel development have used tetracyclines. One approach is to inject them at known intervals into laboratory animals. The other is to use the collections of extracted human teeth made by dental hospitals for training student dentists, which at one time contained a considerable proportion showing the characteristic markers of treatment with tetracycline during childhood. In the latter case, the interval between tetracycline treatments is not known, but the lines mark out parts of the teeth that mineralised at the same time, so these can be matched up. In dentine, the brightly fluorescing tetracyline bands follow the spheritic mineralisation pattern of dentine (page 102), rather than the flat layering of collagen fibres (Kawasaki and Fearnhead, 1975). This confirms that they primarily mark the front of mineralisation, but Simpson (1981) found that the fluorescing tetracycline markers were retained in a less pronounced form even when the dentine was demineralised, so it appears that both mineral and organic components are marked in some way.

Tetracyclines do not greatly affect the development of dentine so long as the doses are relatively small. Kawasaki and Fearnhead (1975) showed in pigs and goats that, whilst there were some signs of dentine disruption associated with fluorescing lines from tetracycline doses above 62.5 mg/kg, doses less than 31 mg/kg produced no signs. In laboratory studies, the dose of tetracycline can be controlled and, in humans, the expected therapeutic dose would be 5–26 mg/kg. In extracted human teeth, they found that few tetracycline bands were associated with evidence of disturbance. Even for those few cases, the disturbance was not more pronounced when associated with broader or more intensely fluorescing markers, which suggests no particular relationship with dose.

Miani and Miani (1971) injected dogs with small doses of tetracycline every 12 hours to show by the regularity of the marker bands that dentine matrix was secreted at a steady rate. Yilmaz et al. (1977) injected young pigs with xylenol orange and tetracycline markers with a 14-day interval between them. Under the microscope, 14 short-period dentine increments could be counted between each pair of markers. Two further studies were based on laboratory pig-tailed macaques from the Regional Primate Research Center of the University of Washington, which had been marked with fluorochromes as part of a study of both pre- and post-natal bone development (Newell-Morris et al., 1980; Newell-Morris and Sirianni, 1982). Bromage (1991) sectioned the teeth of two monkeys from this study. In the permanent first molars, the fluorescent bands were well marked in dentine and some were faintly evident in enamel. They were also defined by sharp increment lines in enamel and counts of prism cross striations between these matched well with the number of days between injections. Smith (2006) found the same in teeth from 17 macaques from the collection. They also identified more closely spaced features in between cross striations, intradian bands, which were more difficult to resolve under the microscope. These features occurred in patches and appeared to represent a 12-hour periodicity. Molnar et al. (1981) investigated four laboratory rhesus macaques which had been injected with low doses of tetracycline. They also identified incremental lines but, with a mean spacing of 16.8 µm and a periodicity of four to six days, these seem to have been long-period lines. Dean (1993) found that short-period dentine lines matched a daily rhythm in the tetracycline marked teeth of a single rhesus macaque.

Studies of human teeth have all been based on markers from tetracycline which was administered therapeutically at unknown intervals. In extracted teeth from dental hospital collections, Kawasaki et al. (1979) used tetracycline bands to match long-period and short-period dentine lines with the brown striae of Retzius and prism cross striations in neighbouring enamel. Dean and co-workers (Dean et al., 1993a; Dean and Scandrett, 1996) examined the teeth of an unidentified forensic specimen in which there were 30 fluorescent lines marking tetracycline treatments

Underlying principles

between approximately 1 and 12 years of age. The intervals could be estimated because some fluorescent bands continued into the enamel, where they were associated with pronounced brown striae of Retzius, between which cross striations could be counted. Long-period lines were prominent in the dentine, at spacings varying between 15 and 30 μm, and formed at a regular rhythm equivalent to eight cross striations. This was the same as the periodicity of the brown striae.

The simplest approach, however, for checking the periodicity of enamel increments is to make total counts of prism cross striations or brown striae, from initiation to completion of the crown, and then to compare these with the crown formation time expected from other studies (Chapter 3). Hans Asper (1916), a dentist from Zürich, sectioned ten human permanent upper canines and counted an average of 197 brown striae in each. From other evidence, he reckoned it would take 3 years 8 months to form the crown, or about 1340 days, which is not far from the estimates given in this chapter. Dividing 1340 days by 197 brown striae, he estimated an average of 6.8 days to form each and, as he had counted five to ten cross striations between regular brown striae in his sections, he felt it reasonable to suggest that these cross striations represented a rhythm of 24 hours.

Another early report is the conversation recorded on a train ride during a tour of the USA by Asper's fellow countryman Alfred Gysi, Professor of Histology at the University of Zürich (Gysi, 1931). Professor Gysi was reported as saying that the number of cross striations in upper premolars varied between about 1200 and 1700, with the average near 1500. If these represented days, then it would imply 3.3 to 4.7 years (mean 4.1) for the crowns to develop, which was close to what he expected for premolar crown development.

The findings of Asper and Gysi still make sense today. Careful microscopy of good sections shows that cross striations and brown striae are regular. If they are counted between the first and last enamel matrix to be secreted, then the total matches reasonably well with other evidence for the time taken to form the crown in that tooth type. When dealing with a regular periodicity over a long time, a small variation in cross striation interval would make a big difference. For example, if an upper canine takes 1300 days to form, or 31 200 hours, then if cross striations represented a 28-hour rhythm (rather than 24), there would be 1114 of them. This is a six-month difference. Dean *et al.* (1993a) and Reid *et al.* (1998a) estimated the age at initiation and completion for permanent tooth crowns in the dentitions of single individuals (Appendix A, Tables 27 and 28), assuming that cross striations represented a 24-hourly rhythm. There were differences between them, but not beyond the range of individual and population variation reviewed in Chapter 3. They can be compared directly with macroscopic observations (Liversidge, 2000) of incisor and first molar crown completion based on 50 children from the crypt at Christ Church, Spitalfields in London, whose ages are independently known from coffin plates and parish records (page 223). Again, they fall well within the range of variation. Antoine *et al.* (2009) tested cross striation periodicity using teeth from five Spitalfields children. The ages-at-death were independently known and the tooth crowns were still forming at the time of death. Cross striations were counted from the first formed enamel matrix or neonatal line under the permanent first molar cusps up to the last matrix being formed when death occurred. The counts matched well with the age when allowance was made for the fact that the most recently formed matrix at death would have been only partly mineralised and therefore was incompletely preserved (page 88).

In conclusion, all the evidence available suggests a circadian rhythm. Most markers used to label development are capable of disrupting normal development, but the periodicity of prism cross striations or short-period dentine lines remains constant within one individual whether or not the count is between markers associated with disturbance. Thus, even if enamel or dentine development was affected, the rhythm represented does not seem to be and all labelling studies so far carried out have suggested an approximately 24-hour rhythm. This is supported by counts of cross striations for the duration of individual tooth crown development which match the timing expected from children whose age-at-death is independently known and whose teeth were still forming at the time of death.

Periodicity of long-period lines

To date, no careful study has yet shown variation in cross striation count periodicity of brown striae of Retzius within any one tooth or between the teeth of one individual. Individuals do, however, vary between one another in the largest studies, which have been on modern human permanent teeth. In 49 lower canines from the Medieval Tirup site in Denmark (Reid and Ferrell, 2006), the modal periodicity was 8 and the mean 8.5 (Figure 5.1), periodicities of 8 and 9 accounted for 68% of the individuals and 95% had periodicities between 7 and 10. FitzGerald's collection of 96 incisor and canine sections showed some population differences, with modal periodicities of 10 in Africans, 9.2 in Europeans and 9 in Native Americans (FitzGerald, 1998). Reid and Dean (2006) found that almost all incisors and canines of Africans had modal counts of 9 or 10, whereas the same teeth in Europeans had modes of 8 or 9. The molars of both populations all had modal counts of 8. Further work with these molars (Smith et al., 2007a) showed that about 72% of individuals had periodicities of 8 or 9, with 95% between 7 and 10 (Figure 5.1). The overall minimum was 6 and the maximum 12.

Reid and Ferrell (2006) noted that in permanent lower canines from one human population the count of perikymata in the imbricational enamel was inversely related to the periodicity. That is, the lower the count of cross striations between brown striae, the more perikymata at the surface and vice versa. At first sight this might suggest that the overall formation time for imbricational enamel remains constant, but this is not the case because the relationship is not linear. With the lowest periodicities, the count of perikymata is disproportionately higher than that associated with the higher periodicities. It is not known whether or not these relationships hold for other hominins, or other primates in general, so at present they do not allow periodicity to be estimated from the perikymata count. This is an important question for fossils which cannot be sectioned. There is, however, too

Underlying principles

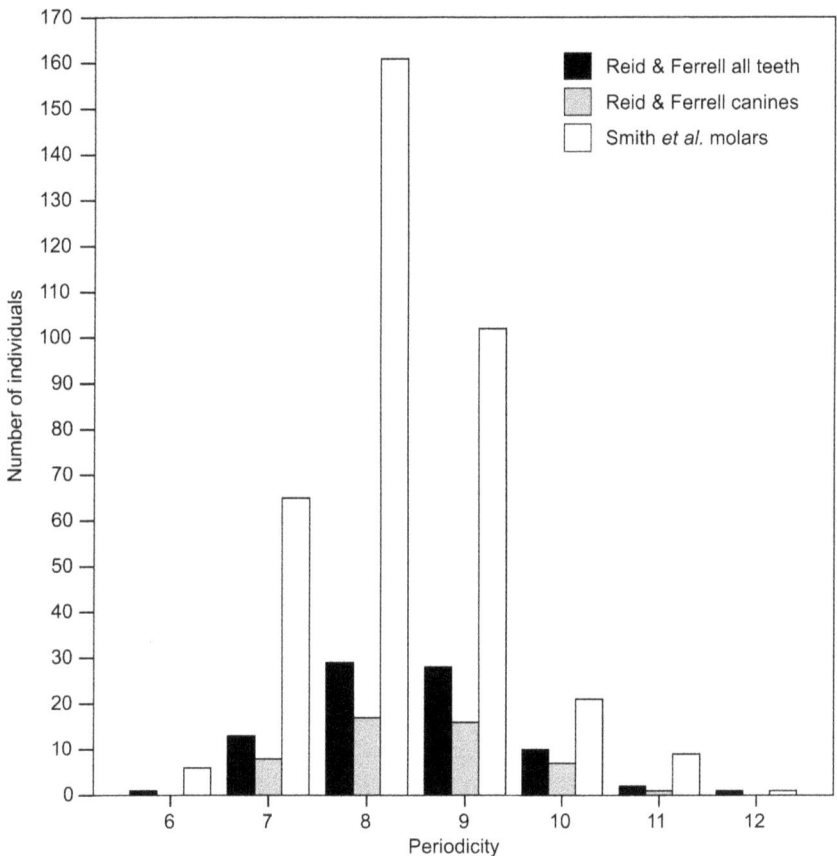

Figure 5.1 Periodicity of brown striae of Retzius in human teeth, assessed by counts of prism cross striations between them. Black and grey bars represent frequencies for 84 permanent teeth of all types from the Medieval Danish site of Tirup and for 49 lower canines only, scaled from the histogram in Reid and Ferrell (2006), Figure 1C, p. 197. White bars represent frequencies for molars from 365 individuals from South Africa, North America, Denmark and the UK, scaled from the histogram in Smith *et al.* (2007a), Figure 2, p. 183.

much variation, between individuals, populations and species, to know what value should be assumed. Fortunately, it is now possible to make cross striation counts without sectioning by using confocal microscopy and synchrotron radiation micro-CT (Appendix B).

Variation in long-period lines

As described in Chapter 4, brown striae of Retzius vary in their prominence and appearance throughout one tooth crown section: relatively more sharply defined, more diffuse, brighter or darker. Several such striae may together form a recognisable pattern which can be seen on both the labial and lingual sides of the same tooth, or matched in the equivalent tooth (antimere) from the other side of the

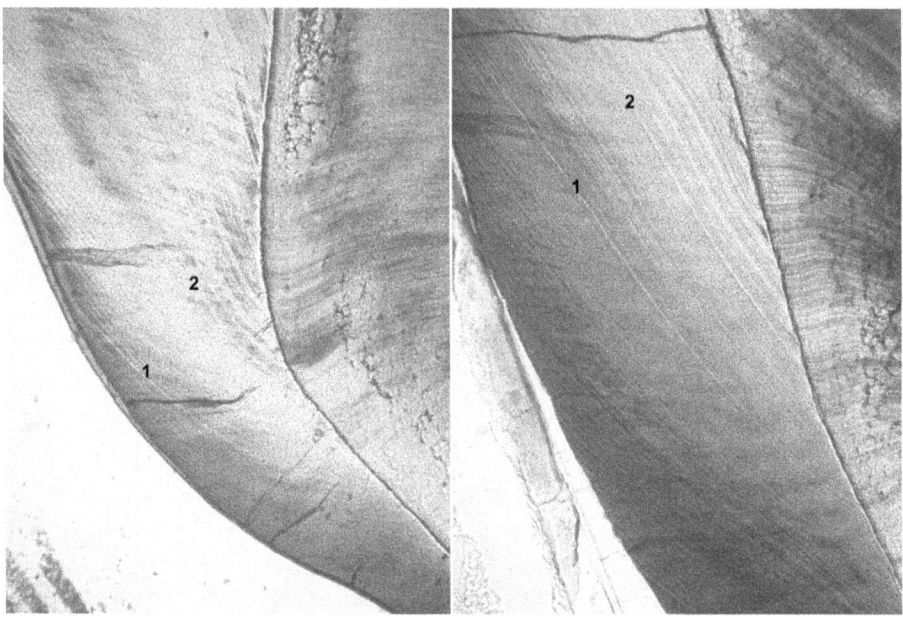

Figure 5.2 Sequences of prominent brown striae of Retzius matched between a human permanent lower first molar (left image) and a lower canine from the same individual. Two of the more prominent striae are marked '1' and '2', but the whole sequence matches well. The first molar crown starts to form earlier than the canine, so the matching striae are further down the crown side. Matches between molars and anterior teeth are generally more difficult than between teeth from the same part of the dentition, so this is a particularly clear example. Polished sections viewed with conventional transmitted light microscopy as described in Appendix B. In each case, the field of view is 1.5 mm wide. The specimens come from the crypt of Christ Church, Spitalfields in London and were kindly loaned by Daniel Antoine, The British Museum.

dentition (Gustafson, 1955). Furthermore, the pattern can be recognised in other teeth from the dentition whose crown formation times overlap (Figure 5.2). From this evidence, it is clear that the variation in striae is due to some systemic factor which does not change the even spacing of brown striae or count of cross striations, but causes variation in the way in which the striae are perceived under the microscope. As the brown striae seem to be generated largely by light-scattering effects from some discontinuity within the enamel (page 95), it seems reasonable to suggest that the discontinuity is more pronounced in those striae which appear more marked. Gustafson and Gustafson (1967) used microradiography to show that most striae were associated with radiolucent lines, suggesting they were less heavily mineralised than the intervening enamel. Some pronounced striae were associated with radio-opacities, suggesting heavier mineralisation. In a similar way, they are visible using synchrotron radiation micro-CT (Smith *et al.*, 2007c; Tafforeau and Smith, 2008; Smith *et al.*, 2010b).

Underlying principles

Long-period lines in the dentine also show variation in prominence and, by following them to the EDJ, can often be matched with prominent brown striae of Retzius. The dentine lines at the junction are always slightly displaced to cervical relative to the enamel lines because dentine development is a little in advance. Groups of dentine lines can also be defined and matched with other teeth from the same individual (Gustafson, 1955).

At the crown surface, there is variation in the spacing and prominence of perikymata, seen in defects of enamel hypoplasia (page 164). Such defects are often associated with prominent brown striae in the underlying enamel. A group of perikymata can therefore be matched between antimeres from one individual, down to the count of perikymata within the group. Matching between different teeth of similar type – for example, incisors with other incisors or canines – is usually clear, but molars are harder to match with incisors because of their different crown morphology. Detailed examination of potential matching regions (Figures 4.6 and 4.7) is required.

What can these matches of lines be used for? Firstly, teeth from one individual can be identified unambiguously. It is most unlikely that an exactly similar pattern of prominent lines and count of intervening perikymata/striae and cross striation periodicity could be found in another individual. There are important applications in fragmentary and commingled burials, from forensic cases or archaeological sites and in fossil assemblages. Development sequences are extended by matching prominent lines to carry the count of cross striations on from one tooth to another. In this way, a full sequence can be built for one individual from the first permanent tooth to initiate development, to the next and so on to the last enamel matrix to be formed. Such extended sequences are the basis of the discussions in this chapter.

The neonatal line

The anchor point for development sequences is the neonatal line (page 100). This is found in all deciduous teeth and one or more of the cusps of the permanent first molar. In essence, it is a particularly prominent incremental line in the dentine and enamel. The distinguishing characteristics are the positioning of this line within the tooth crown and the observation that the prenatal enamel and dentine are more homogeneous than the post-natal tissues. It is also often quite sharply defined, but not necessarily so, and this may partly be related to the thickness and orientation of section. Occasionally, there may be faint incremental lines in prenatal enamel, but they are never so strongly developed as post-natal brown striae. In spite of all this, the neonatal line can be difficult to identify, particularly in permanent first molars where a relatively small amount of tissue has been formed at the time of birth. If the neonatal line is not strongly developed, there may not be enough prenatal enamel and dentine formed to contrast with the post-natal tissues. In addition, the centring of the plane of section is of vital importance. It must be exactly over the dentine horn underlying the mesiobuccal cusp (and ideally also the mesiolingual).

The cut is decided on the surface morphology of the crown and it takes considerable experience to position the tooth in the saw (Appendix B). Difficulties with the plane of section are probably the most common cause of problems in identifying the neonatal line in permanent first molars.

But how strong is the evidence that the neonatal line does represent the day of birth and not some other growth-disrupting event near that time? The idea originally came about as part of 'tooth ring analysis', a series of papers by Isaac Schour and co-workers (page 191) suggesting that different phases of growth were marked by variations in the quality of enamel which recorded physiological and pathological changes in metabolism. Schour (1936) described a distinctive line in both the enamel and dentine of human and macaque deciduous teeth. Such lines were found in all deciduous teeth and their position matched the expected state of development for different teeth at birth. They were *not* found in the deciduous teeth of full-term but stillborn babies. In addition, they were absent from permanent teeth except the mesiobuccal cusp of first molars. Schour therefore suggested that the distinctive lines were formed at birth or soon after it. As proof, he referred to his experiment (Schour and Poncher, 1937) in which a child was given injections of sodium fluoride markers at known intervals after birth (page 114). He identified a line in the dentine and enamel in the deciduous teeth of this child which, from a daily dentine apposition rate calculated between the markers, corresponded with the birth day of the child. Schour and Kronfeld (1938) showed similarly marked enamel and dentine lines in the deciduous teeth and permanent first molars of a child who had suffered a severe brain injury at birth. The deciduous tooth crowns were marked on the surface by hypoplastic defects (page 192) and enamel was missing entirely from the mesiobuccal cusp of the permanent first molars. Once again, from the known age-at-death of the child, it was possible to show that these lines and defects corresponded with the timing of birth.

Talmi and co-workers (Talmi *et al.*, 1986; Eli *et al.*, 1989) collected teeth from children with normal births, abnormally difficult births and delivery by caesarean section. They measured the apparent width of the line and found it significantly wider in the difficult birth group and narrower in the caesarean group. It should be pointed out that, like a normal incremental line, the neonatal line seems to be a light-scattering phenomenon for which the prominence is affected by orientation of the section, its thickness and the way the microscope is adjusted. The apparent width of the fuzzy dark band seen in the microscope does not represent the actual width of a structure in the section, but presumably in some way represents the degree of disruption.

The position of the neonatal line remains the strongest proof of its timing. Skinner and Dupras (1993) collected exfoliated deciduous teeth from children who were known to have been born preterm, full-term or post-term, and found that there were consistent slight differences in the positioning. The lines were closer to the first dentine and enamel formed in children born preterm, further away in those born full-term and furthest in children born post-term, as would be expected if they were truly neonatal in origin.

Prenatal enamel thickness will depend on the gestational age at which the permanent first molar was initiated and the timing of birth relative to that (Antoine et al., 2009). It would, however, be expected that 8 to 12 weeks of matrix secretion would have occurred under the mesiobuccal cusp for a normal full-term birth (preterm would be less and post-term more). That should amount to 220–330 µm thickness over the dentine horn. If a neonatal line is not seen, it is necessary to count from the first-formed matrix and bear in mind the potential error.

Methodological issues

Cuspal or appositional enamel

The cuspal enamel poses particular problems. This is the region of the crown that underlies the tip of the cusp (or central mamelon of incisors), above the dentine horn. All the enamel matrix increments are buried under the surface and are not represented by perikymata. It grows in an *appositional* way, each increment being laid down over another until the full crown height is formed. Sometimes the cuspal enamel is distinguished as *appositional enamel*, as opposed to the *imbricational enamel* of the crown sides (which has imbrications or perikymata). The problem with this terminology is that, in fact, the imbricational enamel is also formed appositionally, although each layer overlaps the previous layer down the crown side so that previous layers are not completely covered.

The centring of the section is very important as it must pass exactly through the dentine horn and cusp tip, not only to expose the first increments of enamel matrix, but also to minimise the difficulties caused by decussation. Under the cusp, all the prism undulations of decussation pack closer together. When seen under the microscope, this produces such a complex patchwork of bright and dark areas that it is known as gnarled enamel (page 90). Hirota (1982) used X-ray diffraction techniques to measure the orientation of the apatite crystallites which, in cuspal enamel, run parallel to the prism boundaries so they follow their orientation. Decussation is most marked in the deepest two-thirds of cuspal enamel and, here, Hirota found that the prisms were wound into two conical spirals, wrapped inside one another with one layer going clockwise and the other anticlockwise. The prisms ran relatively straight in the outer one-third of the cuspal enamel. This is an extremely complex structure which means that prisms are rarely parallel to the section surface, so that both the cross striations and brown striae tend to be poorly defined. These difficulties are minimised if the section plane cuts through the centre of the prism spirals, immediately under the cusp, but it is always difficult to make cross striation counts from the first-formed tissue to the tip of the cusp.

Decussation is not, however, the only problem with cuspal enamel. Another is tooth wear. All fossil hominins and, until recently, all modern humans lived a lifestyle which caused their teeth to wear rapidly. The dental wear rate was rapid even in children so, as soon as it emerged into the mouth, tissue would start to be lost from the cusp tips. This is frustrating, because a tooth in which the root is still developing and which still seems at first sight to have intact cusps may nevertheless have lost a substantial number of the final increments of cuspal enamel. It might seem logical to exclude such teeth from study, but the reality of research with rare primate fossils is that we must work with what we have, so many studies examine the trend of brown striae of Retzius to estimate which of them could reasonably be said to mark the last cuspal increment.

The first requirement in estimating the time taken to form the cuspal enamel is to establish the *mean enamel formation rate* or *daily secretion rate*, which is estimated by dividing a measured length of prism boundary by the count of prism cross striations along it. It is slower near the EDJ and faster at the tip of the cusp. In the human molar shown in Figure 5.3, it varies from 2.5 to 6.4 µm per day (Dean, 1998a). Following this, six strategies have been used to estimate cuspal enamel formation time:

1. Cross striation count along an apparent prism track. It is often possible to follow the course of prisms from the EDJ at the side of the dentine horn tip through the enamel at the side of the cusp. Given the intensity of decussation, it is unlikely that this is a single prism, but it presumably represents the course along which prisms deviate from a straight line and thus yields a valid count. In Figure 5.3 the count is 460.
2. Cumulative prism lengths, stepped down along the EDJ. Brown striae are often clearer next to the EDJ than they are in the cuspal enamel above. At the dentine horn, a length of prism can be measured out to a prominent brown stria. This can be followed down to the point at which it meets the EDJ. From there, another prism length can be measured out to the next prominent stria and so on, in zigzag fashion down to the stria that represents the first perikyma and thus the edge of the cuspal enamel. All the prism lengths from each step are added together and then divided by the mean enamel formation rate near to the EDJ. In Figure 5.3, the cumulative prism length is 1383 µm and the enamel formation rate is 2.9 µm per day, which gives a time of 477 days.
3. Cumulated EDJ extension time. As the enamel over the cusp grows in height above the dentine horn, it also spreads outward along the EDJ from the dentine horn. Shellis (1984) devised a formula (Figure 5.3) for calculating the *enamel extension rate* – the rate at which this spread took place – from the mean enamel formation rate and the angles that the prism boundaries and brown striae make with the EDJ. This extension rate decreases down the crown (Dean, 1998a), so the EDJ must be divided into lengths for which the extension rate is estimated separately. From this, it is possible to estimate the number of days required to complete each length, along to the point at which the brown stria marking the edge of the cuspal enamel meets the EDJ. These lengths are then summed. In Figure 5.3 the extension rate varies from 13.6 to 3.0 µm per day and the overall sum for EDJ extension is 474 days. By following the EDJ, this approach avoids the difficulties of the overlying cuspal enamel.

Methodological issues

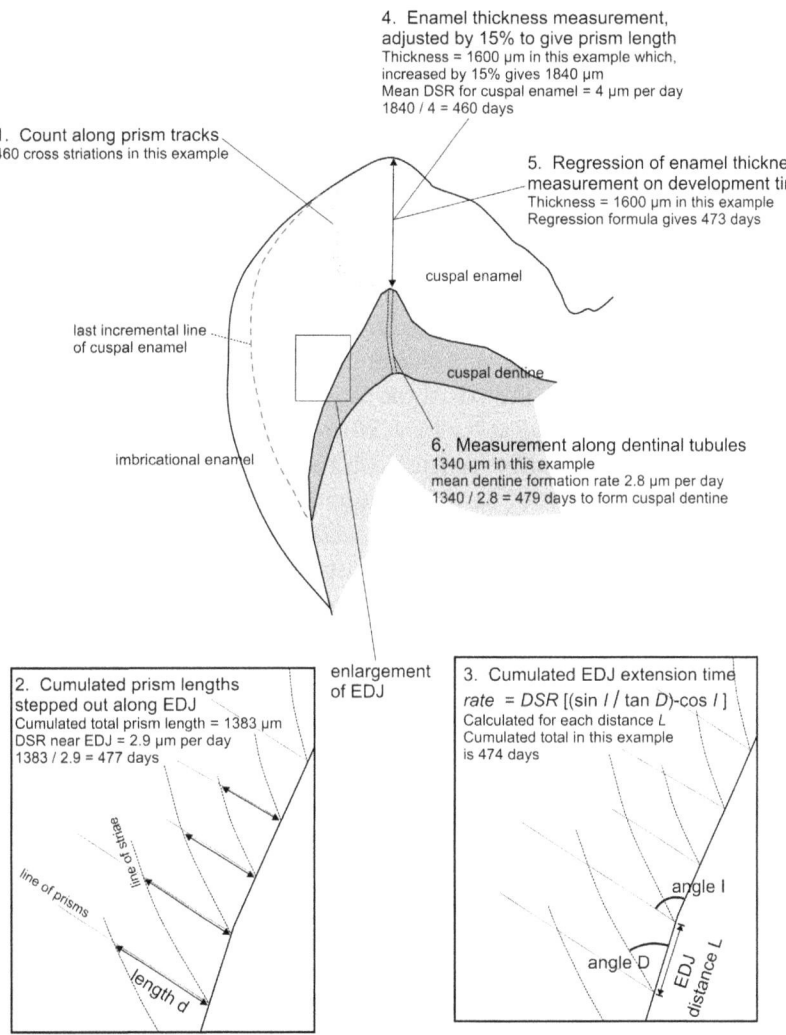

Figure 5.3 Six methods for determining the time taken to form cuspal enamel. The numbering of the methods refers to the text. Section of a human permanent second molar, redrawn from Dean (1998a), Figure 4, p. 456. DSR is the daily secretion rate or mean enamel formation rate determined by measurement along the line of the prism boundaries.

4. Enamel thickness measurement at the cusp tip, converted to prism length by 15% rule. Risnes (1986) estimated the effect of decussation on the length of a prism passing from the tip of the dentine horn to the tip of the cusp. Reid *et al.* (1998a) suggested from this that the prisms might be 15% longer than if they had run a direct course from the EDJ to the tip of the cusp. For Figure 5.3, the enamel thickness measurement scaled from the drawing is 1600 μm which, increased by 15%, gives an estimated prism length of 1840 μm. The overall mean enamel formation rate for the full thickness of cuspal enamel is 4.0 μm per day (Dean, 1998a), which yields 460 days.

5. Enamel thickness measurement at the cusp tip, converted to formation time by regression formula. Schwartz and Dean (2001) found that there was a strong relationship between cumulative cross striation counts out from the EDJ and cumulative cuspal enamel thickness in chimpanzee, gorilla, orangutan and human canines. They established polynomial regression equations with which the counts could be estimated from thickness measurements. The same approach was used by Dean *et al.* (2001) for a range of fossil taxa and teeth (Appendix A, Table 26). Taking the 1600 μm thickness for Figure 5.3, the appropriate regression formula gives 473 days.
6. Estimation of the time taken to form cuspal dentine at the horn. The trend of the long-period lines in dentine makes it possible to identify the point directly under the dentine horn in the axis of the tooth which corresponds to the last brown stria of the cuspal enamel at the EDJ. Where short-period dentine lines are visible, the rate of dentine matrix secretion can be determined. Dean and Scandrett (1995) found that the rate of development remained relatively constant in the main axis of the tooth and so it is possible to make a reasonable estimate of the time taken to form the dentine underlying the cuspal enamel (Dean, 1998a). In Figure 5.3, the measurement of cuspal dentine thickness at the dentine horn axis is 1340 μm and the dentine secretion rate is 2.8 μm per day, which gives a timing of 479 days.

In the example given here, these different methods therefore vary from 460 to 479 days; a difference of about 4%. Given the difficulties of the cuspal region, this is encouraging and Dean (1998a) found reasonable agreement for chimpanzee and orangutan molars, as well as for human. Method 4 works well in Figure 5.3, but there are some uncertainties. Reid *et al.* (1998a) quoted Risnes (1986) as the source of their 15% factor, but it is clear that he was writing about lateral or imbricational enamel and suggested that the deviation for cuspal enamel would be greater. Risnes (1986) proposed instead a mathematical model for the length of prisms in the cuspal region. The parameters for the equation are difficult to estimate, but it is possible to use it to make an adjustment for the increased length of a prism. In Figure 5.3 the cuspal enamel thickness of 1600 μm gives, from Risnes' formula, estimates for adjusted prism length ranging from 1926 to 2193 μm depending on the parameters used. With a 4 μm per day mean enamel formation rate this gives 482–548 days formation time, which compares badly with the other estimates, including the 460 days based on a 15% adjustment. The conclusion that a factor derived from lateral enamel works better than a formula designed for cuspal enamel does give some pause for thought. In any case, as Smith (2008) has pointed out, the 15% factor cannot be used for non-human hominoids, particularly because some taxa seem to have a much smaller degree of decussation than modern humans.

Lateral or imbricational enamel

If the tooth is unworn and the periodicity of its long-period lines has been securely established, then there is usually little difficulty in counting the sharply defined brown striae in the section, just below the crown surface, to provide an estimate of lateral enamel formation time. Similarly, perikymata can be counted at the surface by scanning electron or optical microscopy. If the tooth cannot be sectioned to determine the cross striation periodicity, then the only reliable way to estimate it is by synchrotron radiation micro-CT or confocal light microscopy. To estimate the full crown formation time, it is necessary to add together the separate estimates for cuspal and lateral enamel. Measurements from a conventional micro-CT scan may be sufficient to yield an enamel thickness from which an estimate for cuspal enamel formation time might be derived, but the cross striations would not be visible and the enamel formation rate could not be estimated.

Where the tooth is worn, or the surface abraded, perikymata cannot be counted reliably and, without sectioning or synchrotron radiation micro-CT, cuspal enamel formation time cannot be estimated because the full height at the cusp tip is not preserved. If the tooth can be sectioned, however, these difficulties can be avoided. Dean developed Method 2 above to provide an estimate of formation time for the whole crown by measurements along the EDJ, without the necessity for considering cuspal enamel separately from lateral enamel (Dean, 2012). The mean enamel formation rate is established near the EDJ in the cuspal, mid-crown and cervical regions. On a montage of microphotographs, at the side of the dentine horn tip to avoid the gnarled enamel, a line is drawn out parallel to the prism boundaries to a scale length of 200 µm. Another line is drawn from the end of this line, parallel to the brown striae, back to the EDJ, making the first of 10 to 15 zigzags down the crown side, measuring out a 200 µm prism length each time, except for the last, which needs to be adjusted so that the zigzag meets the cervical margin exactly. The appropriate mean rate of matrix secretion is applied to each of the prism boundary lengths and then the sum provides the total crown formation time. If the relative timing for particular points in the sequence needs to be determined (for example, the completion of the cuspal enamel, or prominent brown striae/perikymata), then extra zigzags can be added at those points.

Groups of prominent brown striae that can be matched between teeth make it possible to transfer the cross striation count between sections of teeth formed at progressively older ages, until the last-formed enamel is reached. In this way, it is possible to build an overall sequence from the permanent first molar, via the incisors and canines, to the second molar, which is completed between 6 and 7 years after birth in humans. In humans, it is not possible to make a direct link with the sequence of enamel formation in the third molar because it does not overlap with that of the second molar (it might be possible in chimpanzees). Similarly, variation in perikymata may be matched between teeth to extend a sequence.

Enamel and dentine matrix being formed at the time of death

Where crowns were still forming at the time of death, sequences can be counted from the neonatal line to the last-formed enamel matrix. In theory, this yields a very precise estimate of the age-at-death. In practice, one difficulty is in the preservation of enamel matrix for which the process of maturation had not yet been completed. Enamel matrix as first secreted is only one-third mineral. The front of maturation, at which there is a pronounced increase in mineralisation, follows behind the matrix-forming surface by about 50 μm (Boyde, 1990). Thus a zone corresponding to perhaps 12 cross striations will be less heavily mineralised than the remainder of the crown. In a fresh specimen, the newly formed enamel matrix is marked by sharply defined Tomes' process pits. These are not generally visible in archaeological specimens (Figure 4.19) because the less mineralised tissue has been eroded and the appearance varies according to the post-depositional history of the specimen. When matching known age-at-death with the total counts of cross striations from the neonatal line to last-formed enamel which could be observed, up to 14 days of enamel matrix secretion may be lost through decomposition or erosion (Antoine et al., 2009). Similarly, a layer about 40 μm thick of unmineralised predentine matrix would have been at the developing dentine surface. After death, this decomposes to expose the developing calcospherites of the mineralising front, which may be seen under the microscope.

Development chronologies for living and fossil primates

Most detailed chronologies have been built from cross striation counts in microscope sections of teeth. This requires permission from curators and the presence of specimens that are not too worn to be useful. For some questions it also requires dentitions at an appropriate stage of development in which the key teeth are exposed to view, but new technologies of micro-CT imaging do provide alternatives. Human teeth extracted during treatment have for many years been collected in dental schools and these form the bulk of the available sections. The difficulty is that they are isolated teeth, rather than complete dentitions. Some substantial archaeological collections of human dentitions have, however, been made available.

Dental sections for monkeys, chimpanzees, gorillas and orangutans mainly come from museum collections. The specimens are much less numerous in the first place and it is more difficult to obtain permission to section them. As fossils, primates in general are rare. For any animal to be preserved as a fossil, it needs to be buried rapidly by sediments which are then not disturbed; for example, in a lake or swamp, fine volcanic ash, or a cave. An arboreal or open country mammal is unlikely to die conveniently in these situations and large accumulations of primate bones and teeth require some very special circumstances indeed. An example is the extraordinary assemblage from Rusinga Island on Lake Victoria in Kenya. The lake clays preserve not only a rich fossil mammal fauna, but also plants and insects due to the chemical environment (Peppe et al., 2009). They are also interspersed with ash layers from nearby volcanoes which allow them to be dated by the potassium-argon method to between 17 and 15 million years ago. The fossil assemblage in general suggests a forested landscape and the primates include members of the genus *Proconsul*: one about the same size as a small female chimpanzee

(*P. nyanzae*) and the other (*P. heseloni*) about one-third that size (Beynon *et al.*, 1998). From other sites in Kenya and Uganda come additional members of the genus (*P. africanus* and *P. major*), which are thought to have been more agile arborealists living in something like today's tropical evergreen forest. The Rusinga species are thought to have lived in a drier, more open forest (Andrews, 1996). Their skeleton combines ape-like and monkey-like features (Begun, 2007) in a way that suggests to many anthropologists that they are near the root of the hominoids (page 3). This makes their dental development pattern of particular interest for the discussion in this book and they are represented by a large collection of beautifully preserved developing teeth. A similar mosaic of ape- and monkey-like features characterises the genus *Afropithecus*, found at a few sites in Kenya dated about 17 million years ago.

The hominoid fossils discussed throughout the rest of this chapter come from the Miocene (23 to 5.3 million years ago), Pliocene (5.2 to 1.8 million), Pleistocene (1.8 million to 10 000) and Holocene (10 000 to Present) geological epochs (Gradstein *et al.*, 2005). One of the more widespread Miocene hominid genera is *Dryopithecus*, for which the classification varies depending on the scheme used (Appendix A, Tables 1 and 2). It is well known from European sites dated 13 to 9 million years ago and includes four species, all of which seem to have been highly arboreal, with teeth similar to those of chimpanzees. Three early Asian hominin genera share many features with living orangutans: *Sivapithecus*, known from sites in Pakistan and India dated 10.5 to 7.5 million years ago; *Lufengpithecus*, with many isolated teeth finds from Yunnan in China dated 9 to 8 million years ago; and *Gigantopithecus*, which is represented by huge teeth and jaws from much later Pleistocene cave sites in South China.

Some of the largest assemblages of Pliocene hominin fossils have been found in limestone caves near Krugersdorp in South Africa: Sterkfontein, Makapansgat, Swartkrans and Kromdraai. Constant dissolution of the limestone has given the caves a complex history, but most of the bones seem to have accumulated as a talus cone underneath shafts that communicated with the surface (Brain, 1981). Animals, including primates, either suffered an accidental fall or were dropped as carcass remnants by predators feeding near the shafts. The underground chambers filled with fallen rock fragments known as breccia, cemented together by water-deposited carbonate, which also protected the fossils. These deposits have yielded evidence of a rich mammal fauna, varying in size from mice and shrews to lions, antelopes and elephants. Representing the primates are several species of baboons, colobine monkeys and two species of hominins: *Australopithecus africanus* and *Paranthropus robustus*. Dates at Sterkfontein for *A. africanus* range from 4.2 to 2.1 million years ago (Partridge, 2005). *Paranthropus robustus* at Swartkrans is dated 1.8 to 0.8 million years (Balter *et al.*, 2008) and at Kromdraai 1.9 million years ago (Thackeray *et al.*, 2002). Both were perhaps a little smaller than a modern chimpanzee (Robson and Wood, 2008). *Australopithecus africanus* is thought to have been capable of upright bipedal walking, but to have been more arboreal than *P. robustus*. In both, the incisors and canines were a little larger than those of modern humans, more so in the case of *A. africanus*, but the cheek teeth were greatly increased in size with thick enamel. This was particularly the case in *P. robustus*, where a permanent lower third molar could measure over 18 mm from mesial to distal, or more than half as much again as the same tooth in a living human.

A large collection of well-preserved Pliocene hominin molars and premolars was found in the River Omo basin which flows into Lake Rudolf in Ethiopia (Ramirez-Rozzi, 1995). They must have been tumbled along like small pebbles and accumulated in sediments where the current dropped. Potassium-argon dating of volcanic ash layers has put them at 3.4 to 2.1 million years BP. The teeth are even larger than those of *P. robustus* and they have been assigned to the species *P. boisei* and *P. aethiopicus*. Altogether, the three large-toothed hominin taxa are labelled 'megadont' (Wood and Lonergan, 2008) and presumably represent a particular adaptation to the requirements of heavy mastication. Another important source of Pliocene hominin finds is the Awash valley in Ethiopia, from which there is a good representation of the skeleton and dentition of *Australopithecus afarensis*, dated four to three million years ago. Like *A. africanus*, this taxon shows evidence of arboreal as well as bipedal locomotion, but differs substantially in morphological details. *Ardipithecus*, on the Miocene–Pliocene boundary from Ethiopia, is dated 5.8–4.3 million years ago and displays a mosaic of features, some like living chimpanzees and some like *A. afarensis*.

One of the most important early Pleistocene hominin fossils is the Turkana Boy or KNM-15000, a largely complete immature skeleton attributed to *Homo erectus* or *ergaster* (Schwartz and Tattersall, 2003; Wood and Lonergan, 2008) and discovered on the banks of the Nariokotome stream not far from Lake Turkana in Kenya. The site is dated from volcanic ash layers to around 1.6 million years ago, early in the Pleistocene geological epoch. Complete skeletons are very rare finds in hominin palaeontology. In life, the body size and proportions would have been similar to those of living people and are seen as the first hominin adaptation for long distance bipedal walking. The teeth were also similar in size to those of living humans. This is therefore a key fossil in the story of hominin development, particularly as the teeth and bones were still developing at death. *Homo erectus* is known mainly from sites in China and Indonesia (but also from other sites in Africa), at dates ranging from over 1.5 million to 30 000 years ago. Stone artefacts chipped on both sides (bifaces or handaxes) have been found in African sites as early as 2.6 million years ago (Semaw, 2000), in some cases from deposits in which hominin fossils have also been found (Klein, 1999). None of the Indonesian *H. erectus* sites has associated artefacts and dating is difficult, but there are some directly associated finds in China. In addition, there is well-established evidence for the use of fire.

One of the most coherent and well-represented Pleistocene fossil hominin taxa is *Homo neanderthalensis*. It includes both adults and children, mostly from cave sites throughout Europe and the western part of Asia, dated between about 200 000 and 28 000 years ago. Whilst the Neanderthal skeleton was similar to that of *Homo sapiens*, it had comparatively short lower arms and legs, with a markedly stout physique and wide ribcage. In the skull, the braincase was as large as that of living humans, but longer, lower and broader. The nose was broader too, within a projecting midface. The palate was wide, with relatively large incisors and canines, although the premolars and molars were no larger than in living humans. In addition, there is a whole list of smaller distinctive dental and skeletal features. The physique and face are often explained by adaptation to cold climatic conditions and the distinctive teeth, palate and lower jaw by adaptation to their use as an addition to the Neanderthal toolkit. Stone artefacts have been directly associated and show evidence of complex manufacture to produce a wide range of forms, presumably with specialist functions. Together they are classified as Middle Palaeolithic or Mousterian technology. There are also graves containing deliberate burials of Neanderthal dead. Several large collections of hominin fossils from the Sierra de Atapuerca in Spain are important in the debate about the origins of the Neanderthals. The earliest, from the Gran Dolina cavern, dated about 780 to 500 000 years ago, are assigned to *Homo antecessor*, a taxon which shows none of the characteristic Neanderthal features. A large collection of fossils from the extraordinary deep shaft of the Sima de los Huesos is more than 350 000 years old and combines some Neanderthal-like features with others shared by Pleistocene hominins from elsewhere in Europe. The question of Neanderthal origins is not straightforward (Tattersall, 2007; Wood and Lonergan, 2008).

Modern humans have a much less coherent fossil record, but it is of similar antiquity to that of the Neanderthals. The earliest finds attributed to *H. sapiens* are from Africa, dated before 180 000 years ago. Larger assemblages come from cave sites in Israel, Mugharet es Skhūl and Jebel Qafzeh, currently dated 100 to 130 000 years ago, and from a limestone fissure at Jebel Irhoud in Morocco, currently dated around 160 000 years ago (Grün *et al*., 2005; Smith *et al*., 2007c). They are associated with Middle Palaeolithic stone artefacts and antedate the Neanderthal sites in Israel. The major change in Europe and western Asia is the disappearance of the distinctive Neanderthal taxon and the appearance of *H. sapiens* with a morphology that overlaps that of living people. The earliest *H. sapiens* finds in Europe are dated to 35 000 years ago. They are associated with a more complex and more rapidly developing Upper Palaeolithic technology (Mellars, 2004). On palaeontological, archaeological and genetic grounds, many anthropologists argue that *H. sapiens* evolved in isolation in Africa and then, from 40 000 years ago, colonised the rest of the world to replace previously existing hominin taxa.

If the slow pace of human development is indeed related to cognition, the fossil record suggests several key places to look for evidence of transition. One is in the Pliocene hominins which, although they have never been directly associated with stone tool technology, are the first to show evidence of bipedalism, brain expansion and reduction in the face and dentition. Another is the Turkana Boy, again without direct evidence of tool use, but with a modern human-like, fully bipedal skeleton. Then there are the Neanderthals, directly associated stratigraphically with a complex stone technology and, finally, the early modern humans.

Periodicity of brown striae and spacing of perikymata in fossil hominins

The periodicity or cross striation repeat interval (page 118) between brown striae of Retzius has been important in a number of estimates for crown formation timing, but it varies within and between taxa. Based on synchrotron radiation micro-CT (Smith *et al.*, 2010b), Neanderthal values varied between 6 and 9, with a mean of 7.4 and counts of 7 or 8 in most crowns. Qafzeh, Skhūl and Irhoud *H. sapiens* teeth also mostly had counts of 7 or 8. This compares with a mode of 8 for modern human molars and a mean of 6.4 for chimpanzees. Confocal light microscopy of 29 fossil teeth assigned to *Australopithecus* and *Paranthropus* yielded intervals varying from 6 to 8, with a mode of 7, whilst teeth assigned to early *Homo* varied from 7 to 9 with a mode of 8 (Lacruz *et al.*, 2008). For sections of Rusinga *P. heseloni*, the count was consistently 5 and for *P. nyanzae*, 6 (Beynon *et al.*, 1998). As discussed in Chapter 4, it is not clear what underlies these periodicities. Smith (2008) has pointed out that there is a relationship with hominoid body mass (known for living taxa and estimated for fossils). Smaller animals tend to have a shorter periodicity and larger animals a longer periodicity.

There is some evidence that the spacing of perikymata (sometimes called 'perikymata packing pattern') down the crown side shows differences between fossil taxa. In his remarkable dental description of *Australopithecus* and *Paranthropus*, Robinson (1956) noted their prominent perikymata, which were wider spaced and more irregular than in recent Greenland Inuit (Pedersen and Scott, 1951), especially in the occlusal half of premolar and molar crown sides. Dean and Reid (2001) found that, for living humans, perikymata spacing decreased gradually from a maximum near the occlusal tip to a minimum near the cervix (Figure 5.4). Chimpanzees and gorillas differed markedly, with less widely spaced perikymata occlusally, decreasing to a minimum at crown mid-height, becoming more widely spaced again towards the cervix. In *Australopithecus* the occlusal spacing was similar to that of modern humans, but decreased less cervically. *Paranthropus* spacing was, however, in general wider and showed even less of a decrease to cervical. Of the *Australopithecus* teeth, *A. afarensis* were most like those of modern humans. The small number of teeth assigned to early *Homo* resembled *Paranthropus* more closely. It may be that those taxa with large crowns developed them in a fundamentally different way to those of modern humans. Lacruz *et al.* (2006) found that, for an *Australopithecus* molar, 61% of crown formation time was taken up by lateral enamel-forming perikymata, whereas for a *Paranthropus* molar it was 39%. The tables in Reid and Dean (2006) suggest that lateral enamel occupies about 60% of crown formation time in modern humans.

There has been a considerable debate about perikymata packing patterns in Neanderthals (Guatelli-Steinberg, 2009). Ramirez Rozzi and Bermúdez de Castro (2009) studied dental impressions of anterior teeth from Gran Dolina (*H. antecessor*) and Sima de los Huesos in the Sierra de Atapuerca (above), and also European Neanderthals and European modern humans from Upper Palaeolithic sites. They divided the crown side into ten equal zones of crown height and counted the number of perikymata in each using a low power stereomicroscope. They suggested that

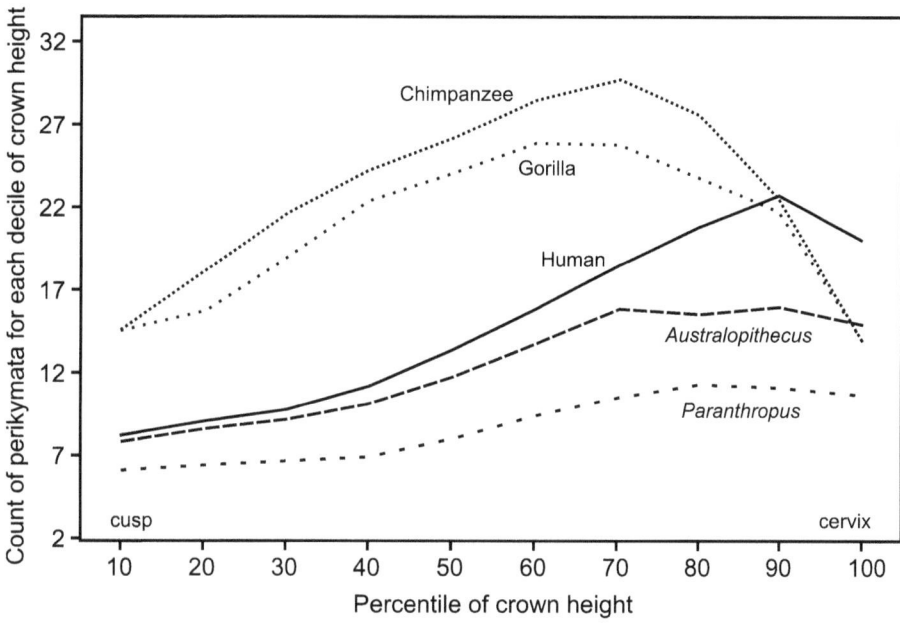

Figure 5.4 Perikymata per decile of crown height in humans, chimpanzees, gorillas, *Australopithecus* and *Paranthropus*. Incisors and canines combined. Redrawn from Dean and Reid (2001), Figure 1, p. 212. Where perikymata are spaced further apart, the count per decile is smaller.

the crowns from the Atapuerca sites and the Neanderthals showed closer perikymata spacing than the modern humans, particularly in the cervical region, and concluded that Neanderthal crowns were formed 15% faster than those of modern humans. Guatelli-Steinberg et al. (2005) made similar counts for Neanderthal anterior teeth (Figure 5.5), but instead compared them with the data of Reid and Dean (2006) for recent modern humans (Ramirez-Rozzi and Sardi, 2007; Guatelli-Steinberg et al., 2007b). Further studies (Guatelli-Steinberg et al., 2007a; Guatelli-Steinberg and Reid, 2008) concluded that, although Neanderthal anterior teeth did not fall outside the modern human range in crown formation time, they differed in their perikymata packing. Neanderthals showed less contrast between occlusal and cervical perikymata. More recent work has now suggested that Neanderthal tooth crowns were indeed formed within a shorter time than those of modern humans (Smith et al., 2010b). It appears, however, that this was not due to a smaller number of long-period increments in the lateral enamel (that is, the number of perikymata), but thinner cuspal enamel, a shorter periodicity of cross striations between brown striae and faster extension rates along the EDJ.

Enamel matrix secretion rate

The mean cross striation spacing gives the daily secretion rate of enamel matrix underneath each ameloblast as the prisms grow in length. This rate increases from

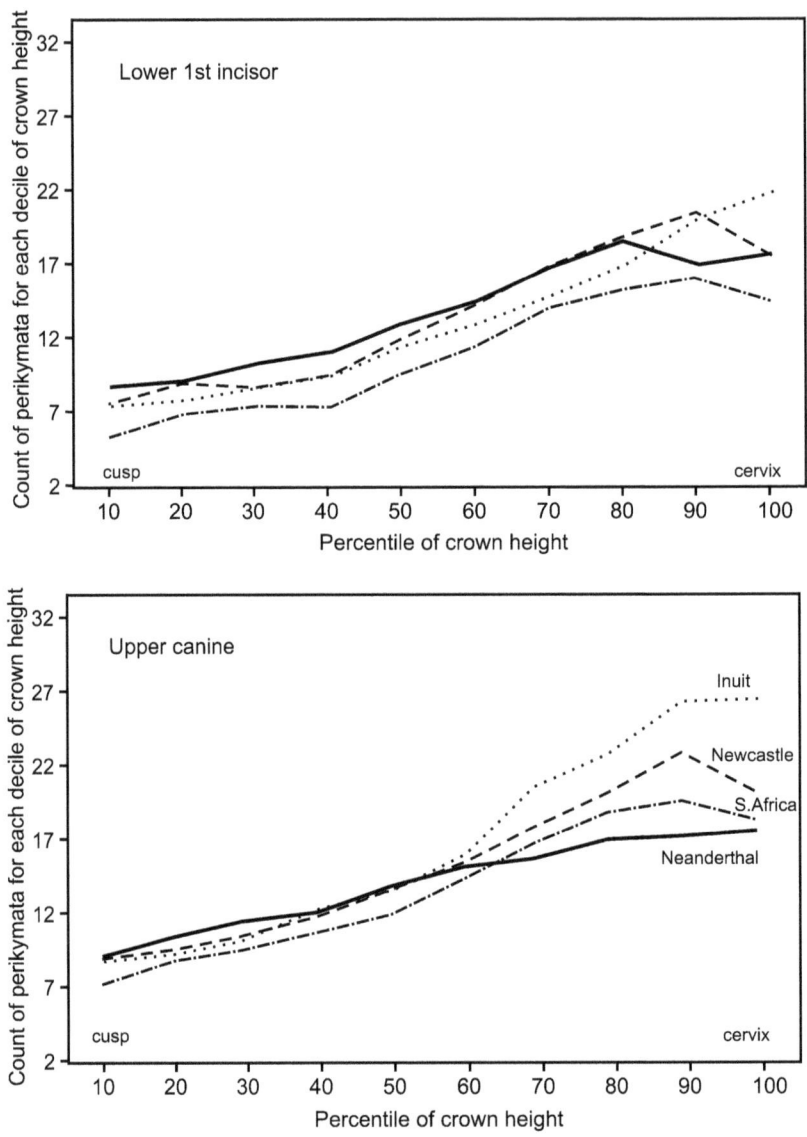

Figure 5.5 Perikymata per decile of crown height in Neanderthals and recent humans including Canadian Inuit, South Africans and English subjects. Permanent lower first incisors and upper canines. Scaled from Guatelli-Steinberg et al. (2007), Figure 3, p. 76. The axes have been made to match those of Figure 5.4.

the EDJ towards the crown surface and is faster under the cusps and slower at the cervix. It is therefore determined separately in enamel sections for inner, middle and outer zones, and in three regions from occlusal to lateral and cervical (Beynon et al., 1991b; Schwartz et al., 2001; Lacruz and Bromage, 2006; Mahoney et al., 2007). Alternatively, the cuspal enamel may be more finely divided into zones of equal thickness; Mahoney (2008) used 60 μm. Most comparisons are made on the

Development chronologies for living and fossil primates

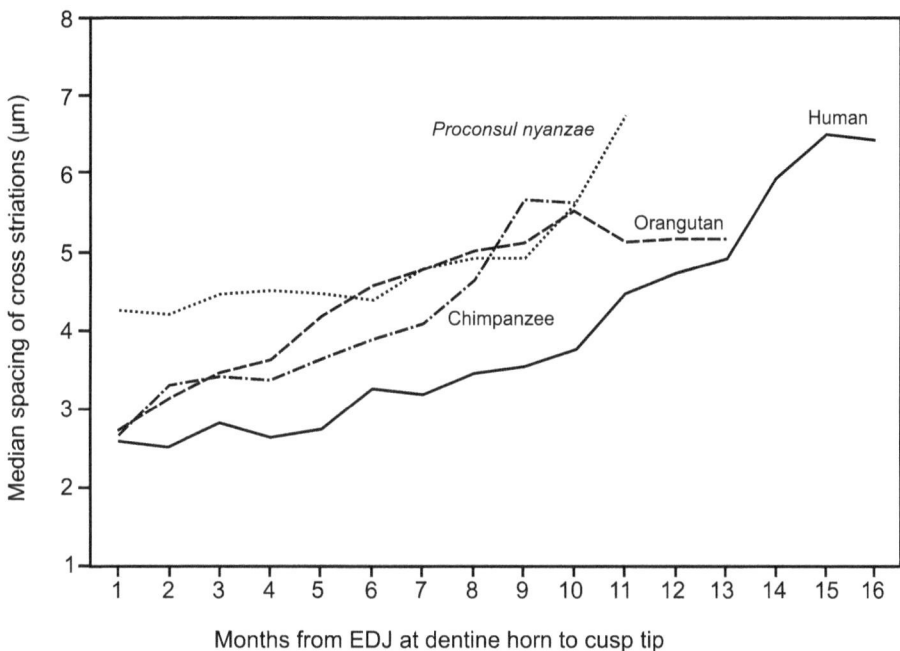

Figure 5.6 Median secretion rate curves for cuspal enamel matrix in molars from a modern human, chimpanzee and orangutan, together with fossil *Proconsul*. The thickness of the enamel at the cusp tip, overlying the EDJ at the dentine horn, is divided into one month (30 cross striation) layers. Plotted from boxplots in Beynon *et al.* (1998), Figures 8 and 9, pp. 181–184.

basis of the cuspal enamel and another way to reckon changes in rate is to plot a continuous record through the thickness of the crown at this point. This is done by repeatedly counting a fixed number of cross striations and measuring the length of prism that they occupy; Dean (1998a) used 30 cross striations to give months of enamel matrix secretion. Such rate curves have a characteristic form (Figure 5.6), as shown by Beynon *et al.* (1998), who compared permanent molars in humans, chimpanzees, orangutans and *Proconsul*. The pattern of rate increase varied. In the human tooth, it increased slowly until the last third of enamel thickness. In the chimpanzee tooth, it rose rapidly to start, slowed and then climbed rapidly nearer the surface. For two orangutan molars, the rate rose steadily for the first half of the enamel thickness, after which it steadied. Enamel secretion in the *Proconsul* tooth started near the EDJ at a higher rate, which was maintained until it increased just under the surface.

It is possible to fit polynomial regression models to the cumulative curves for mean cuspal enamel secretion (page 124). Schwartz and Dean (2001) did this for permanent lower canines in chimpanzees, gorillas, orangutans and humans. Dean *et al.* (2001) applied a similar technique to sections of teeth from fossil *Proconsul*, *Australopithecus*, *Paranthropus*, early *Homo* and a Neanderthal. Chimpanzees and

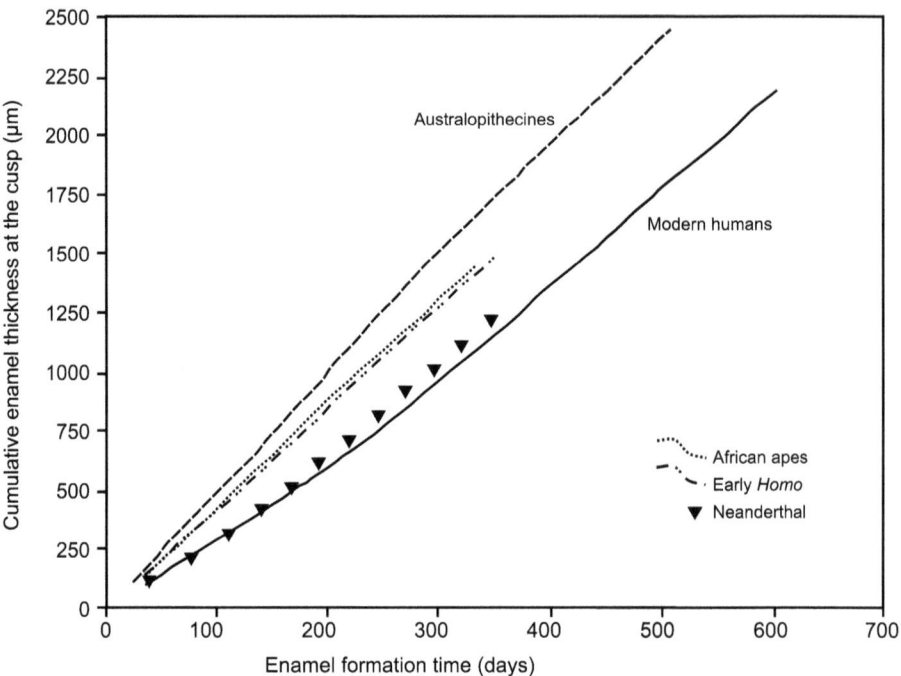

Figure 5.7 Curves of cuspal enamel matrix secretion time versus cumulative matrix thickness for modern humans and African apes, *Proconsul*, australopithecines, early *Homo* and a Neanderthal. Redrawn from Dean *et al.* (2001), Figure 1, p. 629. For regressions, see Appendix A, Table 26.

gorillas showed a faster gradient of cumulative increase in cuspal enamel thickness than living humans (Figure 5.7). The curves for all fossil hominins except the Neanderthal fitted best with chimpanzees and gorillas. Only the Neanderthal fitted well with living humans (Macchiarelli *et al.*, 2006).

Crown and root extension rate

The rate at which the crown grows by extension along the EDJ (Figure 5.3) is a different concept to the rate at which enamel matrix is secreted along the prisms. It was first proposed by Shellis (1984; 1998). In modern human and chimpanzee permanent molar crowns, extension rates may be 15 to 30 µm per day or more along the EDJ under cuspal enamel, but decrease to 4 to 8 µm per day under lateral enamel (Dean, 2009; 2010). Deciduous tooth crowns have a considerably greater extension rate than permanent crowns (Shellis, 1984).

The crown extension rate curve varies between individuals in relation to the variation in overall formation time (Figure 5.8). At one extreme there are crowns which took longer to form, with a low peak extension rate under the cusp and a relatively long period of slower extension rate lateral enamel. At the other extreme are crowns

Development chronologies for living and fossil primates

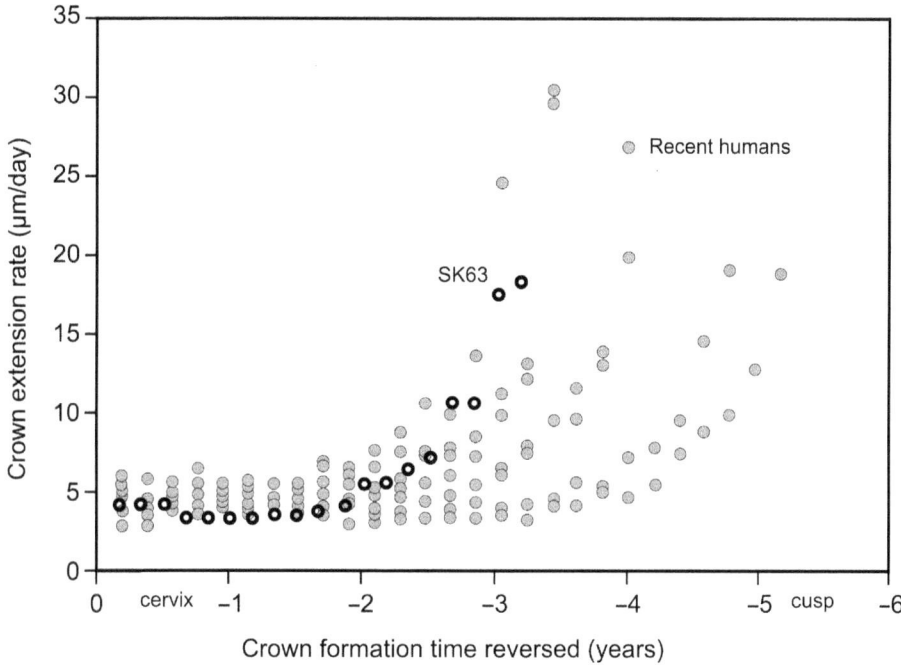

Figure 5.8 Crown extension rate along the EDJ for recent human canines and the SK63 *Paranthropus robustus* canine. The horizontal axis is plotted with the cervix on the left, so completion of the crown is at zero and the scale is in years before this stage. This shows differences in completion time more clearly. Redrawn from Dean (2009), Figure 3, p. 71.

which formed in less time overall, with a higher peak cuspal enamel extension rate and a relatively short period of slower extension rate lateral enamel. It is apparent that the balance between the two is complex. A canine from the *P. robustus* fossil SK-63 (page 145) followed the latter trend, but was still within the modern human range of variation (Dean, 2009). Another way to display extension rate data is to plot the cumulative extension along the EDJ with each month (30 cross striations) of development (Figure 5.9). This produces a gentle curve in which a steeper slope representing the more rapid cumulative extension under the cusp grades into a shallower slope representing the slow lateral crown extension. For one tooth type, in one taxon, the overall slope of these curves varies between individuals. The ranges of variation for modern humans and chimpanzees overlap extensively (Dean, 2010), as did the SK-63 lower canine, the La Chaise Neanderthal, and the S7–37 *H. erectus* first molar from Sangiran (Dean, 2009).

The extension rate concept was applied to the dentine of roots by Beynon *et al.* (1998) in their study of *Proconsul* discussed earlier. With reference to Figure 5.3, in place of cross striations they used the short-period line spacing; angle I was taken between the dentinal tubules and the CDJ and angle D between the long-period lines and the CDJ. Initially, only the root itself was studied (Dean, 2007; Dean and Vesey, 2008) but, given the continuity of EDJ/CDJ, there is little point

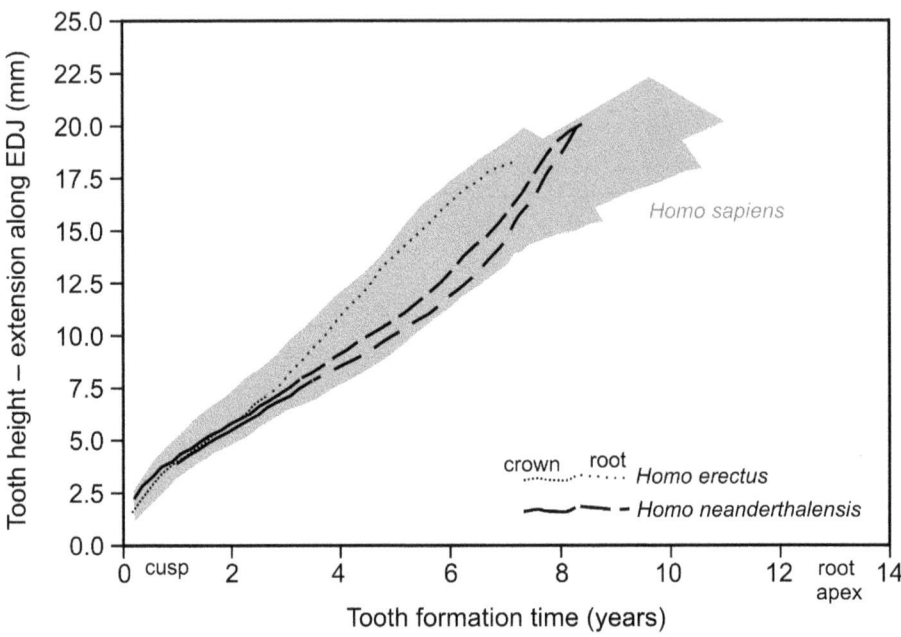

Figure 5.9 Curves of cumulative crown and root (whole tooth) extension along the EDJ for permanent first molars. The area of the plot covered by a number of curves for recent humans has been shaded grey to reduce complexity. Curves are plotted separately for a *Homo erectus* molar from Sangiran in Java, and a Neanderthal molar from La Chaise-de-Vouthon in France. Two curves are shown for the Neanderthal; one assuming a daily rate of 2 μm and the other a rate of 2.5 μm. Redrawn from Dean (2009), Figure 2, p. 70.

in distinguishing between crown and root, so more recent studies have included both together (Dean, 2009; 2010). Cuspal crown extension rates are much higher than anything that follows. At the transition to the root (Figure 5.10), there is little change from the low lateral crown extension rate but, in most teeth, the rate rises to a peak a short way further down the root, before falling again towards the apex. Dean (2010) called this the *peak height velocity* and showed that it occurred in a group of chimpanzee first molars 3 to 4.7 years after the first dentine was deposited (at around birth). This matched well with the eruption age for this tooth (Appendix A, Table 7). This is a crucial point because it might yield a way in which age at gingival emergence of the first molar could be estimated in fossils (page 144).

Smith *et al.* (2010b) used synchrotron micro-CT to estimate mean crown extension rates in a different way, as the length of the EDJ from dentine horn to cervix, divided by the overall crown formation time. On this measure, Neanderthal permanent tooth crowns showed faster extension rates than living humans, amongst which early *H. sapiens* crowns fitted better. On the other hand, Macchiarelli *et al.* (2006) were able to plot curves of cumulative root extension rate for the La Chaise Neanderthal first molar and showed that it matched well with curves for modern human molars. Some evidence therefore suggests a difference in extension

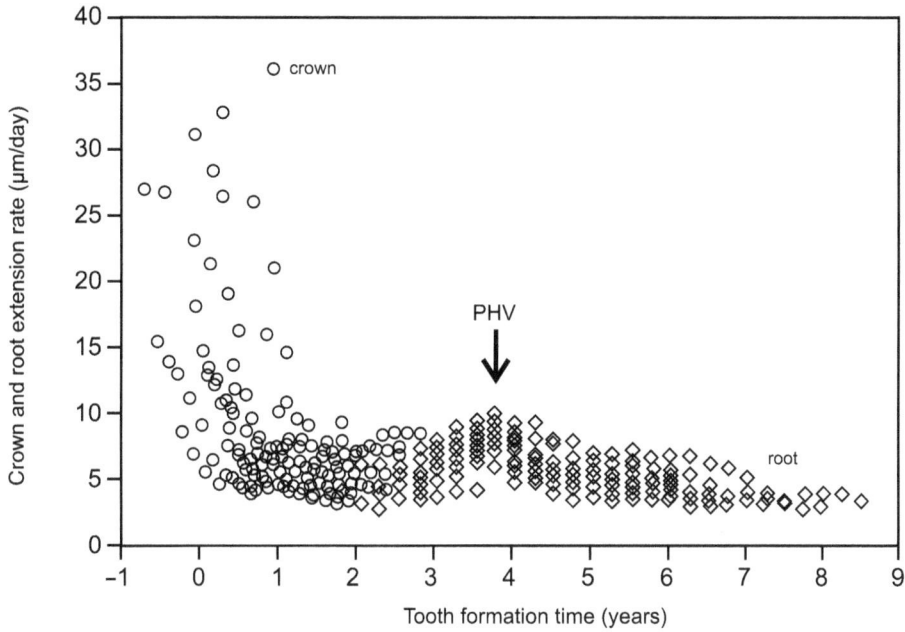

Figure 5.10 Crown and root (whole tooth) extension rate along the EDJ for recent chimpanzee first molars. The curve of points for each tooth has been aligned at the peak extension rate for the root (the peak height velocity, PHV) for which the mean age is 3.8 years. This makes the correspondence of the peak between different teeth clearer. Redrawn from Dean (2010), Figure 5, p. 3403.

rate between Neanderthals and modern humans, and some does not. Bermúdez de Castro et al. (2010) claimed that the state of development in a fossil dentition assigned to *H. antecessor* matched what would be expected for modern humans for their estimate of age-at-death. This was based on root length in the growing first molar root, applied to published extension rates in modern humans and chimpanzees. If so, this would represent the earliest evidence for a modern human-like dental development pattern.

Histological age at initiation and completion of the crown

For some specimens, it is possible to section a permanent first molar and use the neonatal line as the basis of a sequence from which the ages of developmental stages can be estimated. Beynon et al. (1991a) examined dentitions from a juvenile gorilla in the collections of the Royal College of Surgeons and a young adult orangutan from London Zoo (Figure 5.11). On sectioning, the gorilla dentition showed a neonatal line in the permanent first molar and a prominent hypoplastic defect allowed sequences in different teeth to be linked. The orangutan dentition displayed tetracycline lines which were used to link the sequence together. Cuspal enamel formation times were estimated by measuring enamel thickness, adding 15% to take account

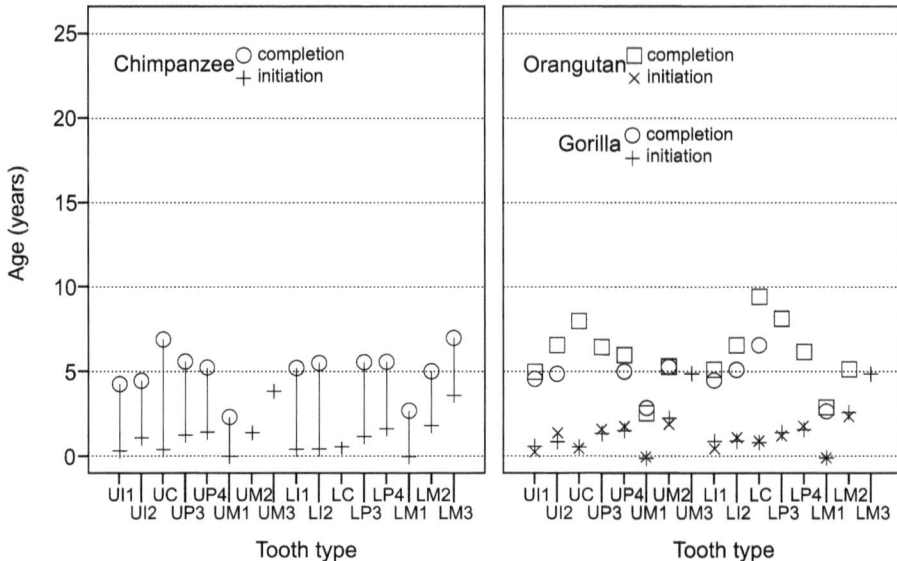

Figure 5.11 Crown initiation and completion in chimpanzees and orangutans, estimated from enamel histology. Data from Appendix A, Tables 27 and 28.

of decussation (page 125) and dividing by the spacing of the prism cross striations. Lateral enamel formation time was estimated from counts of the regular brown striae of Retzius near the crown surface multiplied by the mean enamel secretion rate. From this it was possible to derive charts of permanent crown initiation and completion (Appendix A, Tables 27 and 28).

Schwartz et al. (2006) carried out a similar study of a young female gorilla that had lived at a zoo and died aged 3 years 3 months. Ages at crown initiation were established for permanent teeth by cross striation counts from the neonatal line. Death occurred before some of the crowns had been completed, which made it possible to confirm the periodicity of cross striations, but also meant that crown completion could only be determined for permanent incisors and first molars. The durations of crown formation were much less than those noted by Beynon et al. (1991a) for their gorilla specimen, perhaps reflecting sexual dimorphism or differing environment.

Reid et al. (1998b) sectioned chimpanzee teeth from comparative anatomy collections at University College London and the Royal College of Surgeons: an adult female, two juveniles and one infant. Very similar methods to those of Beynon et al. (1991a) were used, producing equivalent charts (Appendix A, Tables 27 and 28; Figure 5.11). The ages at crown initiation overlapped with those of radiograph-based studies (Appendix A, Tables 17 and 18), but the ages at completion were older by as much as 1 or 2 years. They suggested that this might be because radiographs show only the mesial and distal ends of the crown, where development is less advanced compared with the buccal and lingual sides through which the sections were cut, but there were also differences in the samples studied (Kuykendall, 2001).

The older completion ages of the study by Reid *et al.* (1998b) suggest considerable overlap in development between molars, and between premolars and molars. This may be a useful distinctive feature for the chimpanzee pattern of dental development. The detailed studies of Smith and co-workers (2007b; 2010a) produced similar results for molars using sections from both wild and captive animals. They found that the wild animals fell into the middle to older half of the captive animal range.

For humans, Dean *et al.* (1993a) used similar methods on a single forensic case with multiple tetracycline bands to match sequences between teeth and establish ages at crown initiation and completion for the whole dentition. They compared reasonably well (Appendix A, Tables 27 and 28) with ages determined for an archaeological dentition from a Medieval monastery site in Picardie, France (Reid *et al.*, 1998a). These histologically derived initiation ages were applied to a larger study of isolated, extracted teeth from dental hospitals in England, North America and South Africa (Reid and Dean, 2006). For these, counts of cross striations provided estimates for the number of days required to complete the crown, which were added to the Picardie initiation ages to estimate ages at completion (Figure 5.12). With the exception of molars, the Africans completed their crowns slightly earlier than the Europeans. As with the chimpanzee study, there are considerable differences between these histologically based estimates and those, based on radiography of living children, reviewed in Chapter 3. Ages at attainment for both the initiation and completion of incisor and canine crowns are similar, but in premolars and molars tend to be younger by up to 1 year for the histological estimates. As discussed above, this is likely to be because radiographs can only record a stage after the matrix is sufficiently mineralised and this will always be some time after the matrix was actually secreted. The histological estimates show more strongly the contrasts between humans and chimpanzees (page 68), particularly in the extent to which the formation of first, second and third molar crowns overlaps. In both species they show that the first molar crowns started to form before birth and were completed between 2 and 3 years after birth. In humans, the second molar crowns were initiated just after first molar completion but, in chimpanzees, they were initiated between 1 and 2 years and thus overlapped extensively. Similarly, in humans, the third molars were initiated long after the completion of second molars, whereas in chimpanzees the third molars were initiated just before that stage. Fourth premolar crowns followed a similar schedule to second molars in both species and thus also showed an overlap with first molar formation in chimpanzees, but not in humans. Third premolar crown initiation was about a year earlier in chimpanzees than in humans, but completion was similar. The incisors were initiated at similar ages in both species, but, whereas upper incisor completion was similar, chimpanzee lower incisor completion was about one year later than human. Finally, canine formation times also overlapped extensively, which is remarkable as chimpanzee canines are so much larger than human canines, although completion of the upper canine was two years later in the chimpanzee than in humans.

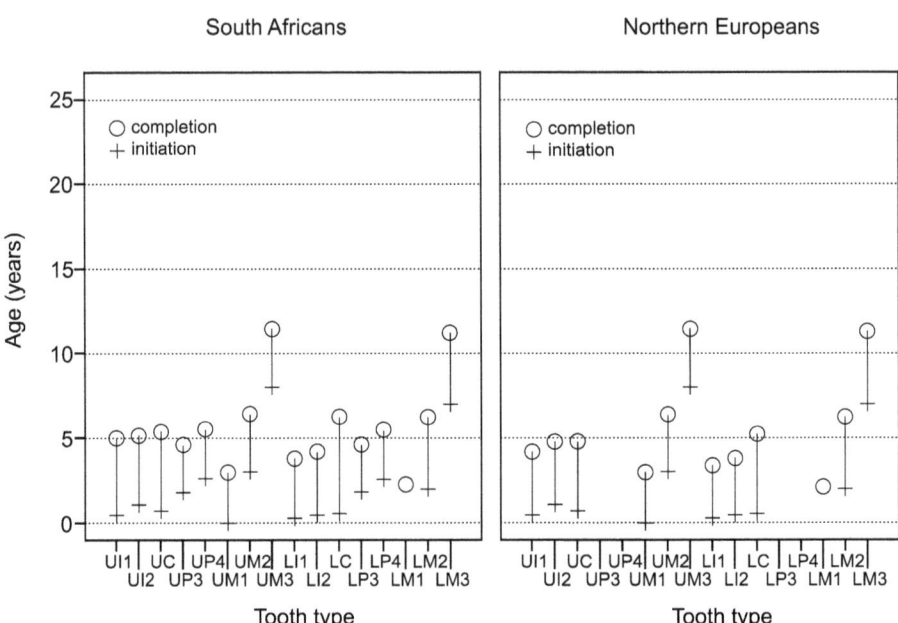

Figure 5.12 Mean crown initiation and completion in recent humans, estimated from enamel histology. Data from Appendix A, Tables 27 and 28.

Crown formation times have also been established for anubis baboons (Appendix A, Tables 29 and 30). For these, the completion times were remarkably delayed in comparison with radiographically derived ages for the same genus (Appendix A, Table 24). The first and second molar crowns overlapped slightly in formation, but the third molars were separated. The incisors were delayed relative to the first molar, as in chimpanzees, and the large canines took a long time to develop, overlapping all other permanent tooth crowns.

Amongst fossils, only in *P. heseloni* has it been possible to section enough teeth (Beynon et al., 1998) to establish crown initiation and completion for all permanent lower teeth. From this it can be seen that molar crown development took place within a remarkably short space of time (Appendix A, Table 35). This might partly be the small size of *P. heseloni* in relation to the living apes, but the estimated body weight was not far from that of a pig-tailed macaque (Appendix A, Table 23), for which radiography suggests the third molar crown was completed at least six months later than *Proconsul*. In addition, the crowns formed in a shorter time than in chimpanzees and there was an even greater degree of overlap in timing between molars.

A combination of micro-CT methods, along with a first molar section, has been used to estimate the ages of crown initiation and completion in a juvenile Neanderthal from Scladina Cave in Belgium, dated 127 to 80 000 years ago (Smith et al., 2007d). The results (Appendix A, Tables 29 and 30) were rather young in comparison with living humans. Histologically based age-at-death estimates for a further six juvenile Neanderthal fossils suggested that the developmental stage they had reached was advanced in comparison with modern human children (Smith *et al.*, 2010b). This faster developmental schedule contrasts with the evidence for slower crown and root extension rate in the La Chaise Neanderthal discussed earlier. By comparison (Smith *et al.*, 2007c), a synchrotron radiation micro-CT examination of permanent lower second incisor, canine and third premolar in early *H. sapiens* from Jebel Irhoud produced estimated ages at crown completion which were older than the mean for recent humans.

Histological estimates for times taken to form crowns

With isolated teeth, unless they are permanent first molars in which the position of the neonatal line can reasonably be estimated, it is not really feasible to estimate ages at crown initiation or completion. It is, however, possible to estimate the overall time taken to complete the crown (Appendix A, Tables 31, 32 and 33). Without sectioning, it may still be possible to estimate lateral enamel formation time from perikymata counts and their associated cross striation intervals. Some naturally fractured crowns expose brown striae of Retzius and prism cross striations clearly enough to estimate the periodicity using a low-power stereomicroscope (Beynon and Dean, 1987; Lacruz and Ramirez-Rozzi, 2010). More recently, it has been possible to use a tandem scanning confocal light microscope or synchrotron micro-CT to image intact fossils. Conventional micro-CT allows enamel thickness over the dentine horn to be measured so that cuspal enamel formation can be estimated, but this involves assumptions about the mean enamel secretion rate (page 124).

The large study of Reid and Dean (2006) using sections of modern human isolated teeth from dental schools shows that incisor and canine crowns take from 3.5 to 5 or more years to form. Premolars and molars are formed more rapidly at between 2.5 and 4 years. There are fewer estimates for chimpanzees (Appendix A, Table 32), but what there are suggest longer formation times for incisor crowns than in humans and shorter formation times in molars. A single estimate for gorilla incisors suggests a much shorter time of 2.7–2.9 years and it may be that the current chimpanzee figures are not typical of living African ape incisors. The large chimpanzee, gorilla and orangutan canine crowns (Schwartz and Dean, 2001; Schwartz *et al.*, 2001) generally take much longer to form than those of humans (nearly ten years in some cases). The Rusinga *Proconsul* fossil assemblage has yielded estimates for crown completion from sections of most teeth (Beynon *et al.*, 1998). An upper first molar and lower second molar crown from *P. nyanzae*, around the same body size as a small female chimpanzee, were formed in about two years, which is within the range of chimpanzee (Appendix A, Table 35). *P. heseloni* was a much smaller creature, perhaps the size of a gibbon, with smaller crowns which took a much shorter time to form than any other anthropoids considered here. For other non-hominin fossil taxa, data are only available for single teeth (Appendix A, Table 36). *Afropithecus, Dryopithecus, Sivapithecus, Lufengpithecus* and *Gigantopithecus* all took a similar time to chimpanzees to form their molar crowns.

For many hominin fossils, the estimate has been made from perikymata counts alone, but this requires an assumption about the cross striation repeat interval. This is problematic, so the crown formation times are calculated for a range of different periodicities (typically 7, 8, 9 or 10). For *Australopithecus* incisors, these values overlap the faster end of the modern human range (Appendix A, Table 33). The small *Paranthropus* incisors formed more rapidly still (Dean et al., 2001). Incisors and canines assigned to early *Homo* species from Africa and southeast Asia also showed relatively short crown formation times. In the few *Australopithecus* and *Paranthropus* molars (and one premolar) for which there is detailed information, crown formation seems to have been relatively rapid, overlapping chimpanzee more than modern human timing. One of the largest collections of hominin teeth is the *Paranthropus boisei* and *aethiopicus* premolars and molars from the River Omo basin (page 133). Their large, thick-enamelled crowns (Appendix A, Table 34) were completed in a similar time to other *Paranthropus* specimens (Ramirez-Rozzi, 1995).

The large incisors and canines of Neanderthals have been a particular focus of discussion. Ramirez Rozzi and Bermudez de Castro (2009) suggested from counts of perikymata that the crowns of these teeth formed more rapidly in dentitions from Gran Dolina, Sima de los Huesos and Neanderthals than in Upper Palaeolithic and Mesolithic modern humans. By contrast, Guatelli Steinberg et al. (2005; 2007a) instead estimated Neanderthal formation time to be within the range of living humans. Many assumptions underlie these estimates, but more recent studies based on sections and synchrotron radiation micro-CT (Smith et al., 2010b) have found shorter overall crown formation times in Neanderthals for all tooth types, both in terms of mean values and ranges, in comparison with recent humans and the early *H. sapiens* fossil from Jebel Irhoud (Appendix A, Table 31).

Age at eruption of the molars

First molar eruption age has formed the centre of discussions of the evolution of human life history because of its relationship with brain size in living primates (page 154). Some parts of the eruption sequence (page 42) are preserved in fossils (alveolar eruption and the initial light wear marking occlusal eruption) but age at gingival emergence is needed for comparison with living primates and this is difficult to estimate in the absence of soft tissue.

There is a small and select group of key fossils in which the permanent first molars were erupting or had just reached the occlusal plane at the time of death. One of the most famous is the Taung *Australopithecus africanus* skull from South Africa. It is the type fossil of the species and, soon after its discovery, Dart (1925) described it as a 'man-ape'; that is, it differed from the living chimpanzee and gorilla to the same extent as they did from one another and at the same time it showed similarities with humans in the morphology of its brain, the cranium and its teeth. Dart observed that the state of dental development corresponded with that of a 6-year-old modern human child and, at that time, it was assumed that the tooth development schedule was similar in chimpanzees and humans. It has been known since the studies of Nissen and Riesen (1946) that chimpanzee dentitions develop to a faster schedule than human dentitions, but it was still widely assumed that a slow human-like pace of development would be one of the distinctive characteristics of *Australopithecus* and *Paranthropus*.

Mann (1975) studied dental development in Swartkrans *P. robustus* using direct observation and radiographs. In the SK-63 mandible, he noted that both the permanent first molar and first incisor crowns were completed and erupting, as would be expected in modern humans, but not chimpanzees (page 68). The logical conclusion was that *Paranthropus* had developed along a similar schedule to that of living humans. There were, however, some dissenting voices. Biggerstaff (1967) believed, on the basis of the small body size of *Australopithecus*, that it would have had a short generation time similar to that of the chimpanzee. This does indeed make sense in a context of recent primate life history theory (page 149). Biggerstaff therefore used chimpanzee dental development timings to suggest an age-at-death of 4 years for the Taung individual. Smith (1986) showed that, when all teeth in developing dentitions were taken into account, fossil taxa did not fit consistently either with a human or a chimpanzee schedule. Bromage and Dean (1985) estimated age-at-death from perikymata counts in incisors for three early hominins with erupting first molars: *A. afarensis*, *A. africanus* and *P. robustus*. Their estimates varied from 3.2 to 3.5 years at death and therefore suggested a schedule closer to that of chimpanzees than modern humans. As noted earlier, perikymata alone are not a reliable basis for estimating developmental ages where the time taken to form the cuspal enamel and cross striation repeat are not known. Dean (1993b) therefore gained permission to section the canine of SK-63 to check the age estimates with direct observation of cross striations. This provided a slightly older estimate of age-at-death at around 4 years. The first molar had been in the occlusal plane long enough for some enamel to have been worn from the tips of its cusps, suggesting that a few months had elapsed since its gingival emergence, which might therefore have been at 3.5 years. Lacruz *et al.* (2005) used assumptions about the crown formation time and rate of root formation from other histological fossil studies to arrive at an estimate of age-at-death of 3.7 to 3.9 years for the Taung skull. Assuming that some months must have elapsed since gingival emergence, this matches well with SK-63, confirming a younger age of permanent first molar eruption in early hominins than in living humans. The confusing factor in *Australopithecus* and *Paranthropus* is therefore the rapid development and very early eruption of the incisors, which thus arrived in the mouth at the same time as the first molars.

Human dental development is also characterised by relatively late eruption of the permanent second and third molars. It is even more difficult to find fossils in which either of these teeth was in the process of eruption at the time of death (Dean, 2010). One important example is the Nariokotome early *Homo* fossil (page 130). Dean and Smith (2009) estimated the age-at-death on the basis of perikymata and root growth in the second molar to have been between 7.6 and 8.8 years. This tooth had just erupted into occlusion, with a little wear, suggesting an age of gingival emergence some months beforehand, which would be young for a modern child. The state of dental development did not fit well with either modern humans or chimpanzees, but Dean and Lucas (2009) found that overall it fitted best with a living child 10 to 12.4 years of age. In addition, the development stage reached by the arm bones would be more typical of a 13- to 14-year-old modern human child. It appears that this individual was following a developmental schedule which has no modern parallels.

There are some estimates for the gingival emergence of permanent first molars in non-hominins. An *Afropithecus* lower jaw yielded an estimate of emergence at 2.4 to 3.6 years (Kelley and Smith, 2003). Kelley (1997; 2002; 2004) also estimated the age of first molar gingival emergence for a lower jaw specimen of *Sivapithecus parvada* at between 2.2 and 4.5 years, and a minimum age at emergence of 2.6 years for an isolated tooth of *Dryopithecus laietanus*. In all three cases, the estimates are not far from the figure for living chimpanzees.

Another approach to estimating eruption timing is to record the extent of molar root development at that stage. Many factors are involved in eruption. Not only is the tooth increasing in overall height within its crypt, but bone and soft tissues are remodelling around the tooth. The crypt and the tooth migrate through the outline of the alveolar process as the cusps of the tooth move towards the occlusal plane, so eruption takes place faster than can be explained by a simple extension of the roots. The gap between alveolar eruption and full occlusal eruption is somewhere between 6 months (premolars) to 2 years (molars) in modern humans (Haavikko, 1970; Liversidge, 2003). Gingival emergence is at some point between them. Dean (2007) measured permanent lower first molar root lengths in radiographs of London children and found that, at alveolar emergence, the mean for the mesial root measured cervix to apex was 7.1 mm.

Kelley et al. (2009) examined jaws of young chimpanzees, gorillas and orangutans in museum and other collections in which the lower first or second molar cusp tips were between the alveolar and occlusal eruption stages. Some had stained enamel and others not, and they reasoned that the stained teeth had been within the mouth for longer. Therefore where some cusps in one tooth were stained and some not, they argued it had been in the midst of gingival emergence at the time of death. The mean mesial root length at this stage in first molars was 4.8 mm (range 4.2–5.3 mm) in their chimpanzees, 5.2 mm (range 4.3–6.1 mm) in orangutans and 6.9 mm (range 4.9–7.9 mm) in gorillas. Second molar means were 7.0, 7.0 and 7.4 mm, respectively. The range of variation in these relatively small assemblages suggests that it would be unwise to rely on such measurements, but they might give some idea about the eruption status of an isolated tooth in a fossil.

Kelley and Schwartz (2010) histologically determined gingival emergence in first molars from two young orangutans and a young gorilla that had been shot in the wild and were kept in museum collections. One orangutan had first molar cusps slightly higher than expected at gingival emergence, but the other orangutan and the gorilla showed that they had just emerged because the position of the gingival margin was marked by a clear stain line. Time elapsed during crown and root formation was determined from counts of incremental structures in enamel and dentine. For the orangutan jaws, they estimated that gingival emergence of the first molar took place at 4.6 years of age and, for the gorilla, 3.8 years.

Sexual dimorphism

One prominent feature of most living higher primates is their large degree of sexual dimorphism. This is seen, for example, in body size and in the canines, but the difference between sexes is much less marked in living humans. In a total of 51 canines, Schwartz and Dean (2001) found that dimorphism (the difference between the sexes expressed as a percentage of female size) in mean crown height varied from 11% in humans and 33% in chimpanzees, to over 60% in gorillas and orangutans. Both males and females varied around the mean, but the males varied more than the females. The level of dimorphism was reflected in crown formation time, with male orangutan and gorilla canine crowns taking on average over half as long again to form as female crowns. The difference in timing was much smaller in chimpanzees and humans. Canine cuspal enamel was thickest in humans and orangutans, with the longest cuspal enamel formation time occupying the greatest proportion of overall crown formation time. In these features, modern human canines were clearly distinguished from those of chimpanzees and gorillas. As early hominin canine morphology is relatively similar to that of humans, it seems reasonable to suggest that development dimorphism might also be similar. This is a crucial issue for interpreting fossil hominins in which the level of dimorphism is difficult to establish.

Summary

Several different lines of evidence confirm that prism cross striations represent a circadian rhythm and strategies have been developed to use them to estimate the time taken to secrete enamel matrix at different stages in the formation of the crown. It is possible to define short-period lines in dentine that are equivalent to prism cross striations and which can be used in a similar way. Regular long-period lines are present in both enamel and dentine and represent a development rhythm of between 5 and 12 days in the higher primates described here, with a periodicity that is constant for each individual. These lines vary in prominence through the tissue to create a sequence which can be matched between teeth from one individual and, in teeth which started to form before birth, which mark the point of birth with a neonatal line.

From this evidence, it has been possible to build development chronologies for the dentition that can be compared with the estimates based on radiography which were described in Chapter 3. The ages for crown initiation and completion are generally earlier, as would be expected given the nature of the radiographs. They confirm the overlap in timing between living chimpanzees and modern humans for the permanent incisors, but have made more clearly apparent the overlap in formation timing between the permanent first, second and third molars in chimpanzees. This contrasts with the lack of overlap in humans. It also seems that molar crowns take less time to form in chimpanzees than in humans, although the ranges overlap. There appear to be only small differences between geographically separated populations of modern humans, and wild chimpanzees overlap with laboratory animals. For fossils, the most detailed chronology currently available is for *P. heseloni*, an animal much smaller than the chimpanzee, somewhere near the root of the Hominoidea. Microscope sections of this exceptionally preserved material showed that crown development was initiated and completed at a very young age in comparison with chimpanzees. It has not been possible to section examples of so many different teeth in fossil hominins, but the revolutionary synchrotron micro-CT technique has made it possible to investigate key Neanderthal and early modern human fossils. From currently published evidence, it appears that the Neanderthals on average initiated and completed their crowns a little earlier than modern humans.

It has been possible to estimate the overall crown formation time in a larger number of fossil teeth from simple counts of perikymata, which represent long-period lines at the surface. Whilst *P. heseloni* molar crowns formed in a remarkably short time compared with those of chimpanzees, fossil hominoids such as *Afropithecus*, *Dryopithecus*, *Sivapithecus* and even *Gigantopithecus* seem to have formed their molar crowns in a similar time. *Australopithecus* molar crowns formed more slowly than the thick-enamelled molar crowns of *Paranthropus*, but both overlap the range of chimpanzees and modern humans. The small *Paranthropus* incisor and canine crowns were completed rapidly.

Part of this variation must be related to variation in the rate at which enamel matrix was secreted. Microscope sections of cuspal enamel have shown that it develops at a slower rate in modern humans than in chimpanzees. Rare sections of fossils can be compared. *Proconsul*, *Australopithecus*, *Paranthropus* and early *Homo* fit well with the fast chimpanzee rate curve, whereas a single Neanderthal specimen fits with modern humans. Was the slow modern human development schedule present in Neanderthals? There has been considerable discussion on this point. It has been claimed that a fossil from Atapuerca, assigned to *H. antecessor*, also had a schedule like modern humans.

Chapter 3 presents a great deal more evidence for the eruption of permanent molars in primates than it does about any of the tooth formation stages. Just a few fossil dentitions represent an individual that died during a molar eruption phase. These include representatives of *Afropithecus*, *Sivapithecus* and *Dryopithecus*, all of which were erupting molars at an estimated age similar to that found in chimpanzees. The same is true of a small group of *Australopithecus* and *Paranthropus* dentitions which show the first molar in eruption and the estimated age is too young for the modern human schedule. The remarkable skeleton from Nariokotome in Kenya discussed in this chapter had a dentition in which the second molar was erupting. Once again, the estimated age-at-death is young for a modern human child, but it is also rather old for a chimpanzee. This dentition was clearly developing to a very different schedule.

6 Human evolution, pace of development and life history

Life history

The life history of a mammal species can be defined by variables such as the mean length of gestation, the size and number of offspring at birth, the interval between births, the age at which weaning takes place and body size at that time, the age at which males and females first breed successfully, their size as adults and the age at death. Some species 'live fast and die young' and, at the other extreme, some species 'live slow and die old' (Promislow and Harvey, 1990; Smith, 1992). A 'fast' mammal might typically have a short gestation time, multiple small offspring, a short birth interval, early age of weaning and of sexual maturity, and a short lifespan. It would be small in body size at each stage. A 'slow' animal would have a long gestation time, after which one or a few large babies would be born, a long interval for each female between successive births, a late age of weaning and maturity, and a long lifespan. Offspring would be large at all stages and born from large mothers. Body size is thus strongly related to the pace at which life history unfolds but, when the mean values for life history variables in different species are plotted against one another, they remain correlated even when the effect of size is removed.

Models of life history emphasise the switch between growth and reproduction and suggest that the age of sexual maturity in a given species represents a balance between the two. Fast mammals, on one hand, minimise their risk of dying without reproducing themselves by maturing young and giving birth to many, but they are vulnerable to predators because of their small size. Slow mammals have an increased risk of dying before they can give birth, but their larger size decreases their overall risk of predation and increases their ability to protect their offspring through a longer period of development.

Primates in general live 'life in the slow lane' (Charnov and Berrigan, 1993; Walker et al., 2006; Mumby and Vinicius, 2008). Their development is slow, with a long juvenile period between weaning and sexual maturity in which they are relatively independent with respect to food, but still require their group for protection. They mature late and give birth at long intervals to a few, large offspring that can live for a long time. Life history theory predicts that primates evolved this slow pace of development because their young were at relatively low risk of predation during growth, which set the balance point between growth and reproduction at an older age. One possible explanation is that most primates are arboreal, which affords a degree of protection to their young. Arboreal species from other orders of mammals also tend to show slow life histories (Kappeler *et al.*, 2003). Another possible explanation is that selection for a large brain in primates, the growth of which requires a large share of resources (page 16), would act as a constraint on the overall rates of development (Barrickman *et al.*, 2008). The benefits of a large brain might include the decreased mortality of adult animals involved in the reproduction and care of offspring due to complex social organisation, foraging strategies and avoidance of predators. This would allow a longer reproductive period, maintaining numbers of offspring per female in spite of the delay in sexual maturity, longer birth interval and longer gestation that form part of a slow life history (Leigh, 2001). Amongst mammals, brain size is correlated with body size, but primates do, on the whole, have relatively larger brains for a given body size than most mammals (see, for example, Figure 'a', p. 111 in Deacon, 1992).

Characteristic features of human life history

Within the primates, the prosimians (lemurs, lorises, galagos and similar species) lie at the faster end of the scale, whereas chimpanzees, gorillas and orangutans are very much characterised by a slow pace of life history. Humans have the slowest life history of all, but they are also distinct from chimpanzees, gorillas and orangutans in more specific ways, as discussed in the following sections.

Human lifespan

Records of modal age-at-death for recent hunter-gatherers and forager-horticulturalists vary from 68 to 78 years (Gurven and Kaplan, 2007). The oldest person with an independently substantiated record of age was Jeanne Calment, who was born in 1875 and died in 1997 in Arles (France) aged 122 years (Robine and Allard, 1998). For wild chimpanzees at Gombe, Taï, Kibale, Mahale and Bossou, the oldest animal recorded was 55 years old (Hill *et al.*, 2001), but few survived into their 40s (Emery Thompson *et al.*, 2007). The maximum age for captive chimpanzees is about 60 (Robson and Wood, 2008). The maximum recorded age for captive gorillas is 52 years and those above 35 are considered old (Atsalis and Margulis, 2006). For wild orangutans at Ketambe in Sumatra, the oldest female was 53 years old and the oldest male 58 (Wich *et al.*, 2004). Other primates live considerably less long: the average lifespan for gibbons and siamangs is in the 30s, Old World monkeys mostly in the 20s and New World monkeys varying from 16 to 46 years (Judge and Carey, 2000).

As lifespan was strongly correlated with mean body size and brain size, Judge and Carey predicted by regression the lifespan that would be expected from these relationships. The actual lifespan comfortably exceeded that predicted in chimpanzees, gorillas and orangutans, some Old World monkeys and the New World monkey genus *Cebus*. Humans, on the other hand, were predicted a lifespan between 72 and 91 years, which fits comfortably with the observed range. This implies that human longevity is part of a broader evolutionary trend within the primates, rather than being a divergent human characteristic.

Menopause

Many women today live more than one-third of their lives after the menopause (Cohen, 2004). This represents a selective advantage because older females who are not reproducing can help to look after the children of younger mothers in the group, who can then maintain shorter intervals between births (Hawkes *et al.*, 2003). This is important when development to maturity lasts for so many years. Some researchers suggest that the human menopause is unique (Pavelka and Fedigan, 1991), whereas others (Cohen, 2004) point to evidence for menopause in non-human primates and other mammals. In modern humans, the average age at which the menopause starts is about 50 years, although it varies considerably. If the modal lifespan is 72, then a little less than one-third is post-reproductive.

Many chimpanzees also show signs of declining reproductive function with age, but the timing varies between studies (Walker and Herndon, 2008). Some show that the menstrual cycle ceases in most females between 35 and 40 years, whereas others record wild chimpanzees that remained reproductive over 40 years. Even if a post-reproductive stage exists, it is only a small proportion of the chimpanzee's lifespan. Little is known about this in orangutans or gorillas. Both macaques and baboons appear to lose reproductive function in their mid-20s, which again makes the post-reproductive period only a small proportion of their average lifespan. Several primate species other than humans do therefore show evidence of a post-reproductive stage in some long-lived females, but this does not necessarily mean that the animals are post-menopausal (Walker and Herndon, 2008). This, of course, is also true of many humans for at least part of the post-reproductive stage, but what does appear to be unique to humans is the length of the stage, which is more a function of the extended lifespan. The ability of humans to find food and contribute to the care of children thus declines remarkably slowly.

Long gestation period and large neonatal size

A full-term pregnancy in humans lasts 38–42 weeks (page 10). In chimpanzees it is 29–37 weeks (Thompson and Wrangham, 2008), in bonobos 34 weeks (Robson *et al.*, 2006), in orangutans 35–37 weeks and in gorillas 37–40 weeks (Harvey and Clutton-Brock, 1985). Amongst the primates as a whole, there is a strong relationship between body size and length of gestation, into which these values fit reasonably well (Martin, 2007). The mean weight of the neonate also shows a strong relationship with adult body weight amongst the primates. This trend is followed by chimpanzees, orangutans and gorillas, but human neonates are particularly large in comparison with adult size. This is due primarily to a rapid increase in body fat during the last ten weeks before birth (Tanner, 1989), which is seen as a peak in the growth rate of body mass (page 13). Human babies are noticeably chubby in comparison with the young of other primates and this presumably represents an energy reserve for the early post-natal period.

Early age at weaning and short intervals between births

Weaning in all mammals is a gradual process, starting with solid food and ending with a complete end to suckling. In humans it lasts from months to years and its timing varies. Clearly it is not driven by biological priorities alone and there is a social element, including the control of reproduction and the availability of alternatives to the mother's breast milk. In recent non-industrialised communities, some solid food was introduced on average at 5 months and most children were fully weaned between 2 and 4 years after birth – the mean age was 2.4 years (Humphrey, 2010). This compares with 3.2 years in gorillas, 5 years in chimpanzees, and 7 years in orangutans. Within the primates as a whole there are relationships between the timing of weaning and the weight of the newborn offspring and of the mother (Martin, 2007). The mean weight of human mothers predicts a weaning age of 2.5 to 3 years, but neonate weights predict over 6 years. Human weaning age does therefore appear to be young in relation to our size as a primate.

Humans are still not able to fend entirely for themselves after they are weaned. They require a diet specially adapted to their needs until 6 or 7 years of age. It seems reasonable to suggest that this is because the permanent dentition is not yet established and the digestive tract is still small (Humphrey, 2010). Bogin (2006) has argued that, for humans, this stage immediately following weaning should be distinguished as *childhood* (page 14).

Long gap between weaning and reproductive maturity

Amongst all primates, the age at which a mother successfully bears offspring is later than menarche. So, for example, in modern human populations the mean age of menarche is around 12–13 years (Tanner, 1989), but amongst most recent populations the average age of mothers when they first give birth is 19.5 years (Robson *et al.*, 2006), 6.5 to 7.5 years after menarche. In chimpanzees, menarche may be at 7 to 10.8 years (Tanner, 1962); the average age of wild chimpanzee mothers when first giving birth is 13.3 years, although in captivity this is usually 11 years (Robson *et al.*, 2006). This is a gap of about 1 to 3 years. It therefore takes humans a good 5 years more than chimpanzees to start reproducing themselves. There is also a long interval of about 17 years between the relatively young age of weaning and old age of first birth in humans. This compares with 8 years for wild chimpanzees, 8.6 in wild orangutans and 6.8 years in wild gorillas.

Life history of fossil primates

All the life history variables discussed in the preceding sections are difficult, or impossible, to estimate directly in fossil primates. Robson and Wood (2008) instead suggested *life history-related variables*; that is, morphological features which might be used to represent life history in fossil taxa. They included adult body size, adult brain size and the pattern of dental development.

Body size and life history

Body size is correlated with most aspects of life history. Throughout the mammals as a whole, including the primates, bigger species tend to have longer gestations and larger babies, a late age of weaning and first birth, and a longer lifespan (Harvey and Clutton-Brock, 1985; Purvis *et al.*, 2003). One of the reasons that chimpanzees, bonobos, gorillas, orangutans and humans all have relatively slow life histories is that we are the biggest primates and, in general, big mammals. As Robson and Wood (2008) have pointed out, chimpanzee, bonobo and orangutan females have a similar mean adult body mass of about 35 kg, and have similarly paced life histories. Human hunter-gatherer females are perhaps 10 kg heavier and have a slower pace of life history. Gorilla females, on the other hand, have a mean body mass of around 95 kg, but develop to reproductive maturity considerably faster. Thus the relationship is not entirely simple.

In fossils, bone measurements must be used instead of body mass. Lengths of the long bones in the limbs are strongly correlated with body size, but complete specimens are rare finds. Fortunately, dimensions at the middle of the long bone shafts and the joint surfaces, which can be measured in fragmentary specimens, act as good estimators for body mass in living primates (Ruff, 2003). Skull fragments are rather more common finds and there are correlations between skull measurements and body weight which allow similar estimations to be made (Spocter and Manger, 2007). Teeth and jaws are the most common fossils, and mean tooth sizes correlate highly with mean body mass in living primate species, so once more estimates can be made (Gingerich *et al.*, 1982).

There are, however, still difficulties in applying these estimations to fossil taxa. One is the question of sexual dimorphism, which is pronounced in many primate species and can only be estimated for fossil taxa. Similarly, there are considerable uncertainties involved in using regressions developed from measurements of living primates to estimate body mass in extinct fossil primates for which the relationships might be very different. Nevertheless, Robson and Wood (2008) gathered from the literature estimates for body mass in different taxa. They found a division in the hominin fossil record around two million years ago. Fossil taxa dated before this, including the genera *Ardipithecus*, *Australopithecus* and *Paranthropus*, all had an estimated body mass of around 35–45 kg, overlapping the range of living chimpanzees. Those dated later, including the genus *Homo*, had body mass estimates overlapping the range of modern humans.

Brain size and life history

Longer lived primates tend to have heavier adult brains, as do those with a longer gestation period and later average age at which females give birth for the first time. Brain mass is also correlated with body mass. As discussed in the preceding section, body mass is itself correlated with most life history variables, but it is possible to isolate brain mass statistically and to show its relationships independently, particularly its relationship with lifespan (Deaner et al., 2003). It may be that brain size places a limit on the rate at which an individual can grow because it is expensive to develop relative to the rest of the body in terms of the energy required (Aiello and Wells, 2002). Another possibility is that a larger brain requires a longer period of development for the matching adult skills to be attained. The implication of this as an adaptation is that a juvenile would not be able to function adequately until maturity was reached, but most living non-human primates are able to forage and process food adequately as soon as they are weaned (van Schaik et al., 2006).

Brain size is estimated in fossil skulls from the endocranial volume (cranial capacity), although this is difficult because it is rare for a skull to be preserved completely. Robson and Wood (2008) found that endocranial volumes for *Ardipithecus*, *Australopithecus* and *Paranthropus* overlapped one another and those of living chimpanzees. Volumes for *H. habilis*, *rudolfensis*, *ergaster* and *erectus* all fell between those of living chimpanzees and humans, whereas *H. heidelbergensis* and *neanderthalensis* were not significantly different from those of *H. sapiens*.

Dental development and life history

The diagram of Schultz (1960, figure 2, p. 3) has an important place in thinking about the stages of life history amongst the primates. It distinguishes five periods: prenatal, infantile, juvenile, reproductive and adult. The dividing line between the infantile and juvenile period is drawn by the eruption of the first tooth in the permanent dentition (generally the first molar) and the line between the juvenile and adult period by the eruption of the last tooth in the dentition (third molar). This connection has been extensively explored by Smith (1989; 1991a; 1991b; 1992) in 21 living primate species. First molar eruption ages were highly correlated with birth weight, age at weaning and female age at first birth, neonate brain weight and adult brain weight. Third molar eruption ages showed similarly high correlations with body weight, birth weight, age at weaning, age of sexual maturity, female age at first birth, neonatal brain weight and adult brain weight. It therefore appeared that, in fossil primates, age at molar eruption might act as an excellent proxy for life history variables.

As pointed out in Chapter 5, fossils at exactly the right developmental stage to show first molar eruption are rare. There are, however, some key specimens from which it is possible to suggest that the age at gingival emergence was young, similar to living chimpanzees, in both non-hominins such as *Afropithecus*, *Sivapithecus* and *Dryopithecus*, and the early hominins *Australopithecus* and *Paranthropus*.

Equivalent fossils showing third molar eruption are not available, but the second molars had just erupted in the Nariokotome early *Homo* skeleton, suggesting gingival emergence at around 8 years of age, compared with about 12 in living humans (page 145). This contrasts in an intriguing way with the state of arm bone development, which was similar to that of a 13- to 14-year-old living human child.

Much more data are now available on the development of permanent tooth crowns in fossil taxa. As explained in Chapter 4, this is much less commonly recorded than gingival emergence for living primates, but it is part of a highly integrated developmental sequence which leads to eruption. It is therefore to be expected that there would be a similar relationship between adult brain weight and age at crown completion. The histological estimates for crown formation in Pliocene fossil hominoids suggest a relatively fast schedule for most of them, closer to that of living chimpanzees than living humans. The examples of *Paranthropus* and Nariokotome, however, show that living primates do not provide directly comparable models. There is, it must be said, no particular reason why extinct creatures should have developed in exactly the same way. The focus of debate, however, has shifted to the Pleistocene and centres on the developmental schedules of the largely contemporary Neanderthals and early modern humans. Whilst all accept that the early *H. sapiens* fossils from Qafzeh, Skhūl and Irhoud developed their crowns following a similar schedule to living humans, some studies suggest that Neanderthal tooth crowns developed on a faster schedule, whilst others disagree (page 142). Both taxa were obligate bipeds, even if the post-cranial skeleton of Neanderthals is highly distinctive, and had brains of a similar size. Premolars and molars overlap in size and, although Neanderthals have, on average, distinctively large incisors and canines, they too overlap with the full modern human range. Both Neanderthals and early *H. sapiens* have been found in Israel and North Africa in association with similar Mousterian or Middle Stone Age stone tool assemblages, which would make it even more remarkable if they differed markedly in their life histories. It is only later that the more complex Upper Palaeolithic technologies appeared. The debate is continuing at the time of writing. Once again, it seems unlikely that the median developmental schedule for living humans will provide an exact model for either of these fossil groups. In all probability the stone and bone tools that survive represent only a part of the technology in use, which might have included wood, plant fibres, skins and even the teeth themselves.

Weaning, giving birth and the expansion of post-canine teeth

When the permanent first molar erupts behind the deciduous premolars, it increases the length of the cheek tooth row. It seems reasonable to suggest that this expansion would increase the individual's ability to process solid food, and that it might even be a prerequisite of this in some primates (Bogin, 1999; Humphrey, 2010). It is notable, nevertheless, that in the study of Smith (1989) of 21 living primate species, the eruption age of the third molar showed a slightly higher correlation with weaning age than that of the first molar.

Godfrey *et al.* (2003) addressed the question in a different way. They measured 'dental precocity' for any given age as the number of post-canine (deciduous and permanent premolars and molars) teeth erupted in a species, expressed as a ratio of the total number of deciduous and permanent post-canine teeth that could eventually erupt. Dental precocity was estimated at 4 months, at 1 year of age and at weaning. They also estimated 'dental endowment at weaning', which was the occlusal area of deciduous and permanent post-canine teeth erupted at weaning, expressed as a ratio of the average adult post-canine occlusal area. They predicted these two dental development measures in 40 primate species from their relationship with cranial capacity, neonatal body mass, adult female body mass, age at weaning and length of gestation. They found that the most important predictor of dental precocity was cranial capacity: primate species with a small brain in relation to body size were likely to be dentally precocious, in comparison with larger brain for body size species. Species with a longer gestation also tended to be more precocious. Age at weaning was less important as a predictor, and it did so in the opposite way to cranial capacity. Species with a young age at weaning tended to have more advanced dental development and a larger post-canine occlusal area. Overall, the state of dental development at weaning was very variable across primate species, partly because weaning varied markedly in its duration between individuals and a single age at weaning was therefore often difficult to define.

In living humans, at the average age of weaning, children have only just completed their deciduous dentition, with the fourth premolars erupted. The permanent first molars do not erupt for another four years. Chimpanzees have a mixed dentition at weaning, with all their deciduous teeth, and the permanent first molars erupted behind the deciduous fourth premolars. Gorillas still have a deciduous dentition, although their permanent first molars are just beginning to erupt. Orangutans have much of their permanent dentition erupted or erupting, with the exception of the canines and third molars. It is difficult to discern a pattern. Only in gorillas and chimpanzees can it be said that the permanent first molar erupts at the time of weaning. In general, the case for weaning being accompanied by an expansion of the chewing area of the dentition by the addition of permanent teeth is not a strong one. There is similarly little strong agreement between the completion of a full adult permanent dentition and the onset of reproduction. In humans, the eruption age for permanent third molars coincides reasonably well with the average age at which females give birth. In gorillas, the third molars erupt a little afterwards. In both chimpanzees and orangutans, the full permanent dentition has been established for two or more years by the time that females give birth for the first time.

It is thus difficult to argue that the state of dental eruption is a good basis for predicting the ages of weaning and first birth (Skinner and Wood, 2006; Robson and Wood, 2008). In the primates as a whole, there does not seem to be a strong relationship once the effects of other factors, such as body size and brain size, are taken into account. Perhaps this is not surprising because it is difficult to see the transition to a permanent dentition as an event that confers for the first time an ability to cope with adult solid food. This must vary hugely with the requirements of the diet. In any case, the molars, which are so important in this discussion, are best seen as part of a continuum in which they follow on behind the deciduous premolars (page 30). The eruption of the first molar can therefore be seen as the next step in the development of the deciduous dentition rather than the first step in establishing the permanent dentition. In addition, weaning is usually a gradual transition for which the timing may well vary according to environmental factors and changes in the dynamics of a group of animals. The ecology, biology and behaviour of different primate species vary greatly and it is hardly likely that weaning has the same significance in each species.

Another check on the effect of weaning is to consider dental attrition. It seems logical to suggest that tooth wear would increase in relation to the consumption of more solid food, so that variation in deciduous premolar attrition at a comparable age in different species of primates might imply a different relative age at weaning. Aiello *et al.* (1991) found that these teeth started to wear earlier and to a greater extent in gorillas than in either chimpanzees or orangutans. The age at eruption for deciduous premolars does not vary greatly between these taxa. They concluded that the relatively early weaning age of gorillas provided the best explanation. In fossils, they found that for *Paranthropus* specimens with an estimated age-at-death of 2.5 to 3.5 years the enamel on deciduous premolars was worn through to expose dentine. In contrast, in *Australopithecus* specimens with estimated ages of 3 to 4 years, there was only minimal polishing of the enamel. The large cheek teeth of *Paranthropus* had considerably thicker enamel than the smaller teeth of *Australopithecus*, and would be expected to wear more slowly. From this they suggested that *Paranthropus* might have been weaned at a younger age.

Skinner (1997) explored a similar idea in comparing Neanderthal and Upper Palaeolithic *H. sapiens* dentitions from western Europe. He found that, whilst the estimated wear rate in deciduous and permanent cheek teeth was similar in both groups, deciduous anterior tooth wear was more rapid in Upper Palaeolithic children. Permanent anterior tooth wear was, in contrast, more rapid in Middle Palaeolithic dentitions. Once more, he suggested an earlier onset to weaning for the Upper Palaeolithic children.

Fast and slow mammals and Schultz's rule of eruption

In the same book chapter in which he introduced his well-known primate development diagram (page 154), Schultz (1960) also tabulated the order of permanent tooth eruption. He showed that, in the fastest developing primates such as the night monkeys (*Aotus trivirgatus*), all three permanent molars erupt before any deciduous teeth are exfoliated and replaced by permanent incisors, canines and premolars. In the slowest developing primates such as living humans, the deciduous incisors are replaced at the same time as the permanent first molar erupts and the remaining deciduous teeth are replaced before the eruption of the second molar. Other primates could be seriated in between these two extremes. Schultz explained that this was a consequence of the longer development period over which the permanent molars emerged, during which the deciduous teeth were worn out. Smith (2000) has called this observation *Schultz's rule*. She not only confirmed it in primates by the addition of new data, but also showed that it could be applied to other orders of mammals.

Life history, development and cognition in primates

The concept of cognition is often used in discussions of hominid development and evolution. Cognition used in its psychological sense relates to knowledge and thought: the ways in which an individual acquires information, processes and uses it. Cognition can be graded within the primates, with living humans showing the most complex cognition, chimpanzees, bonobos, gorillas and orangutans showing an intermediate grade, and other non-human primates showing the lowest grade of complexity (Russon, 2004; Russon and Begun, 2004). The living great apes share many cognitive features and abilities with humans including symbolism, the manufacture and use of tools, cooperative hunting and food sharing, but they are not developed to the same extent or the same level of complexity. There are several models for the development of human cognition through childhood. They include the ideas of Jean Piaget (Piaget and Inhelder, 1969), who divided the process into four periods: sensorimotor (birth to 2 years), preoperations (2 to 6 years), concrete operations (6 to 12 years) and formal operations (12 to 15 years). Taylor Parker (2002) applied Piaget's definitions of these periods to living great apes and macaque monkeys. She found that great apes only reached the early preoperation period and that macaques reached just the sensorimotor period.

Cognition amongst the primates is related in complex ways to brain size, body size and life history. Most studies have linked brain size with the level of cognition and this makes sense in terms of the larger numbers of additional neurons (over and above those required for simple functions) which are more extensively interconnected (Russon and Begun, 2004). In addition, brain structures such as the neocortex and cerebellum, which have been linked with cognition, tend to be a larger proportion of the total brain size in larger brains. Body size is strongly correlated with brain size in mammals (Martin, 1981) and primates as a whole show somewhat larger brain sizes for a given body size than other groups of mammals. Living hominids (the great apes and humans) are the largest primates and they have the largest brains. In the case of chimpanzees, bonobos, gorillas and orangutans, brain size in general follows the same scaling with body size as the smaller primates, but the human brain is more than twice what would be predicted from the relationship between body and brain size in mammals (Ward *et al.*, 2004).

The overall relationship may partly be a matter of simple scaling. A larger body has more neural receptors sending inputs to the brain and has more elements to control and therefore more motor outputs, together leading to greater complexity. In addition, a brain is expensive in terms of energy costs both to develop and to maintain. A large brain in a small body would consume proportionally a larger share of the total energy resources than a large brain in a large body. So body size creates a metabolic limit on the size of brain that can be supported. Smaller mammals have a higher mortality and their 'live fast, die young' life history can be seen as an adaptation to the requirements of faster reproduction. In those circumstances, there is little advantage for them to invest in the higher energy costs of a larger brain. By contrast, most large mammals develop more slowly, which allows them to develop larger brains more slowly, and provides more time for more lengthy learning with which to take advantage of their larger brain.

None of these hypotheses, however, explains the observation that there is not a perfect allometric relationship between brain size and body size in mammals. There is variation between fairly closely related taxa in the relationship, and some groups of taxa (such as the primates) have a different relationship to others. Most primates are arboreal. Does this provide challenges for control of locomotion that require a proportionally larger brain? Is it the nature of the diet, which is related to body size through the pace of metabolism (smaller mammals have a faster metabolism)? The smallest primates are insectivores and most of the largest are folivores (leaves require fermentation for digestion, so body size may in this case also be related to the size of the gut). Most of the medium-sized primates are primarily frugivores and fruit can present particular challenges in gathering and processing. Some taxa take a wider variety of foods than others, including meat. This might lead to some variation in the relationship between body and brain size. The size of the social group shows a relationship with brain size in primates. The number of individuals in a group may represent the need for protection against predators, competition within and between groups, requirements for collaboration in gathering food, and so on. This again could cause variation in the body/brain size relationship. There is therefore a whole complex of different potential factors and relationships that need to be taken into account when considering the evolution of great ape and human cognition. As Ward and co-workers (2004) suggest, it seems best to consider that cognition, pace of life history, tempo of development, brain size and structure and body size and shape all co-evolved together. It will be difficult to tease out the different factors from the fossil record and, in particular, difficult to isolate life history variables.

Summary

In comparison with chimpanzees, humans give birth to large, fat babies after a long gestation period. The infants grow slowly, but are weaned young, before they are able to feed themselves without help. This allows their mothers, who typically had their first child at an older age than is the case for chimpanzees, to become pregnant again sooner. Such reduced birth intervals allow human mothers, on average, to give birth to a greater total number of children (Bentley *et al.*, 1993) than chimpanzees (Nishida *et al.*, 2003), in spite of becoming non-reproductive at a similar age (Emery Thompson *et al.*, 2007) and thus having a shorter period over which they can bear children. It also means that the energetic costs of lactation for the mother are substantially lower per child (Key, 2000). The risk for humans is that children must pass through a vulnerable interval between weaning and full independence in feeding themselves. During this interval, they are looked after by their family and the wider social group, including older siblings, parents and grandparents. The longevity of humans allows a long post-reproductive life for women in which they take part in the gathering of food and care of the children of their daughters and other relatives. These human adaptations for reproductive success place an emphasis on social interactions.

Humans are also large primates with large brains for their body size. These are expensive to develop and maintain in terms of the energy required from food, so human life history also entails an emphasis on the gathering and utilisation of resources through mobility, complex social organization and tool use. In turn, all of these are enabled by the development of the large brain, a body adapted for bipedal locomotion and arms and hands adapted for manipulation (Rolian *et al.*, 2010), all combined with a long period for cognitive development as well as physical growth. Being a human female requires the input of more energy from food than being a chimpanzee, but the shorter inter-birth interval means that the cost of producing each unit of reproductive success, a child, costs less energy.

The dental histological studies outlined in Chapter 5 suggest that all extinct Miocene primates so far investigated had a more rapid pace of development than living humans, with a sequence similar to that of living chimpanzees, if not necessarily directly equivalent. The Pliocene genera *Australopithecus* and *Paranthropus*, some of the earliest hominins, similarly had a more rapid tempo of growth than humans today, although their sequence of dental development has no exact parallels in living primates. This is not surprising, given the large, thick-enamelled molar and premolar teeth, particularly of *Paranthropus* dentitions. In comparison with humans, both genera were also relatively small-bodied, small-brained and seem to have had only a partial adaptation to bipedal locomotion.

Robson and Wood (2008) have argued that the point of change for body size in the hominin fossil record was around two million years ago, not long in geological terms before the start of the Pleistocene epoch. The *H. erectus* or *ergaster* skeleton from Nariokotome, dated 1.6 million years ago, is the strongest early evidence for both a body size and fully bipedal limb anatomy equivalent to that of living *H. sapiens*. The relative stages of different teeth in its developing dentition, however, do not fit well into the sequence expected for a human child. The best fit would be around the 11 years stage and yet its histologically estimated age-at-death is about 8 years. This suggests a much more rapid development, which is even more remarkable when it is considered that bone development was around the 13–14-year stage

Summary

of a living human child. In truth, the Nariokotome skeleton does not fit well with either humans or chimpanzees. Care needs to be taken because this is a single specimen and, as Chapter 3 shows, there is considerable variation in growth within living humans. In addition, early *Homo* fossils are rather variable in form and their taxonomy is disputed (Lieberman *et al.*, 2008). There is, however, another example of a *H. erectus* tooth with a rapid rate of crown development from Sangiran in Indonesia, dated perhaps a few hundred thousand years later. Both stone tools and cut-marked animal bones are known in African archaeological sites from 2.6–2.5 million years ago (Semaw, 2000; Domínguez-Rodrigo *et al.*, 2005), although none has been directly associated with hominin fossils and it is not known which taxon made them. It does therefore seem that at least some of the behavioural and technological adaptations were already in place which might have allowed the gathering of more resources.

Lieberman *et al.* (2008) have argued that the appearance of *Homo* in the fossil record marked an adaptation to persistence hunting. This is the running down of prey animals until they are exhausted. Whilst, for example, antelopes may be able to gallop faster over short distances than living hunters, humans are anatomically and physiologically adapted to running for longer periods: endurance running. Reduction in premolar and molar size, thinner dental enamel and stone tools might similarly be adaptations to a more carnivorous diet. Meat, kilo for kilo, provides more calories and protein than the bulk of plant foods that could have been gathered in most parts of the world before the development of cultivation, although Lieberman *et al.* (2008) also suggested the digging of starch-rich tubers. In terms of energetic efficiency (the balance between calories expended in the hunt and calories available for consumption as a result), this strategy would confer an advantage. Increased carnivory could therefore supply the energetic needs of a larger body size, larger brain, longer development and shorter birth interval.

The more complex stone tool assemblages labelled Mousterian, Middle Palaeolithic or Middle Stone Age, starting perhaps 250 000 to 200 000 years ago, have been found in direct association with fossils of both *H. neanderthalensis* and *H. sapiens*. Whilst all studies agree that the dentitions in these early fossils of *H. sapiens* developed on a slow, modern human-like schedule, some have suggested that Neanderthals developed more rapidly. Others suggest a slower schedule. At the time of writing, this is the frontier of the debate, as can be seen by the number of papers reviewed in this book. Neanderthals are in any case intriguing hominins from many points of view, such as their distinctive skeletal and dental anatomy, geographical distribution, overlap in the fossil record with *H. sapiens* and their ultimate disappearance from the record. Similar histological studies have not been carried out on later fossils assigned to *H. sapiens* and associated with the far more complex, variable and rapidly changing Upper Palaeolithic technology which replaced the Mousterian. It is, however, assumed that they would have had a slow, modern human-like pace of dental development.

7 Dental markers of disease and malnutrition

As described in Chapters 4 and 5, enamel and dentine are marked by microscopic features which preserve evidence of their formation sequence, including a record of disturbances to their development. On the surface are the defects of enamel hypoplasia (Definition Box 7). In microscope sections, there is variation in the prominence of the brown striae of Retzius (but see page 174), often distinguished as *accentuated lines* or *Wilson bands*. In the dentine, there may be poorly mineralised bands marked by interglobular spaces (page 104) that are sometimes associated with a deviation of the EDJ and with defects of enamel hypoplasia.

One possible source of confusion is inherited enamel defects, some of which form part of an inherited syndrome affecting several parts of the body and some which are due to factors which affect the teeth alone, such as mutations in the gene which codes for enamel matrix protein. The latter are grouped together as *amelogenesis imperfecta* and they include both hypoplastic and hypocalcification defects. They yield no information relevant to this book and are so rare (no more than 1.4 per 1000 people) that they are unlikely to be encountered in fossils. In addition, they are recognisable because they tend to affect the whole dentition.

Hypoplastic defects

Defects of enamel hypoplasia are divided into three types: furrow-form, pit-form and plane-form (page 86). A wide range of different defects can be seen, sometimes combining features of all three. They are dynamic and vary depending where in the crown formation sequence they occur. One defect may also vary around the circumference of the crown, so that it is different on lingual and labial sides. Close examination is often needed to understand what went on during enamel matrix secretion.

Definition Box 7. Enamel hypoplasia, DDE, LEH or AI?

The defects examined here were originally described by the pioneers of French dentistry as *érosions* (page 185) and that was how they were known for the next century, even though it was accepted that they were not the result of mechanical or chemical erosion. The term *atrophy* was also unsuitable as it implied a wasting away, rather than a lack of development in the first place. Other more descriptive names were also used, such as *wavy enamel, furrowed teeth* or *honeycomb teeth*. It was in 1896 at an international dental conference in Chicago that Otto Zsigmondy, a dentist from Vienna, suggested: 'Where individual organs or part of organs are defectively developed because of external or internal noxae, pathological anatomists are wont to employ the term hypoplasia to express that condition. We may accordingly speak of a hypoplasia of the enamel' (Zsigmondy, 1893, p. 714). The word *hypoplasia* describes the condition and not the individual defect itself. Many authors write 'hypoplasias' to imply more than one defect, where strictly speaking this means more than one type of hypoplasia, so it is clearer to write 'hypoplastic defects'.

Developmental defects of dental enamel (DDE) was the term adopted by the Fédération Dentaire Internationale in 1982 when developing an index for recording them in epidemiological surveys of living patients. They meant the term to include hypoplastic defects (those due to disrupted enamel matrix secretion), opacities (equivalent to hypocalcification – page 71) and discoloured enamel. DDE is therefore not really a good term for the defects described in this chapter, which are specifically of the hypoplastic type.

Linear enamel hypoplasia (LEH) is commonly used in anthropological literature (Goodman and Rose, 1990) to mean what are called furrow-form defects in this book. The term was first used clinically (Sweeney *et al.*, 1971; Infante and Gillespie, 1974) to describe neonatal defects which make a sharp line across the deciduous tooth crowns (page 192), but this might be a furrow-form or a plane-form defect. For this reason, LEH is not used here.

Amelogenesis imperfecta (AI) is a term used clinically (Weinmann *et al.*, 1945) to distinguish the rare defects which are inherited directly, as opposed to those which are caused by environmental disruption during development. There are several different types of AI which can be distinguished by a range of diagnostic signs (Bäckman, 1989). Many dentists might assume that 'hypoplasia' means such inherited defects, but they can in fact involve either hypoplasia or hypocalcification, or both.

Localized enamel hypoplasia as used in anthropology (Skinner and Hung, 1986; Lukacs, 1991) implies the presence of very large pit-form defects on deciduous canines which cannot be found on any other crown in a deciduous dentition. The cause is unknown and it is a very specific phenomenon.

Furrow-form defects

These defects (Figures 7.1 and 7.2) are by far the most common and superficially might seem the most simple, but when considered in detail it is apparent that they could in fact be the most complex. The furrows range from sharp, narrow lines which look almost as if they have been engraved into the crown surface, to broader,

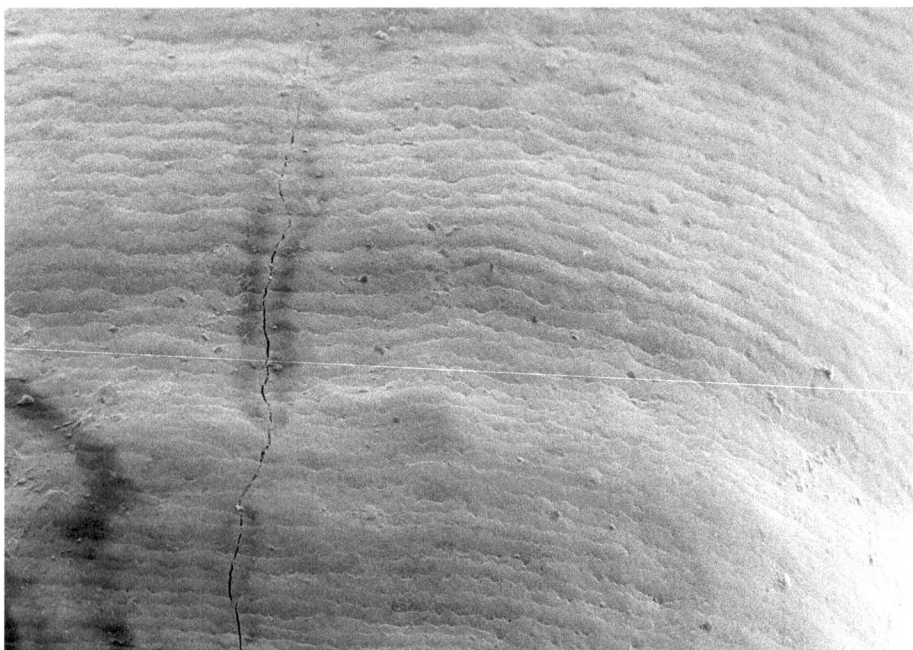

Figure 7.1 Mid-crown perikymata in the human permanent lower second incisor from Figures 4.1 and 4.3, 4.4 and 4.5, with the incisal edge of the crown above the top of the image. There are many small variations in the spacing of the perikymata, but the most prominent feature is a furrow-form hypoplastic defect, occupying a little less than 0.5 mm of crown height, in the centre of the field of view. Uncoated specimen under low vacuum in a scanning electron microscope using an ESED (see Appendix B). Field of view is 2.7 mm wide.

less sharply defined indentations. Several furrows are often combined together to give a washboard effect. In incisor and canine crowns the defects are most well defined on the labial sides, where the perikymata spacing is optimal and the surface is smooth and relatively flat. They are harder to see on the concave lingual sides of these teeth. Defects visible in the cervical region of the labial side will not be visible on the mesial and distal sides because the strong curve of the CEJ to occlusal means that this part is lacking. Furrow-form defects are less pronounced in molars and premolars, but there is less difference between different sides of the crown. This is because a much greater part of their crown height is composed of occlusal-type perikymata, which are wider spaced, or cervical-type perikymata, which are narrowly spaced (page 81).

Furrow-form defects are mainly formed through variation in the spacing of the perikymata. The simplest may be made by an increase in spacing between just two neighbouring perikyma grooves. In the occlusal to mid-part of the crown this may even be visible with the naked eye as a sharp line if the illumination is right. A slightly larger defect might include two or three neighbouring perikyma grooves which are more widely spaced than normal. They form an *occlusal wall* to the defect. The *floor* and *cervical wall* of the defect may be formed by normally spaced

Hypoplastic defects

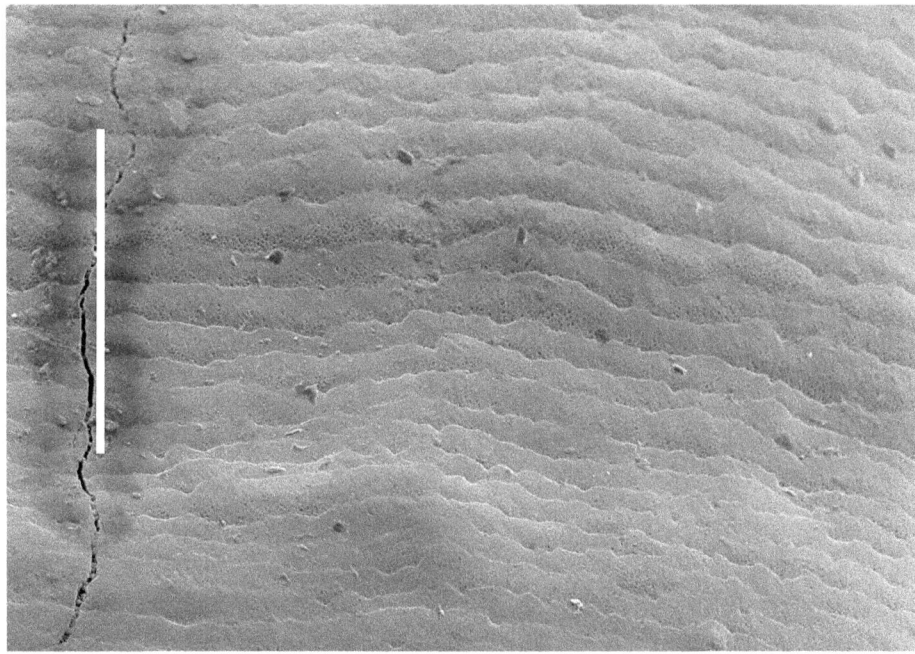

Figure 7.2 Closer view of furrow-form defect shown in Figure 7.1. The occlusal wall of the defect is marked with a white line and comprises eight widely spaced perikymata, some of which expose a broader band of more sharply marked Tomes' process pits. The cervical floor of the defect has more closely spaced perikymata, but it is difficult to be sure where the cervical margin of the defect lies. Field of view is 1.4 mm wide.

perikymata, or perikymata that are more closely spaced than normal (Hillson and Bond, 1997; Hillson *et al.*, 1999; Witzel *et al.*, 2008). Generally speaking, the occlusal wall is more sharply defined and there is a gradual return to normal spacing through the cervical wall. Larger furrow-form defects might involve 10, 20, or more, abnormally widely spaced perikymata in their occlusal wall.

Two main factors seem to underlie the formation of furrow-form defects. One is the number of rings of ameloblasts that cease matrix secretion at each perikyma groove (page 98). This decreases progressively down the crown but, in the more widely spaced perikymata of defects, the number of rings is larger than normal for that part of the crown (Hillson *et al.*, 1999). One effect of this is that more sharply defined Tomes' process pits are seen at the surface (Figure 7.2). The other factor is a slowing of the secretion rate of enamel matrix just before the ameloblasts cease at a perikyma. This is a normal feature, apparent from the curved outline and decrease in spacing of the brown striae of Retzius just below the crown surface. Witzel *et al.* (2008) described a furrow-form defect associated, not with an increase in perikymata spacing, but instead with a further reduction in secretion rate. Another defect, however, had both an increase and a subsequent decrease in perikymata spacing, but without any change in the secretion rate.

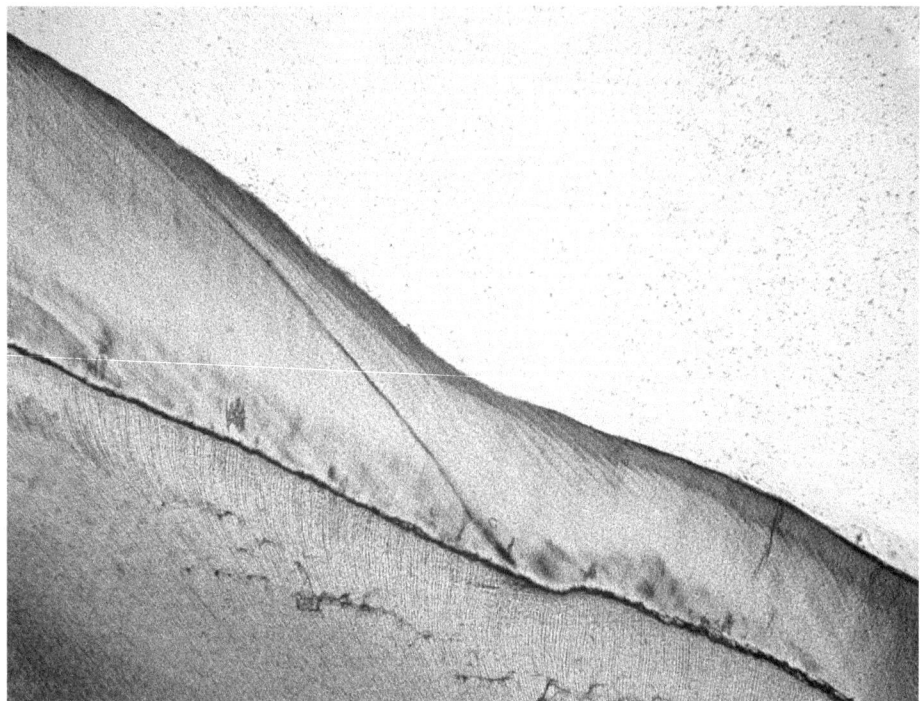

Figure 7.3 Section of furrow-form hypoplasia in a human permanent lower first incisor. This is a prominent furrow defect which deeply indents the crown side. The brown striae of Retzius remain relatively evenly spaced throughout and angle up to meet the perikymata in the normal way. There is one accentuated stria, either at the start of the defect, or slightly beforehand. As often happens with really marked striae, it is associated with a deviation of the EDJ and there is perhaps a suggestion of a matching line in the dentine. Polished section, imaged in a conventional light microscope as described in Appendix B. The specimen comes from the crypt of Christ Church, Spitalfields in London and was kindly lent by Daniel Antoine, The British Museum. Field of view is 2.1 mm wide.

There does, therefore, seem to be some variation in the way in which these defects are formed.

In a microscope section through a furrow-form defect (Figure 7.3), it is apparent that, whatever is happening at the crown surface, the overall daily prism cross striation rhythm continues unabated, with the same periodicity count between brown striae (Hillson et al., 1999). The brown striae of Retzius are sharply defined near the surface and it is apparent that each perikyma groove in the defect is associated with a regular brown stria. There is no evidence for additional or missing striae, although it is fairly common for some of those involved to be more prominent than those not associated with the defect (page 175). These observations imply that the disruption causing the defect is not specific to a particular day or days in the sequence, but is a more general effect which alters in some way the amplitude of the long-period rhythm rather than its periodicity. This fits with the nature of factors

Hypoplastic defects

disrupting development (discussed later in this chapter). If someone is ill with an infectious disease, they are not instantly unwell one day and better the next. The physiological effects take time to develop and recede. The same is true for malnutrition or psychosocial stress.

If furrow-form defects are caused by factors which disrupt growth in general, then they should be matched in any other tooth of the same dentition for which the crown surface was forming at the same age. Antimeres, upper and lower representatives of the same tooth, should match exactly, with the same counts of perikymata within and between defects. Human permanent upper first incisors should match both lower incisors, as their crown surfaces start to form at the incisal edge around 1 year of age (Figures 4.6 and 4.7). Upper second incisors and canines start to form their crown surfaces six to eight months later, but the later part of the sequence should be matched with the other anterior teeth. Matches between premolars or molars and anterior teeth are more difficult because their crown surfaces are formed in a rather different way (page 85). There are relatively few of the medium spaced mid-crown perikymata which make furrow-form defects so clear on incisors and canines. With a microscope and care, however, it is possible to match defects by checking with perikymata counts. It is possible to build an extended sequence of defects, potentially going from the formation of the permanent first incisor incisal edges at around 1 year after birth, to the completion of the canine and second molar crowns around 6 years of age. This kind of matching has a useful application in finding the teeth from one individual within a commingled burial.

It might seem logical to link the 'size' of a furrow-form defect to either its duration or its severity. Size in absolute terms is difficult to measure because the edges of the furrow are rarely sharply defined. In any case, the 'defect width', or the vertical height of the crown surface separating these two edges, will vary for the same defect measured in matched different tooth crowns because it will be found in different parts of the crown, with different spacings of perikymata. The same 'width' of crown surface in different teeth may actually include different numbers of perikymata. Size therefore cannot be used as an index of duration. Similarly, it cannot indicate 'severity' in any simple way. Is a defect involving more perikymata than another a more severe defect because it disrupts a larger proportion of the long-period enamel matrix secretion sequence? Is a defect which includes perikymata which are particularly widely spaced more severe than one in which the abnormal spacing is less pronounced? The answers to these questions are simply unknown.

Plane-form defects

A plane-form defect can be seen as an extreme variation of a furrow-form defect in which the plane of a single brown stria of Retzius is exposed over a much wider band (Figures 7.4, 7.5 and 7.6). The plane varies from 100 µm or so wide to several millimetres. It may be isolated between otherwise normal perikymata or be part

168 Dental markers of disease and malnutrition

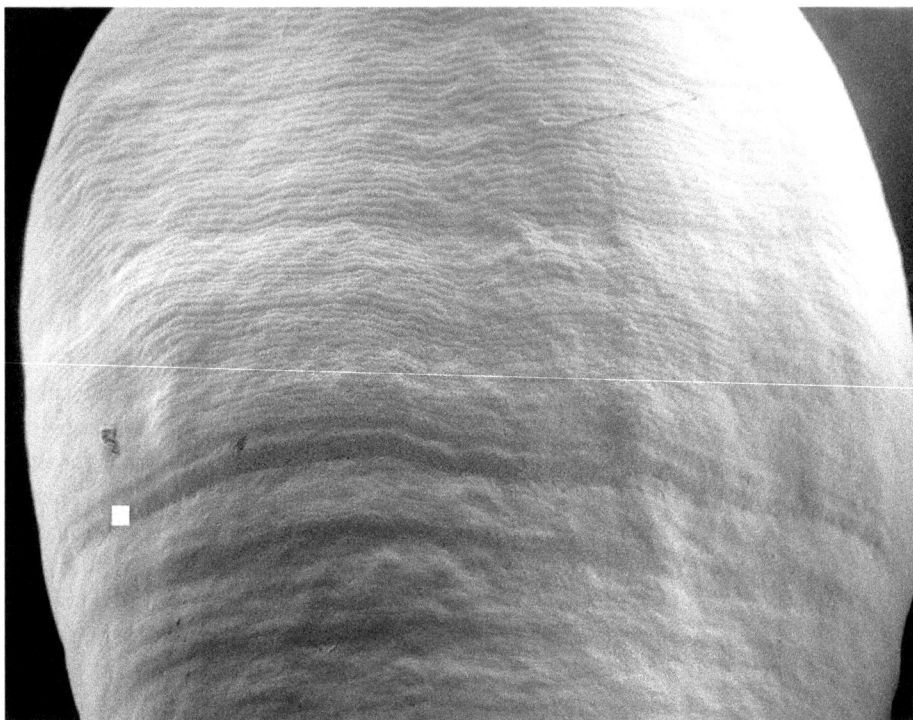

Figure 7.4 Mid-crown perikymata in the permanent lower canine shown in Figures 4.10 to 4.17, with the occlusal tip of the crown above the upper margin of the image. Many small variations in perikymata spacing are shown, from a single pair of increment margins wider apart than those nearby, to more substantial furrow-form defects. The most prominent feature, however, is a narrow plane-form defect, marked with a white square (see Figure 7.5). Field of view is 5.7 mm wide.

of a furrow-form defect, especially when it is lower down the crown side. There may even be two or more planes, separated by normal enamel. When the disruption occurred before completion of the cusps, all the overlying layers of matrix are missing, so that the exposed plane defines a greatly reduced nodule of enamel (see mulberry molar, page 186). Sometimes, particularly around the cusps, there is a pit-form defect at the cervical edge of the plane. Occasionally, the whole plane of the brown stria is exposed down to the EDJ, making a spectacular, very deeply incised defect.

The surface of the plane is marked by sharply defined Tomes' process pits, indicating that matrix secretion was interrupted abruptly for most of the ameloblasts (Boyde, 1970; Hillson and Bond, 1997). There may be some irregular continued matrix formation in patches. Where matrix secretion resumed at the cervical edge of the defect, there are often crowded brown striae which are not represented by perikymata at the crown surface, but a substantial thickness of enamel has often built up at the base of the plane. Witzel et al. (2008) described a plane-form defect

Hypoplastic defects

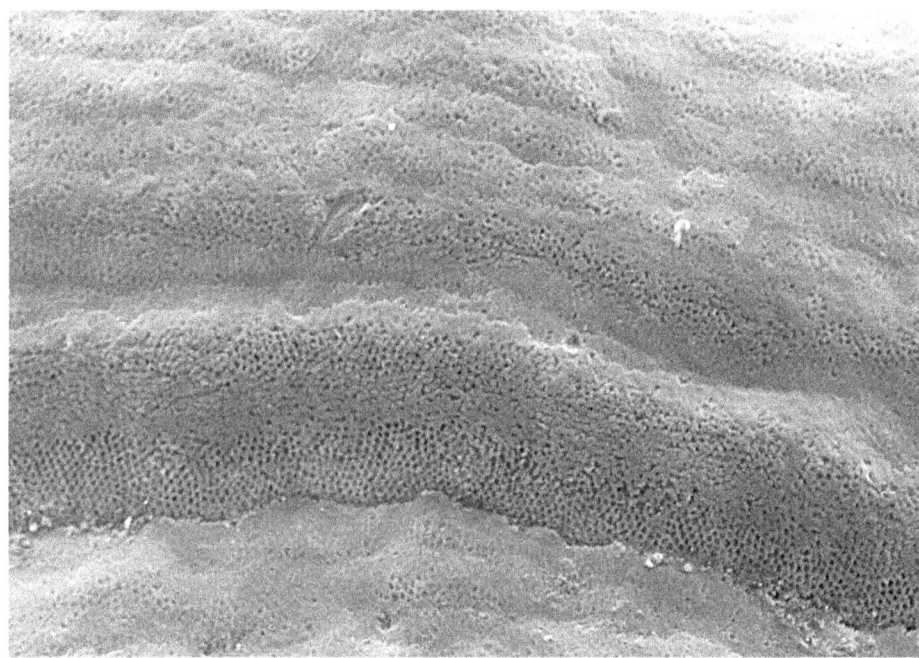

Figure 7.5 Closer view of plane form defect shown in Figure 7.4. Increment margins in this part of the crown are around 100 µm apart, but the defect exposes a brown stria of Retzius plane 500 µm wide. It is marked with sharply defined Tomes' process pits showing that enamel matrix secretion ended abruptly, although irregular secretion did continue over some parts of it. Together these features are what distinguish it as a plane-form defect rather than a furrow-form defect. Field of view is 741 µm wide.

associated with a single accentuated line (page 165). Either side of this, deep in the enamel, the brown striae of Retzius were evenly spaced and it is apparent that the defect represented a single, short disruption. It is not entirely clear what 'short' means in this context, but, as the rhythm of the brown striae marched on unaffected before and afterwards, there is no particular reason to suggest that it lasted more than a single beat of the rhythm.

If abnormally spaced perikymata represent growth disruption, the extreme expansion of spacing in the plane-form defect should represent a very large disruption. There is, however, no clinical or experimental proof that this is so. If the disruption took place in the cuspal enamel, it is also difficult to estimate the age at which the disruption took place from the surface because the defect represents the exposure of a deep layer. Tables of crown surface development are therefore not of much use. The cervical step of a plane defect does not represent the point in the sequence at which the disturbance occurred, but the position of the first row of ameloblasts that was sufficiently unaffected to continue matrix secretion. The width of the exposed brown stria plane therefore has no relationship whatsoever with the duration of the disruption. If, however, the defect occurred in the lateral enamel,

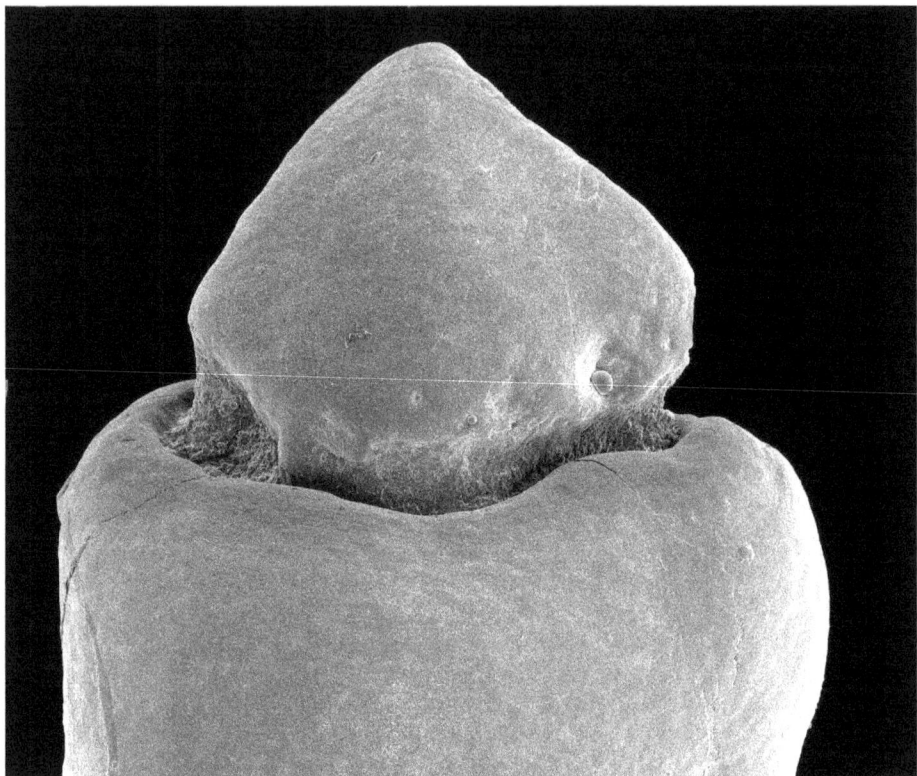

Figure 7.6 Pronounced plane-form defect in a human lower canine. The exposed brown stria of Retzius plane is in the lateral enamel of the crown, in spite of being so near the cusp tip. Deposits of dental calculus obscure the deepest parts of the defect, but it is clear that most, if not all, of the brown stria plane was exposed. Normal perikymata are observable both above and below the defect. Epoxy resin replica, cast from a dental impression of a specimen in the Odontological Collection at the Royal College of Surgeons of England. Sputter-coated with gold and imaged in a scanning electron microscope as described in Appendix B. As commonly happens, some bubbles were trapped in the original dental impression and appear as tiny spheres in the replica. Field of view is 9.5 mm.

then the last normal perikyma which is just to occlusal of the exposed plane could be used to estimate the initiation of the defect.

Pit-form defects

The defining feature of furrow-form and plane-form defects is that they extend continuously the whole way around the circumference of the crown, following the perikymata. In contrast, pit-form defects are discontinuous. A scatter of isolated pits spreads around the crown in a band. Individual pits vary from several millimetres in diameter down to a few tens of micrometres (Figures 7.7, 7.8 and 7.9). In the floor of each is the exposed plane of a brown stria of Retzius, with sharply

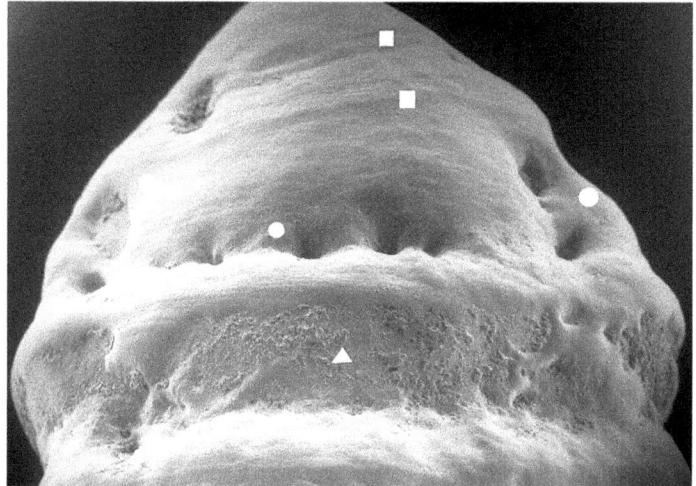

Figure 7.7 Pit-form and furrow-form defects from the human permanent lower canine shown in Figure 4.18. All the smaller pits, marked with white circles in the upper half of the image, seem to represent a single growth disturbance. The pits expose what appears to be a single plane of a stria of Retzius buried within the cuspal enamel, which does not reach the surface of the crown as the increment margin of a perikyma. More detail is given in Figure 7.8. Superimposed over the pit-form defect and marked with white squares are furrow-form defects which represent a later phase of crown development during the secretion of the lateral enamel matrix. Another pit-form defect, marked with a triangle, extends like a band around the lower part of the image. Superficially, it looks like a plane-form defect because some of the pits are so large and occupy such a width of crown. There are, however, areas of relatively normal enamel matrix between them, which marks it clearly as a pit-form defect. Detail is shown in Figure 7.9. Cleaned, uncoated specimen, imaged in a scanning electron microscope under low vacuum using an ESED (see Appendix B). Field of view is 6.3 mm wide.

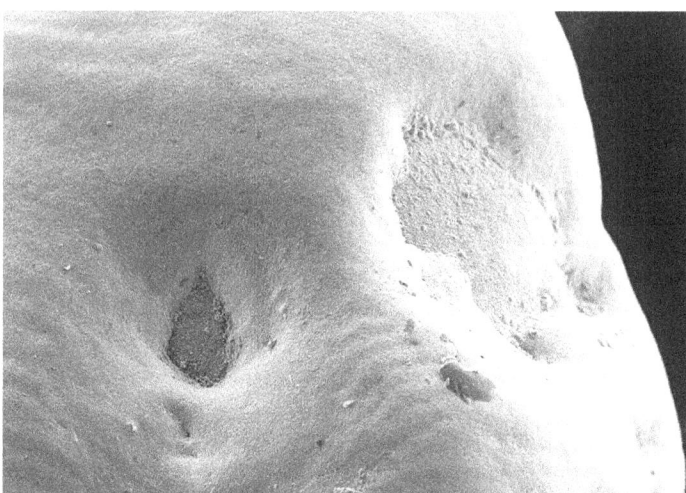

Figure 7.8 More detail of pit-form defect shown in Figure 7.7. The floors of the defect pits, bearing sharply defined Tomes' process pits, are part of a single brown stria of Retzius plane, deep inside the cuspal enamel. In between the defects the enamel surface is marked by poorly defined occlusal perikymata which show that intervening ameloblasts were not strongly affected by the growth disruption and went on to complete cuspal enamel secretion and continue with lateral enamel. Field of view is 2.3 mm wide.

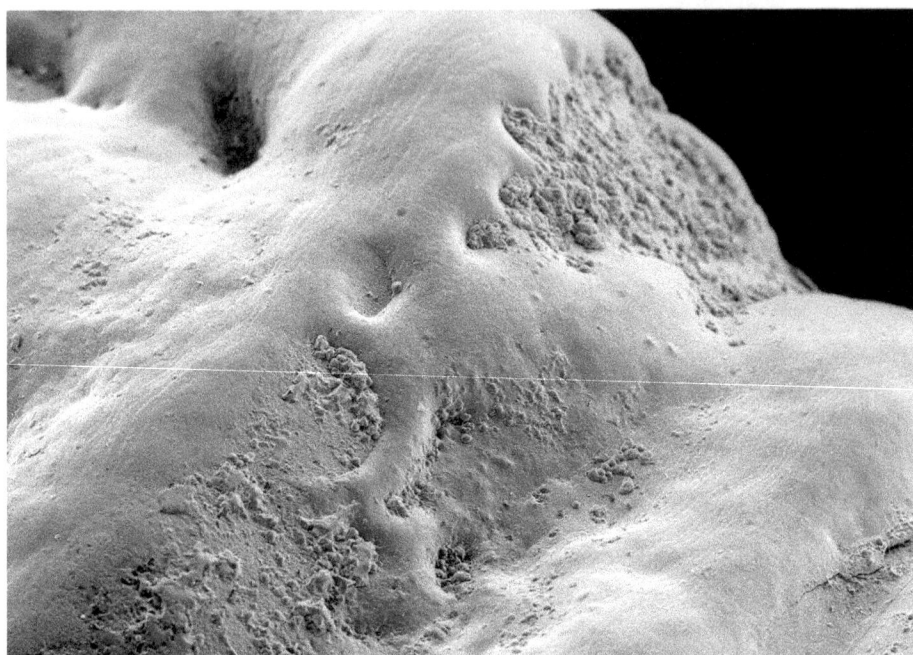

Figure 7.9 Closer view of large pit-form defect shown in Figure 7.7. Some of the pits in it communicate with perikymata at the surface. The plane of the brown stria which is exposed was therefore part of the lateral enamel. Field of view is 3 mm wide.

defined Tomes' process pits denoting a sudden end to enamel matrix secretion, although there may be evidence that this continued irregularly for a short while. There may be some tens or several thousand Tomes' process pits, depending on the size of the defect. The sides of each pit defect are usually smoothly curving, sometimes with perikymata on them, and denote a relatively normal continued matrix secretion of ameloblasts in between the groups which were so suddenly interrupted. This is what distinguishes pit-form defects from plane-form defects: openings in normal enamel matrix that expose the brown stria plane are scattered around the circumference of the crown as opposed to one continuous exposed plane. In some cases, the pits are so large and irregular that they look like planes (Figure 7.9).

Pit-form defects are commonly seen in the occlusal part of the crown side (Figure 7.10). Here, even a broad band of them may represent a disturbance to just a single brown stria. The individual pits within the band vary in depth. Near the cusp tips they may be relatively shallow because the affected brown stria plane is not far below. Lower down the crown sides they tend to be deeper because the brown stria plane angles down deeper into the enamel. One of the principal problems lies in determining when the disruption to enamel matrix secretion took place. When the disruption was in the cuspal enamel it is not

Hypoplastic defects

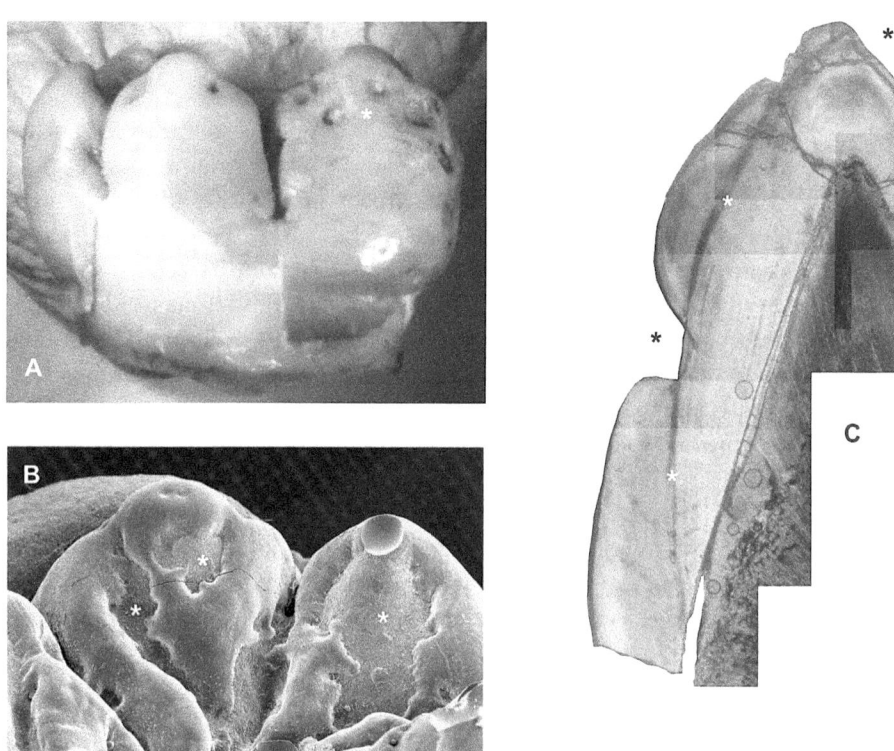

Figure 7.10 Pit-form defect in permanent lower first molar from a child buried at Christ Church, Spitalfields in London. He is known from parish records to have died on 15 May 1782, aged 3 years, 4 months 15 days. (A) Photograph of the buccal crown surface, with the scatter of pits around the cusps marked '*'. Width of field of view 1 cm. (B) Scanning electron microscope image of an epoxy replica showing the occlusal surface with larger, more irregular pits. Width of field of view 6 mm. (C) Ordinary light microscope image of a section through the mesiobuccal cusp showing sectioned pits and a strongly accentuated line, both marked '*'. The line is shown from counts of cross striations to have been initiated when the child was 375 days old, during February 1779. Width of field of view 6 mm. Specimen and section from the work of Daniel Antoine, British Museum, London (Hillson and Antoine, 2011).

possible to tell from the pit-form defects seen at the surface whether it was just one single brown stria plane that was affected, or more. The only solution is to make a section of the tooth. If the stria plane is deeply buried like this, it is clear that the ameloblasts involved were nowhere near their normal limit for switching from matrix secretion to maturation because the unaffected ameloblasts around the defects continued to secrete matrix. It is not at all clear how or why this happens. Perhaps pits around the cusps represent a form of disruption that would have been marked by a plane-form defect in more lateral enamel. Plane-form defects can, however, also be found in the cuspal enamel as, for example, in mulberry molars (page 186).

Wilson bands, pathological striae or accentuated lines

Gustafson and Gustafson (1967) divided brown striae of Retzius into *rhythmic incremental lines* and *pathologic Retzius lines*. Their illustrations make it clear that the former describes the appearance of the brown striae in the zone of lateral enamel where they angle up towards the crown surface and are clearly regularly spaced. They considered pathological lines to be marked largely by variations in mineralisation. In polarising microscopy, some striae were associated with lines of *negative birefringence* (a double refraction phenomenon that makes them appear bright in this type of microscope), signifying heavier mineralisation than other lines. Other lines were associated with *positive birefringence*, which represented less heavily mineralised enamel. They saw the neonatal line (page 121) as a good example, but did not provide evidence that other pathological lines could be matched with episodes of disease.

Wilson and Shroff (1970) took this idea further, distinguishing between *line striae*, sharply defined and regular near the crown surface, and *band striae*, less clearly defined and broader and in deeper enamel. In addition, they defined *pathological bands* as those associated with a sudden change in direction of the prism boundaries and with atypical prism forms. They saw these, together with neonatal lines, as separate entities from the striae of Retzius – merely running parallel with them – and they interpreted the sharp change in prism boundaries as evidence of a sudden developmental disruption. Rose and co-workers (Rose, 1977; Rose et al., 1978; Rose, 1979) published a methodology for identifying them. He decided to rename them *Wilson bands* to avoid calling them 'pathological', because there was no direct evidence that they represented disease. Firstly, he set up the microscope with oblique illumination (Jerome Rose, personal communication), which makes his published images look similar to dark field microscopy (Appendix B). Thick sections were cut and then etched with hydrochloric acid. Wilson bands were defined as 'any striae of Retzius exhibiting abnormal prism bending and absence or distortion of prism structure' (Rose et al., 1978, p. 513). Goodman and Rose (1990, p. 93) defined three more criteria as follows:

1. they should be continuous through at least three-quarters of the enamel thickness (those up to half the thickness were distinguished as *cluster bands*);
2. they should be observed in matching portions of the enamel on both labial and lingual sides of the crown;
3. they should appear with oblique illumination as a trough or ridge, representing the sharp deviation of prism boundaries.

FitzGerald and Saunders (2005), however, argued that this deviation could be seen in both accentuated lines and regular brown striae.

Are Wilson bands simply accentuated brown striae of Retzius, that is, a part of the long-period rhythm of enamel matrix secretion, or can they occur in between striae at any point in the sequence? Logically it might seem that, if there is an episode which disrupts growth, it would affect the secretion of enamel matrix immediately, without waiting for the next beat of the long-period rhythm. From this argument, if the event happened to coincide with that beat, then an accentuated brown stria would result. If not, then an extra accentuated line in between regularly spaced brown striae would result. In a review article, Risnes (1998) distinguished between regularly spaced rhythmic Retzius lines and a different category of Retzius lines, of different origin, which were superimposed over the rhythmic pattern. His main example of this was the neonatal line, but he suggested that other disturbances to growth might have a similar effect. FitzGerald and Saunders (2005) hypothesised that, when a disruption to enamel matrix secretion coincided with the normal brown stria rhythm, the combined effect would produce a Wilson band which, if it reached the crown surface through the lateral enamel, would be associated with a hypoplastic defect. Conversely, when the disruption occurred between beats of the brown stria rhythm, it would have a less pronounced effect which might not be distinguishable as a Wilson band. These discussions lead to several questions.

1. Are there published examples of additional accentuated lines in between regular brown striae of Retzius?
2. If such additional lines appear to be present, are they associated with an additional perikyma groove in between the regular grooves?
3. Are there examples of accentuated lines associated with hypoplastic defects?
4. What is the evidence that any of these accentuated lines do actually represent an external event, for example the impact of some environmental factor or an episode of disease?

The dark fuzzy line that is the main feature of both regular striae and accentuated lines in conventional light microscopy is not a directly observed structure, but an optical effect due to light scattering from a plane of interruption within the enamel (page 93). It becomes even less distinct under higher magnifications, so it is difficult to be sure whether or not the accentuated line coincides with a regular brown stria of Retzius exactly. One possible solution might be to examine a thinner plane of section, such as that provided by a confocal microscope, or simply by carefully polishing a thinner section. The problem is that a thinner section makes the striae much less distinct and instead it is possible to see many minor layerings formed by cross striations lining up in zigzag fashion. Witzel *et al.* (2008, figure 6) made particularly thin sections of human teeth in which accentuated lines are sharply delineated, but it is not possible in their published images to determine whether or not they match precisely the regularly spaced brown striae near the crown surface. So far, there is also no published evidence that extra perikymata occur in between regular striae.

The same paper investigated the relationship between accentuated lines and hypoplastic defects. It showed that plane-form defects were clearly associated with a single strongly marked and sharp accentuated line, but furrow-form defects showed a more variable association. In some, an accentuated line underlay the widely spaced perikymata at the more occlusal edge of the defect, whereas in others there were no associated accentuated lines. In addition, there were accentuated lines not associated with hypoplastic defects. Goodman and Rose (1990) illustrated a number of cases in which defects were associated with Wilson bands or cluster bands, but also some that were not. Hillson et al. (1999) showed a slight increase in prominence of regular brown striae associated with the more widely spaced associated perikymata in a small furrow-form defect.

Only one paper has matched accentuated lines with a known history during development. Schwartz et al. (2006) studied the teeth of a captive gorilla (page 140) for which a detailed diary had been kept, including the ages at which episodes of a potentially stressful nature occurred, such as an eye injury, hospital visits and change to a new enclosure. These were closely matched by accentuated lines for which the age could be established by counting prism cross striations from the neonatal line. Dirks et al. (2002) compared accentuated lines in the teeth of two female baboons from the Awash National Park in Ethiopia. One had been born approximately two years after the other, and they had both died in the same year at a known date. From this information they were able to derive a developmental sequence including both the ages and dates at which accentuated lines were initiated. Many, but not all, of these dates fell within periods of high rainfall or droughts.

In summary, it has been suggested that a separate category of lines can be defined. This category is in general appearance similar to the brown striae of Retzius, but differs from them in the prominence, darkness and breadth of the associated fuzzy band and in showing a particularly marked deviation of the prism boundaries. Some workers suggest that this is simply a marked variation of the regularly spaced brown striae. Others suggest that such accentuated lines can occur independently, which must mean that they can be found in between the regular striae and separated from them by a count of cross striations that departs from the periodicity seen in the rest of the enamel for that individual. Although this might be true, it would be a difficult observation to make and there is currently no published evidence. It is further suggested that accentuated lines are associated with a range of growth-disrupting factors, including disease, nutritional deficiency and psychological stress. At the time of writing there is, however, only a little published evidence for this.

Recording enamel hypoplasia by simple surface observation

Most studies of enamel hypoplasia have been carried out by simple examination with the naked eye, sometimes aided by the tactile test of running a dental probe or thumbnail down the crown. Defects have been recorded as present or absent, or scored according to schemes such as the DDE (see below). Sometimes the defect location is noted simply by the tooth affected, although sometimes the crown side has been subdivided into zones: two (occlusal and cervical); three (occlusal, mid and cervical); or nine or ten. Standard tables have been used to estimate age limits for the zones or for caliper measurements between the cervical margin and defect. This multiplicity of approaches causes its own problems of comparison between studies, but there are other more serious underlying issues.

Threshold for diagnosis of defects

How big must a defect be before it is recorded as present? The international standard for dentists, the epidemiological index of developmental defects of dental enamel (DDE index), is purely descriptive. It requires the defect to be associated with a reduced thickness of enamel, including shallow or deep pits, grooves, or a complete absence of enamel that exposes dentine, but gives no definition of the range of normal. 'Diagnosis will usually be readily evident where a defect is obvious. However, in other instances the most difficult decision will be deciding whether or not an abnormality is present. That is, the examiner may be unsure whether the enamel is defective or falls within the range of normal. When in doubt the tooth surface should be scored normal' (Commission on Oral Health, 1982, p. 161). This leaves the decision up to the observer, but, if the default diagnosis is normal, hypoplasia will be under-reported in those parts of the crown side in which perikymata morphology means that defects are rarely prominent (see below). Goodman and Rose (1990, page 62) cautioned against using a microscope or lens for routine identification because 'if a defect can be seen only under magnification then it is likely to be too small to reliably record and is probably not a true hypoplasia'. As the discussion of enamel hypoplasia in this chapter shows, the defects are part of a continuum which runs from microscopic to visible with the naked eye from a distance. All sizes of defect are 'true' in the sense that they represent a disruption to dental development and there is no evidence to suggest that any size is particularly more important to record than another. Some studies suggest reasonable agreement in intra- and inter-observer reliability studies (Goodman and Rose, 1990), but the defects recorded as present still do not necessarily represent a more substantial growth disruption than those which are recorded as absent. In truth, the only way in which the problem of threshold of diagnosis can be addressed is by the use of careful microscopy for *all* observations of the defects.

The central difficulty lies in the misconception that enamel hypoplasia is a well-defined pathological condition which can be diagnosed. This is discussed in Chapter 8, but hypoplasia is not a disease in the usually accepted sense of the word. It may be initiated by infectious disease or malnutrition, or form part of a syndrome, but most dental defects themselves have little impact on health. They fit at the end of a scale of normal variation and the point at which they depart from normal so much that they can be classed as abnormal remains a matter of debate.

Variation in prominence between different parts of the crown

Different teeth and different parts of even a single tooth crown are not equal in the prominence of the defects found on them. This is a function not of the disruption or form of defect itself, but of the normal pattern of crown development. As a result, defects are not commonly identified on deciduous teeth, except for a few particular cases. In permanent teeth, defects are most commonly recorded in the middle two-thirds of incisor and canine crown sides, and less commonly recorded in premolars and molars. In effect, there is a different threshold for diagnosing a defect in different parts of the dentition. It is meaningless to record hypoplasia merely as 'present' or 'absent' in different individuals. It will depend crucially on which teeth are preserved and the timing of the growth disruption in relation to the development schedule of the anterior tooth crowns. It is not even much help to record its presence or absence in single tooth types. For example, if the unit of recording is the permanent first incisor, defects for which the timing places them in the cervical region will have to be very much more prominent, relatively speaking, to be recorded as present than defects for which the timing places them at mid-crown height. Therefore the proportion of first incisors with defects recorded as 'present' will depend on the timing of the defects as well as their prominence. For this reason, valid comparisons can only be made between the same part of the crown side in the same teeth.

Effects of tooth wear

Tooth wear removes and diminishes defects of enamel hypoplasia in two ways. Tooth-on-tooth contact on occlusal, mesial and distal surfaces produces attrition facets which remove entire sections of the crown, taking the defects with them. More general abrasion of the crown side polishes away the details of the surface, rounding the outlines of defects and perikymata. The teeth of modern people in forensic cases, for example, tend not to show much damage from attrition, but they show heavy toothbrush abrasion which makes it very difficult to study enamel hypoplasia.

Archaeological and fossil teeth show much heavier attrition, with extensive facets even in the teeth of children and young adults, but the more general surface abrasion of the crown side is usually less strongly marked. Nevertheless, the longer the tooth has been in the mouth, the more the general loss of surface detail. Minor defects and those in the occlusal or cervical parts of the crown can be polished away relatively quickly. Larger defects may be apparent even in very worn teeth, but it will not be possible to make out their finer details. Tooth wear therefore exacerbates the problems of recording. The more prominent defects in the middle part of incisor and canine crowns are less rapidly affected by abrasion than those in the occlusal or cervical parts. In addition, attrition facets relatively rapidly remove evidence of defects in the occlusal region. Older individuals, with heavier tooth wear, will show fewer defects in general and a smaller range of defect sizes than younger individuals. Once more, this renders meaningless a simple summary of the proportion of individuals or proportion of teeth affected by defects. Detailed studies of enamel hypoplasia at the crown surface require the little-worn teeth of children. This is frustrating, because it is often possible to see a well-defined defect on a worn tooth crown. In such worn teeth, the only way a detailed study can be made is by sectioning them and examining the incremental structures under a microscope (Witzel et al., 2008), or by using one of the non-destructive alternatives (Appendix B).

Tables for estimating age of growth disruption in furrow-form defects

Estimation of the timing and duration of defects by simple observation of the crown surface can only be very approximate. It is done on the basis of standard charts for the development of the crown surface. Most are presented as diagrams in which shaded bands across the teeth represent different periods of development. Berten (1895) published an early chart based on measurements of the height of developing teeth dissected from the jaws of children. As discussed on page 58, such measurements show a simple increase in height for the tooth crown as a whole. There is no consideration of the internal organisation of the enamel into cuspal and lateral parts (page 123). For this reason, Berten's diagram gives the impression that crown formation starts at the tips of the cusps and spreads down the crown sides and is faster in the occlusal part of the crown and slowing to cervical. This is an early example of a problem that has plagued crown development charts for the past century. The highly influential diagram of Massler et al. (1941, their figure 7) summarised in a simple way the positioning of different developmental periods which they called 'tooth rings' (page 191). This diagram also appears to suggest that enamel formation starts at the visible tips of the cusps and, if interpreted literally, ignores the cuspal enamel to suggest that the crown surface starts to form in each tooth around one year earlier than it actually does. This error is important because influential hypoplasia studies appear to have used the diagram (Sarnat and Schour, 1941; 1942; Schultz and McHenry, 1975).

Counting cross striations and brown striae in enamel sections under the microscope is the only practicable way in which a crown surface formation chart can take account of the cuspal enamel. Hillson (1979) made an initial attempt to do this, dividing permanent crowns into ten equally spaced units of development. The most well-supported chart available at the time of writing is that of Reid and Dean (2006), again with crown surfaces divided into ten segments. The mean ages at which the dividing lines between segments were achieved were reported for dental hospital collections of extracted teeth collected in South Africa and England (Figures 4.6 and 4.7). The mean ages for the South African subjects were younger by up to six months than for the English subjects. If such large differences exist, it cannot be assumed that the charts are representative for all populations. It is better to use the charts as a guide to matching defects in different teeth. The segments were drawn simply to divide the curving crown surface of an unworn tooth equally into ten parts. In a similarly unworn tooth it would be possible to use measurements to draw similar dividing lines but, in a worn tooth, the original crown height is unknown.

Caliper measurements to estimate age of growth disruption

Swärdstedt (1966) used the Massler, Schour and Poncher diagram described above (Massler et al., 1941) to derive a chart of measurements to estimate the age of disruption from caliper measurements between defects and the cervical crown margin. The chart gives a six monthly scale of development next to a scale of measurements in millimetres for each permanent tooth. In this way a defect is assigned to one of the six units. The same chart was published by Goodman *et al.* (1980) with slight modifications. The chart has several problems, however:

1. It perpetuates the error in the original Massler, Schour and Poncher diagram which ignores the cuspal enamel and therefore gives ages which are up to one and a half years too young in the occlusal part.
2. The small size of the original diagram and its approximate nature provide a poor basis for estimating the ages of the units.
3. It assumes a single mean value of crown height, whereas unworn height varies considerably both between and within populations.
4. Even fine point calipers are a coarse instrument to measure so tiny a feature as a hypoplastic defect (although it is possible to take measurements using an eyepiece graticule in a microscope).

The chart has been extended to deciduous teeth (Blakey and Armelagos, 1985) using figures for the initiation and completion of the crown taken from published data. Once again, this ignores cuspal enamel development which, for deciduous teeth, occupies an unknown interval.

Hodges and Wilkinson (1990) showed that variations in crown height did indeed make significant differences to estimates of the age of growth disruption and they suggested a method of adjustment. Goodman and Rose (1990) published a series of regression equations based on Swärdstedt's crown heights and Massler *et al.*'s standards for crown development, but did not address the problems of the initial chart. Ensor and Irish (1995; 1997) proposed the *hypoplastic area* approach. In this, the measurements of the location and width (that is, the difference between the upper and lower edge) of defects relative to the CEJ were used to estimate the proportion of crown height occupied by hypoplastic defects. In order to take account of the effects of dental attrition, they did not record the occlusal onethird of the crown, which added a difficulty because this needed to be estimated. Hypoplastic area avoided the problems of reliance on the Swärdstedt table and allowed previous records of defect measurements to be used, but assumed a uniformity between defects in different parts of the crown side. As discussed above, the same size of defect in terms of perikymata count has a different width if it is in the mid-crown region than in the cervical region; this means that a mid-crown defect in a lower canine, for example, would involve a larger crown height proportion than the same defect in the cervical region. It is therefore clear that hypoplastic area cannot properly represent the proportion of total crown formation time taken up by defects.

Goodman and Song (1999) acknowledged that it was necessary to take account of the cuspal enamel and variation in crown height and they proposed methods for correcting the regression equations. They suggested that the difference in estimated age was relatively small at no more than a few months. Ritzman *et al.* (2008) made a comparison between the Reid and Dean ages for deciles of crown height discussed earlier and the values estimated using the Goodman and Rose regression equations. As might be expected from the discussion in this chapter, they found substantial discrepancies, particularly in the occlusal to mid-crown regions. They then estimated modal disruption ages for defects in a small group of teeth using the Swärdstedt table and Reid and Dean deciles for Europeans and South Africans. The mode estimated from the Swärdstedt table was 2 years younger than that estimated from the Reid and Dean Europeans and 1 year 6 months younger than the South Africans. These differences are so large that they call into question the results of any published studies using the Swärdstedt table or its derivative regression equations.

One alternative might be to re-calibrate existing measurements of defect locations to the deciles of Reid and Dean. Cucina *et al.* (2006) did this for a collection of dentitions from the Roman necropolis of Vallerano in Italy, based on a mean crown height calculated from unworn teeth at the site. The other shortcomings of measurements remain, however.

Building sequences of defects

The methods outlined in Chapter 5 can be used to build sequences of accentuated lines and cross striations, hypoplastic defects and perikymata. The accentuated lines are, in any case, the basis of making matches in sequence between different teeth from one individual. Light microscope sections can be used to investigate the timing and duration of hypoplastic defects seen at the surface of the crown (page 224). This is the only way in which large plane-form and pit-form defects can be investigated. Without sectioning, it is still possible to match furrow-form hypoplastic defects between teeth, as well as minor variations in the spacing and prominence of perikymata within them. In practice, a sequence of small variations is often more straightforward to match than a single big defect. The clearest images for this purpose are achieved with a scanning electron microscope, using either epoxy replicas made from dental impressions of the crown surface, or the tooth itself which can successfully be imaged without coating in the microscope at near atmospheric pressure (see Appendix B). The matches are confirmed by counts of perikymata within and between defects.

Using several crowns from one dentition, an extended sequence of perikymata counts and defects can be built up, from the first perikyma on the cusp tip of the earliest forming tooth to the last perikyma at the cervix of the latest forming crown. It is usually most straightforward to match defects in permanent incisors and canines, but with experience it is also possible to match molars. If the cross striation repeat interval (page 118) is not known because teeth cannot be sectioned, it is necessary to base the chronology on perikymata counts alone. For example, a sequence might start with the first perikyma on an unworn permanent upper first incisor and continue to the last perikyma at the cervix of a lower canine. Even if the actual age of these events is not known, the relative positions of defects between them can be determined as ratios of perikymata counts to the sequence total. Comparisons can be made between individuals simply by aligning the starting and ending points of their sequences.

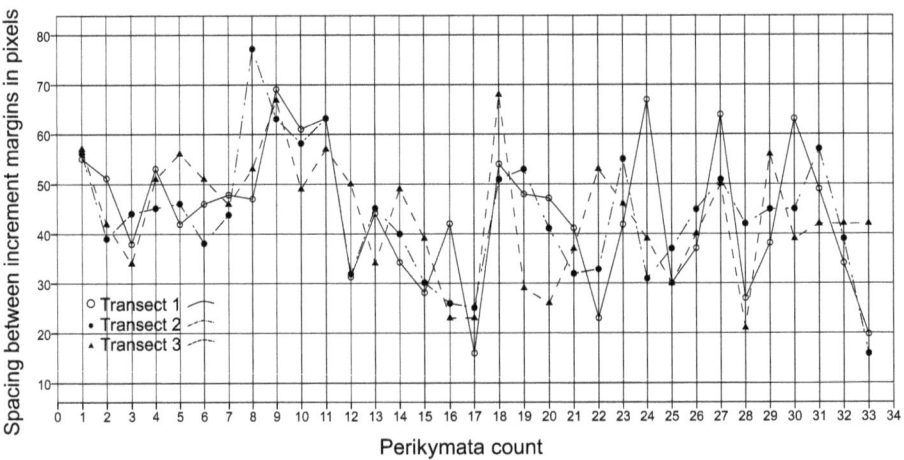

Figure 7.11 Perikymata spacing transects for the crown surface shown in Figure 7.2. The measurements were scaled from the image using ImageJ software (see Appendix B) along three different transects including the same perikymata. All three transects show a gradually decreasing general trend, but considerable variation about the mean value for each part of the transect. In addition, the pattern of variation does not exactly match between transects. This can be confirmed by examining Figure 7.2 where the increment margins undulate up and down relative to one another.

Study of furrow-form defects by measurement of perikymata spacing

Spacing of perikymata can be measured between their increment margins following a transect down the crown perpendicular to the line of the margins. Scanning electron microscope images have a scale embedded in them and measurements can be calibrated using simple image analysis software, but the difficulty is the slope of the crown surface. If the surface is horizontal, then the measurement scaled from the image will be correct. If it is inclined, the foreshortening in the image will affect it and simple trigonometry shows that a slope of 25° will produce a error of roughly 10%. It is always necessary to tilt the crown surface in the microscope to see clearly the very low relief features of the perikymata and it is difficult to determine the exact angle. If absolute measurements in micrometres are required, it is necessary to use photogrammetry software based on stereopairs of scanning electron microscope images (Appendix B). Over a limited area of the crown surface, however, the inclination will vary less and it is possible to use image analysis software to take direct measurements in pixels, which can then be compared without adjustment between perikymata within the same image (Figure 7.11); that is, the measurements will be relative rather than absolute.

Another way to obtain absolute measurements is confocal light microscopy, which produces a stack of images at different levels of focus (Appendix B). These are then combined to generate a three-dimensional surface model and profiles on which measurements can be taken (Bocaege et al., 2010). The limitation of this technique is that, with the high magnification objective lens required to make clear

Building sequences of defects

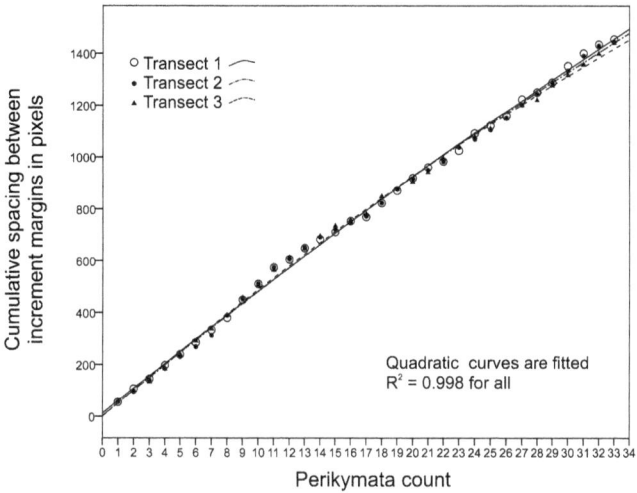

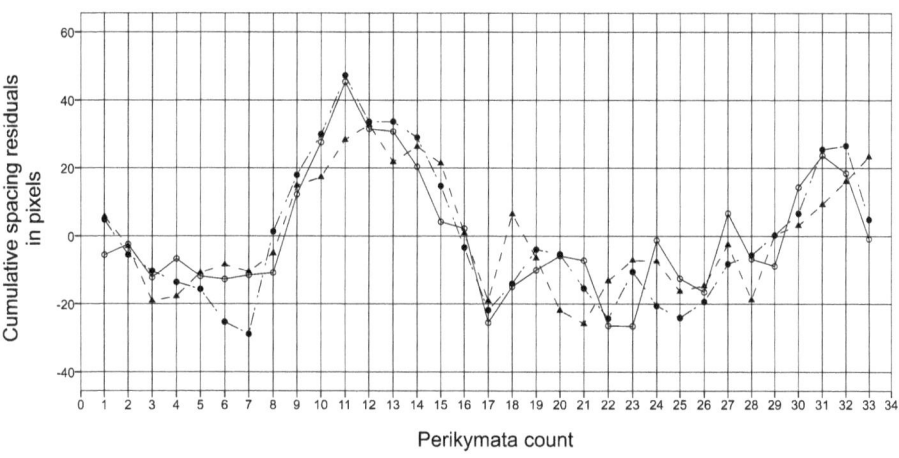

Figure 7.12 Cumulative perikymata spacing. The upper plot shows cumulative values based on the three transects shown in Figure 7.11. There are differences between transects, but they track one another reasonably well. This is also shown in the fitted quadratic curves. The lower plot shows residuals for the three transects. Again there are differences, but they closely match one another with the increase in cumulative spacing at the occlusal wall of the furrow-form defect seen in Figure 7.2 which corresponds to perikymata 8 to 16. The 90th percentile of the residuals is 12.1, 15.6 and 13.9 for transects 1, 2 and 3, respectively. The defect perikymata exceed these values, which suggest it might act as a reasonable rule for recognising abnormal spacings.

images of perikymata, there is a relatively small field of view, so that a whole tooth crown requires multiple separate models to cover it. The simplest method is to use an engineer's measuring microscope (Hillson and Jones, 1989), which is fitted with a mechanical stage and micrometers to measure the position of the specimen in relation to the cross-hairs in the microscope eyepiece. In addition, a micrometer attached to the focussing mechanism measures the vertical displacement of the focussed point. In this way, the microscope scans a transect down the crown whilst the cross-hairs are carefully focussed on each increment margin in turn. This is best done with an epoxy replica of the crown, sputter-coated with gold. It takes practice to achieve these profiles and the limitation of the method is in the depth of focus (Appendix B) of the objective lens.

Perikymata spacing transects of this kind were first made by Hillson (1992a). As described in Chapter 4, the normal geometry of development causes this spacing to decrease down the crown side, but there are minor variations. Figure 7.11 shows the characteristic 'wiggly' line produced in simple relative measurements. It also shows the variation between different transects measured from the same field of view. These effects make it difficult to pick out the variation caused by furrow-form defects. Figure 7.2 shows a well-defined defect in the scanning electron microscope image in which it is clear that the perikymata in the occlusal wall are more widely spaced than in the neighbouring crown surface. This is not apparent in Figure 7.11 because of the variation superimposed over it. If, however, the spacing is measured cumulatively – that is, by adding each successive spacing onto the previous spacing in the transect – these variations have a reduced effect (Figure 7.12). This is because the general trend of perikymata is horizontal around the crown and any local departures from horizontality must eventually be evened out. A curve can be fitted to the cumulative spacings and then residuals clearly show departures from the main trend of spacing. When this is done, the increase in perikymata spacing in the occlusal wall of the defect shown in Figure 7.2 is seen in all of the transects. Even where absolute measurements of perikymata spacing are available, it is still best to convert them to cumulative measurements and use the same curve-fitting and residuals approach. Definitions for departures in spacing that represent a hypoplastic defect can be made on the basis of arbitrary rules. For example, they can be based on the percentiles of the residuals (Figure 7.12).

Although a full crown transect of absolute measurements, from cusp tip to cervical margin, would be the best basis for recognising furrow-form defects, simpler relative measurements could be used to examine potential defects within a smaller field of view. Where a scanning electron microscope image shows what appears to be a defect, relative measurement transects can quickly be made for the defect and its neighbouring perikymata in several places within the image. The residuals will then allow the visual impression of a defect to be tested.

Causes of enamel hypoplasia

The Salpêtriere

Pierre Fauchard (1678–1761) was the great French pioneer of dentistry and his textbook *Le Chirurgien Dentiste* (Fauchard, 1728) formed the heart of dental education for the following century. It inspired Robert Bunon (1702–1748), who wrote *Essay sur les maladies des dents* (Bunon, 1743) and *Expériences et demonstrations faites à l'Hôpital de la Salpêtriere* (Bunon, 1746). Bunon was dentist to *Mesdames*, the eight daughters of the king, Louis XV, and the flowery language in his introduction makes it clear that this was a world of aristocratic patronage. Bunon (1746, p. 24) was particularly interested in a passage in Fauchard (1728, volume 1, p. 95):

La partie émaillée des dents, est encore sujette à une maladie qui ressemble fort à la carie; mais qui cependant n'est point une carie. Leur surface extérieure devient quelquefois inégale et raboteuse, percée de plusieurs petits trou, quasi en forme de rape; mais disposés plus irrégulierement. Je nomme cette maladie érosion de la surface émaillée, ou disposition à la carie. Elle dépend de ce que l'émail est usé pas quelque matiere rongeante; qui a produit en lui le même effet en cette occasion, que la roüille produit sur la surface des métaux.

[The enamelled part of teeth is also subject to a disease which strongly resembles caries, but nevertheless is not exactly a carious cavity. Its exterior surface sometimes becomes uneven and rough, pierced by many small holes, somewhat like a rasp in form; but spaced more irregularly. I call this disease erosion of the enamelled surface, or disposition to caries. It requires the enamel to be worn away by some corrosive substance; which has produced on this occasion the same effect as rust produced on the surface of metals. (translated by S.W. Hillson)]

Fauchard did not expand on the way these *érosions* came about, but Bunon did not agree with the implication that the enamel had been eaten away in some fashion. In his studies of children at the Salpêtriere hospital, he instead linked *érosions* with rickets, rubella, smallpox, venereal diseases or scurvy, particularly rickets. He claimed (his p. 59) that with experience it was possible to tell from the teeth affected which of these diseases was responsible and at what age. His crucial observation (his p. 74), however, was that, when a child had died of smallpox, rubella or neonatal scurvy and the deciduous teeth had not yet been shed, it was possible to remove those teeth to reveal permanent successors marked by *érosions*. From that evidence it was clear that they were a condition of development and not of some factor that affected the teeth after they erupted. Bunon's account is fundamental to understanding the defects. His tone is defensive as he laboriously sets out evidence in front of his peers to support his claims: it was clearly no small thing to disagree with the great Fauchard. The Salpêtriere was not a hospital in the modern sense, but more a place of confinement for the poorest women and children of Paris, those in need of 'correction' and the insane. Doubtless childhood diseases and dental defects were extremely common among its inmates, giving Bunon opportunities for study which can rarely have been repeated.

Congenital syphilis

Syphilis was one of the great scourges of late nineteenth century London and Paris. It was particularly frightening because of the way in which the sexual behaviour of the parents was visited on their unborn children as congenital syphilis. This disease is caused by transfer of the pathogen *Treponema pallidum* across the placenta *in utero*. Sir Jonathan Hutchinson (1828–1913) in London and Alfred Fournier (1832–1914) in Paris were pioneers of syphilology and both were concerned with the diagnosis of congenital syphilis: there is Fournier's sign (mouth scars) and Fournier's tibia, and Hutchinson's incisor and Hutchinson's triad. The latter refers to three diagnostic signs: dental deformities, keratitis (inflammation of the cornea) and neural deafness (Hutchinson, 1857; 1858; 1887).

Hutchinson's incisor is a deformity of the permanent first incisors. It is a hypoplastic defect of the central mamelon, the first part of the crown to start forming (page 57), which either fails to develop altogether or is greatly decreased in height to make a small tubercle. In effect, it is a pronounced plane-form defect in an unusual place, which creates a prominent notch in the incisal edge. The mamelons either side are drawn together, shortening the incisal edge and giving the labial aspect of the crown an oval outline (Figure 7.13). Soon after eruption, the notch of the incisal edge quickly wears and it can be difficult to identify a Hutchinson's incisor with confidence. Just the labial outline and a shallow groove in the crown surface below the notch may survive (Putkonen, 1962). The defect is seen in the permanent upper first incisor, but not the second. This is because the second incisor does not start to form until the end of the first year after birth and the central mamelon of the first incisor starts to develop soon after birth. Sometimes lower incisors, which start to form at a similar age, also show the defect, but this is not so common. Deciduous teeth do not show it. Hutchinson merely described the defect without exploring the mechanisms, but it is possible to localise the disturbance to tooth development closely within a few months following birth (Hillson *et al.*, 1998). This is the point at which the maternal immunity is lost to the baby, which then must build its own. Alfred Fournier (1907), working in Paris during the time of the *belle époque*, gained a vast experience of syphilis. It is confusing, because there were two of them: Jean Alfred Fournier and his son Edmond Alfred Fournier (1864–1938), both of whom wrote books on congenital syphilis. In particular, the Fourniers championed the cause of women and children innocently infected by syphilitic husbands (Fournier, 1881). Jean Alfred knew Hutchinson well and coined the phrase '*le triad de Hutchinson*' (Waugh, 1974). He gave a more detailed description of Hutchinson's incisor and, in particular, distinguished it from more general hypoplasia of the incisors (Fournier, 1884).

If congenital syphilis initiates this characteristic hypoplastic defect specifically during the first few months after birth, what other teeth are developing at the time? Permanent first molars are developing their cusps during the first year after birth and Fournier (1884, pp. 19–20) gave a clear description of a strongly marked plane-form defect around the cusp tips, so that each had a nodule of enamel set into a circular groove almost as though 'it was a smaller tooth growing out of a larger one' (Figure 7.13). Sometimes there was a 'stump of dentine emerging from a normal crown'. These rapidly broke away once the tooth erupted into the mouth and the occlusal surface wore down. This defect has become known as a *mulberry molar* and the term seems to come from the influential paper of Karnosh (1926), presumably as a description of the small nodules, which do look a little like the small

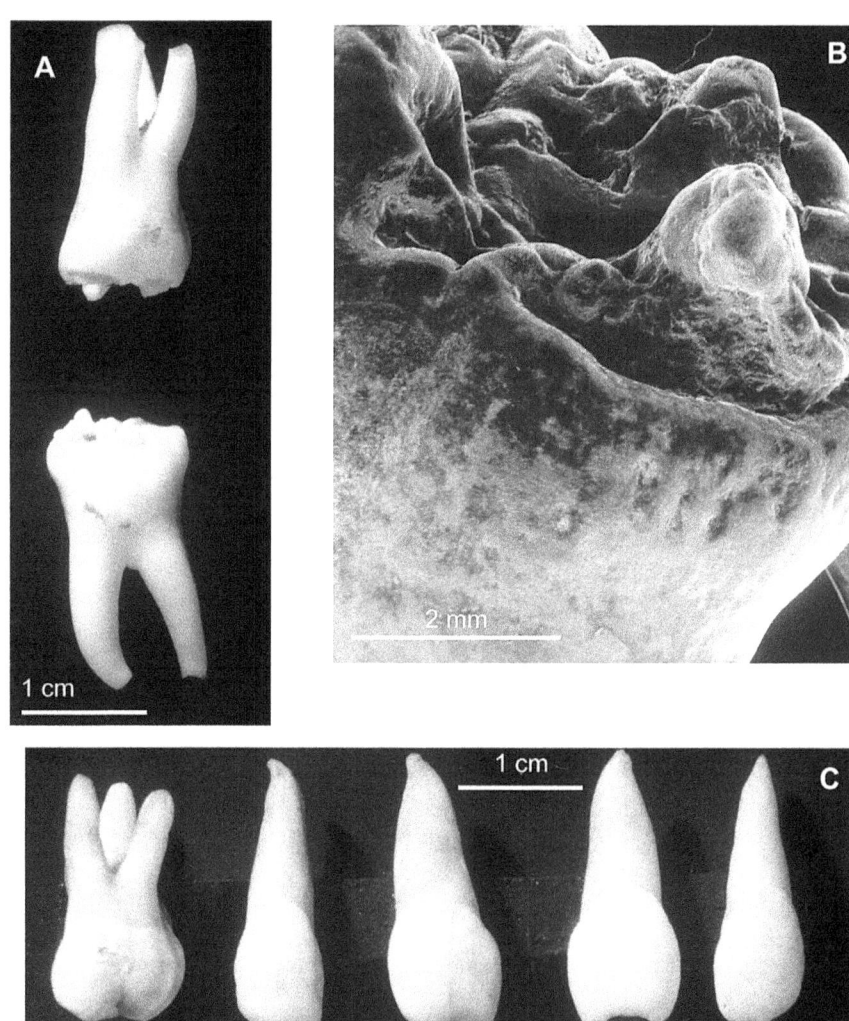

Figure 7.13 Dental stigmata of congenital syphilis. (A) Mulberry (Fournier's) molars. Permanent upper and lower first molars from one individual. (B) Scanning electron microscope image of epoxy replica made from the upper molar and showing the prominent plane-form defect surrounding a cusp which is represented by a small nodule of irregular enamel. (C) Moon's molar and Hutchinson's incisors from one individual. Permanent upper first molar with characteristic 'rosebud' form. Upper first incisors show the characteristic notch on the incisal edge, which is drawn in to give the labial surface of the crown a 'pumpkin seed' outline. The notch has been further modified by tooth wear. The upper second incisors on either side are unmarked by any defect. Specimens from the Odontological Collection at the Royal College of Surgeons of England.

multiple lobes of a mulberry fruit. There is, however, confusion because Karnosh (his p. 33) described it as 'a mulberry molar of Fournier or Moon'. Many years earlier the English dental surgeon Henry Moon (1877) had published a diagram (his plate IV) of first molars 'malformed through inherited syphilis' in which the cusps were closer together than normal, giving a rounded appearance. This is not a good drawing, but Moon also illustrated them in a textbook on surgery (Bryant, 1884, figure 240), showing this dome-like form more clearly. Pflüger (1924, p. 606) described a *bud-form molar* which appears to be the same defect:

Während der normale Molar seinen kleinsten Durchmesser am Zahnhals und seinen grössten im Bereich der Kauhöcker har, ist es bei der Knospenform gerade ungekehrt: die Zahnbasis bildet hier den grössten Durchmesser.

[While the normal molar has its smallest diameter at the neck, and its greatest width at the cusps, in the bud-form these are reversed: the base of the tooth has the largest diameter. (translated by S.W. Hillson)]

Putkonen (1962) was of the opinion that this dome-shaped or bud-shaped crown (Figure 7.13) was the characteristic shape in congenital syphilis and not the marked hypoplastic defect of the Fournier or mulberry molar.

The plane-form defect of the mulberry molar is, however, easier to explain in terms of the timing of the growth disruption (Hillson *et al.*, 1998). It is possible to locate this fairly clearly in a Hutchinson's incisor, in which the central mamelon is either entirely missing or shows a plane-form defect, but the mesial and distal mamelons develop relatively normally. Reid *et al.* (1998a) estimated just over 4 months for the initiation of the first incisor crown (Figure 4.7) and the incisal edge is completed at about 1 year after birth. That means the defect of a Hutchinson's incisor must be confined to around the middle of the first year after birth. The cusps of the first molar are also completed at just over 1 year after birth, having been initiated just before birth, so the position of the plane-form defect also suggests a disruption during the middle of the first year. There is no visible hypoplastic defect in a typical bud-form molar, whereas in the same individual there may be a defect in the incisor (Putkonen, 1962; Hillson *et al.*, 1998). This is a puzzle and at the time of writing there was no published histological description of a bud-form molar.

Many babies with congenital syphilis die during their early development. Rawstron *et al.* (1993b) found that, of 75 babies who tested positive serologically for syphilis, 53% were stillborn. Today, careful monitoring of mothers identifies babies at risk before birth so that they can be tested and treated with antibiotics. In the past, 25–30% of infected babies died soon after birth (Rawstron *et al.*, 1993a) but, of those who survived the perinatal period, many survived into adult life. The dental signs do not appear until the eruption of the first permanent teeth around 6–7 years of age. Putkonen (1962) found that, of 177 patients in whom both the upper first incisor and first molar could be examined, only 58.8% showed the characteristic defects in either tooth, with over half the patients showing Hutchinson's incisors and just one-third showing Moon's molars.

Colorado brown stain

When Frederick McKay set up as a dentist in Colorado Springs in 1901, he was intrigued by patients with brown and white spots marking their tooth crowns. In 1908 he managed to interest G.V. Black of Northwestern University Dental School in Illinois (Black and McKay, 1916). The Colorado Springs dentists commonly replaced the affected incisor crowns with porcelain crowns and McKay sent some originals to Black for histological examination. They were normal in form (the surface was smooth and hard), but the translucent enamel was mottled by irregular white or brown opaque areas underneath the surface. In section, Black found that this mottling was confined to the outer third of the enamel thickness and that prisms and brown striae ran through uninterrupted. Black did not know what caused the mottling, but it was common in other parts of America, Europe and South Africa.

The answer was found in a classic blend of clinical observation, fieldwork and experiment by Smith *et al.* (1932). They recognised that the dark stain was secondary because it followed the line of the gingival margin, and it probably came from dental plaque and food. The primary defect was an opaque white mottling and in some cases is associated with surface pitting defects. Smith and co-workers studied the small community of St David, near Tucson in Arizona. Every child born and raised in St David had mottling in all deciduous and permanent teeth. Children who had come into the area from outside at an older age only showed mottling in those teeth which had formed after their arrival, so it was clear that some feature of the environment was responsible. There was no evidence for a change in diet or a difference in the prevalence of childhood diseases, so suspicion naturally fell on the water. Water from the settlement was concentrated by evaporation and given to laboratory rats, whose permanently growing incisors developed similar defects. Experiments eventually identified high fluoride levels in the water as the cause.

The mottled patches in enamel are poorly mineralised, with a larger than normal proportion of tiny spaces in between the crystallites. The light-scattering effect makes them look white when viewed in daylight. The increased porosity also means that they may take up a stain from plaque and food. It appears that high concentrations of fluoride ions interfere with the maturation of enamel in a way that is not properly understood (Fejerskov *et al.*, 1994). The enamel matrix secretion stage seems to be unaffected, except in experiments where enormously high concentrations of fluoride are administered. Crown surface pitting is produced, not by disruption to matrix secretion, but where the poorly mineralised enamel flakes away under abrasion in the mouth. The result may look a little like enamel hypoplasia when the edges are rounded by abrasion, but it is *not* enamel hypoplasia.

The Mellanbys and vitamin D

In 1914, the young Edward Mellanby was invited by the Medical Research Council of Britain to carry out an investigation into the causes of rickets. From experiments with laboratory dogs, he found that some diets produced the characteristic bone deformation (Parascandola and Ihde, 1977). Once he had established a 'rachitic diet' he could test the effect of adding foods (Mellanby, 1919) and found that cod liver oil, butter and egg yolk had a protective effect. In 1913, McCollum and Davis had discovered two factors that were required for growth in laboratory rats (Combs, 1992): fat-soluble A and water-soluble B. Fat-soluble A was present in cod liver oil and was Mellanby's first choice as the agent responsible, but McCollum suspected that another factor, also present in the oil, was responsible for protection against rickets. He isolated this and it eventually became known as vitamin D (fat-soluble A became vitamin A). We now know that relatively few foods provide an abundant supply of vitamin D and that it is mostly made in the skin through the effect of sunlight (Mellanby's dogs were raised indoors). Rickets is therefore mainly a disease of children who have very little exposure to sunlight rather than a dietary deficiency. It depends where on the planet the children are living and on social conditions which cause them to be covered by clothing.

May Mellanby (they married in 1914) was also a physiologist. She had by chance noticed that her husband's experiments produced not only the classic bone deformities of rickets, but also dental anomalies (Mellanby, 1934). There were delays in eruption, malocclusions, enamel hypoplasia and zones of interglobular dentine. She conducted a series of experiments (Mellanby, 1918; 1929) to show that these anomalies were produced in dogs fed a diet poor in vitamin D, but with adequate resources of other kinds. If cod liver oil or egg yolk were added, the teeth were normal. It was also necessary to have an adequate supply of calcium. She used interglobular dentine as her main index of defective structure, although she also recorded enamel hypoplasia. Klein (1945) carried out similar experiments with rats and pigs, which were fed with a rickets-inducing diet and killed whilst their teeth were still developing. He found evidence of disruption to the ameloblasts of the internal enamel epithelium and to the developing enamel matrix, similar to those pointed out by Kreshover and discussed later in this chapter.

Between 1923 and 1926, the US Children's Bureau carried out a study of rickets in New Haven, Connecticut, with regular examinations of 480 children receiving anti-rachitic treatment through cod liver oil and sunbaths, from birth to 3 years of age. In 1931–2 these children were examined again, when they would have been aged up to 9 years and therefore with a mixed dentition in which permanent incisors and first molars would have been erupted. Eliot *et al.* (1934) divided them into three groups:

1. Children whose frequent examinations during the first 1–2 years after birth showed no development of rickets;
2. children whose examinations had shown moderate rickets, even if in 1931–2 there was no sign of the condition;
3. children whose original examinations had shown severe rickets.

Defects were recorded for 63% of the children with severe rickets, 25% of the children with moderate rickets and 12% of the children with no rickets. In the group as a whole, rickets or not, children with a history of severe infectious disease showed defects more frequently. It thus appears that disruption to enamel development was most pronounced when infectious disease was combined with rickets.

Tooth ring analysis, defects and clinical histories

Isaac Schour has already been mentioned for his chart of dental development (page 51). With his co-workers, he developed the concept of 'tooth ring analysis' (Schour and Massler, 1940a; 1940b), the idea that the quality of enamel varied during different periods or 'rings' of development:

1. Prenatal (excellent quality);
2. birth to 10 months post-natal (very poor quality);
3. 10 months to 2.5 years (good quality);
4. 2.5 to 5 years (poor);
5. 6 to 10 years (fair);
6. 10 to 13 years (poor).

They illustrated these rings in diagrammatic sections of deciduous and permanent teeth (Massler et al., 1941, their figure 6), which match reasonably well with more recent sequences. Unfortunately, in the same paper they included the simplified diagram (their figure 7) which has proved to be so misleading (page 179).

This diagram seems to have been used by Sarnat and Schour (1941; 1942) in a study of 60 child patients at dental hospitals in Illinois, whose teeth showed enamel hypoplasia in either deciduous or permanent dentition. They wished to match the defects with detailed medical histories including chickenpox, diarrhoea, diphtheria, measles, rubella, pneumonia, whooping cough and rickets. Defects were assigned to four different phases: neonatal (birth to 2 weeks); infancy (3 weeks to 12 months); early childhood (13 to 34 months); and late childhood (35 to 80 months). The infancy category appears to have relied on permanent incisors and canines and first molars, which would be incorrect for furrow-form defects of the lateral enamel because this does not start to form in these teeth until 1 year of age (Figure 4.7). It might, however, have been correct for plane-form defects representing deeper increments in the cuspal enamel. The other phases do not seem to be so much in error, but it is a concern that by far the largest number of defects were in the infancy phase. When they tabulated the estimated ages for the initiation of defects against the conditions in the medical histories, they found little or no match. It is hard to escape the thought that this might partly have been due to errors in assigning defects to an age. It does, however, seem in any case to be difficult to match defects and medical histories. Suckling and co-workers (Suckling et al, 1985; 1987) attempted this in a large group of New Zealand schoolchildren, but only 15% of the children showed hypoplastic defects pronounced enough to record. There were no statistically significant associations with episodes of disease, exception for chickenpox.

Lindemann (1958) examined 141 child patients admitted to Queen Louise Children's Hospital in Copenhagen with gastrointestinal conditions. The ages of defect initiation seem to have been estimated using similar methods to Sarnat and Schour and in only ten of the children did the estimates fall within the range of ages during which gastrointestinal symptoms were recorded. In another 11 there was a partial match and, in the remaining 12, no match at all. This is not a very strong indication that the defects were related to gastrointestinal conditions, but it is possible that less marked defects were overlooked or that there were errors in estimating the age of disruption.

Goodman *et al.* (1991) worked with the Mexican National Institute of Nutrition in a study of children from the rural community of Tezonteopan. Their diet fell well below recommended nutritional levels and the project monitored the effect of calorie, protein and vitamin supplements on pregnant mothers and their children. One group received supplements whilst the other did not. Enamel hypoplasia was recorded in a much greater proportion of the group of children who did not receive supplements. The crowns of upper first incisors were divided into nine zones of equal crown height, to which were attached ages following figure 7 of Massler *et al.* (1941). These are in error, but it is possible to revise them (Appendix A, Table 37) using the tables of Reid and Dean (2006) (Figures 4.6 and 4.7). In all zones, a larger proportion of the children who did not receive supplements showed defects than those who did receive supplements, in which proportions were low except in the central part of the crown side. It does therefore seem that the nutritional supplementation was an important factor in the distribution of defects.

Neonatal hypoplasia

Kronfeld and Schour (1939) described the cases of two children who had suffered an injury at birth. In these children, the neonatal line was sharply defined in deciduous tooth sections and was associated with strongly developed plane-form defects. The permanent first molars are not described in detail, but did not show a defect. They called the condition *neonatal dental hypoplasia*. Archaeological human deciduous tooth crowns rarely show hypoplasia, but occasionally there is a sharp plane-form defect at the point crown development would be expected to reach at birth (Figure 7.14).

Few studies of hypoplastic defects in the deciduous teeth of living children have located the defects clearly enough to be sure whether or not they were neonatal in timing, or post-natal. There is little evidence to suggest that hypoplastic defects are formed *in utero* so, if they are recorded on a deciduous first incisor, which completes its crown around 1 month after birth, they are likely to be at least perinatal. Similarly, the deciduous second incisor crown is complete at 3–4 months of age. The deciduous canine, on the other hand, would have just the tip of its cusp formed at birth and the crown completed at around 10 months of age. Stimmler *et al.* (1973) described 12 cases of hypocalcaemic neonatal tetany in newly born infants (page 13). This condition manifests itself only in the first two weeks after birth and all the children had marked defects on the incisal third of the deciduous canine crowns and occlusal surfaces of the third premolars, which would

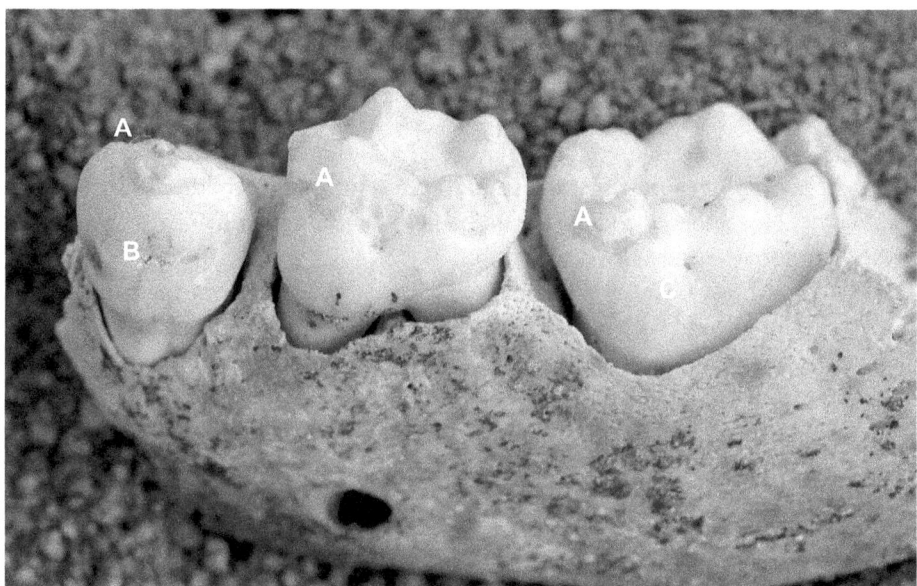

Figure 7.14 Neonatal hypoplasia in human deciduous lower canine, third and fourth premolar showing a state of dental eruption equivalent to approximately 2 years after birth in modern children. Neolithic burial from the site of Çatalhöyük in Turkey. A prominent plane-form defect on all three teeth, marked 'A', represents the state of development expected at birth. There are also associated pit-form defects in the third premolar. The additional defect marked 'B' is a 'localised hypoplasia of the deciduous canine' or LHPC (see page 214 and Figure 7.15). 'C' marks a pit on the fourth premolar which could be associated, or it may simply be a feature of normal morphology, i.e. a buccal pit. Width of field of view 25 mm.

correspond to this period. Purvis *et al.* (1973) sectioned shed deciduous teeth from a large group of children with neonatal tetany and found the characteristic neonatal defect in a little over half of them. Most of the children with pronounced defects were born in winter or spring, suggesting that the condition was related to the lower production of vitamin D in the skin of their mothers. Other studies (Johnsen *et al.*, 1984; Seow *et al.*, 1984) have found that children born very preterm display neonatal hypoplasia more frequently than controls. Children with defects are more likely than controls to have had respiratory distress syndrome requiring supplementary oxygen.

Sweeney and co-workers (Sweeney and Guzman, 1966; Sweeney *et al.*, 1969; 1971) studied children in Guatemala. They were scored as positive for neonatal hypoplasia if there was a marked linear defect in the incisal two-thirds of a deciduous first incisor crown, or the incisal one-third of a second incisor crown. The location is not exactly where a histologist would place it, but with a visual inspection of teeth in the living mouth it must in any case have been difficult to judge. The proportion of children recognised with this defect varied between villages, from 22 to 53%. Amongst extremely malnourished children hospitalised in Guatemala City, the proportion showing the defect was 73%. The strongest relationship was with infectious diseases in the first month after birth. Another large study (Infante and Gillespie, 1974) in Guatemala found similarly high proportions of children with the defect in their deciduous first incisors. The siblings of children with the defect showed it more commonly than the study group as a whole, suggesting that there were shared family factors. In addition, the defect was most common in children born during August and least common in children born from September to November, the latter period being the months when diarrhoea amongst young children was also least common.

Enwonwu (1973) compared Yoruba children from the rural village of Osegere in Nigeria with children of a similar ethnic background, but from affluent families living in the city of Ibadan. There were strong socio-economic differences between them, reflected in larger weight for age in the Ibadan children and relatively delayed eruption of the deciduous teeth in the Osegere children (page 41). Only prominent hypoplastic defects, characterised as grooves which could be felt, were recorded on the deciduous teeth. None of the Ibadan children showed such defects, whereas they were present in 21% of the Osegere children. Only 28% of these defects were in the expected position for neonatal or post-natal disruption. The rest were recorded in positions that implied they were initiated at an earlier stage of development. This could imply premature birth, although Enwonwu observed that many of the women in Nigerian villages were malnourished, with few reserves to fight infectious disease or parasite infestation. Many had iron-deficiency anaemia. He referred to experimental work with pregnant laboratory animals (discussed in the following section) in which both the unborn young and their mothers showed disruption to enamel formation.

Kreshover's experiments

Seymour Kreshover (1913–2006) was a professor of oral pathology and became the Director of the US National Institute of Dental Research. He carried out a series of important experiments on the effect of different factors on dental development in laboratory rodents and rabbits (Kreshover, 1960a; 1960b). These animals have continuously growing incisor teeth, in which there are always some ameloblasts and odontoblasts actively secreting enamel and dentine matrix. In one experiment, he inoculated mice and guinea pigs with tuberculosis, leaving others as controls. After their death, he made decalcified preparations of the jaws, sectioned them with a microtome and stained them with haematoxylin and eosin (Kreshover, 1944). In the animals with tuberculosis there was irregular formation of enamel matrix and the internal enamel epithelium was greatly disrupted. Another experiment involved keeping pregnant rats in an incubator at a continuous temperature of 38°C to simulate fever (Kreshover and Clough, 1953). There was disruption to the internal enamel epithelium and enamel matrix in both the mothers and their young, but the dentine matrix and odontoblasts were not disturbed. Yet another experiment involved the induction of diabetes through injections of alloxan into pregnant rats. In this case, both enamel and dentine matrix secretion were disrupted. Finally, pregnant rabbits were inoculated with vaccinia virus (Kreshover *et al.*, 1954). Marked defects were recorded in the incisor enamel of the mothers, but not their young.

Syndromes

In a well-known review of the aetiology of dental defects, Pindborg (1982) summarised a number of syndromes in which enamel hypoplasia is one of several symptoms. Such defects are associated with, for example, the rare inherited skin disorder epidermolysis bullosa, the inherited metabolic disorder phenylketonuria and with the hormone disorder hypoparathyroidism. They have also been found in the children of mothers with diabetes. Coeliac disease is an immune response to proteins found in cereal-based foods which leads to an inflammation of the small intestine. This interferes with the absorption of calcium and vitamin D and, for that reason, it is to be expected that there might be effects on tooth development. Aine *et al.* (1990) found that 85% of coeliac patients had some form of enamel defect, including hypocalcification as well as hypoplasia, although the latter was much more common. Overall, there is variation between studies (Pastore *et al.*, 2008), but further work does seem to confirm a strong relationship. The permanent incisors and first molars are more commonly affected than later developing teeth, which may reflect diagnosis at a relatively young age and the effects of subsequent treatment. The prevalence of coeliac disease in Europe is around 1%, so it is perhaps not a likely explanation for the majority of defects in the archaeological record.

Another common association is with neurological conditions such as cerebral palsy and some hearing defects. The association is not consistently established, however, and the aetiology is obscure (Bhat and Nelson, 1989).

Summary

Defects of enamel hypoplasia are complex, variable and difficult to record by simple observation of the crown surface. Everyone can agree when a prominent defect is present, but many much smaller perturbations can be seen and the boundary between normal and abnormal has never been defined successfully. It is also difficult to make a reliable estimate of the age at which a growth disruption took place based on the position of a defect on the crown. This is particularly the case for plane-form and pit-form defects, the position of which on the crown surface may bear little relationship to the timing of the disruption. In the case of furrow-form defects, new charts of crown surface formation at least offer a chance to make an approximate estimate of timing. Even then there is the problem that the same defect, in terms of its timing and the number of perikymata it includes, is larger and more prominent in the mid-crown than near the cervix. Added to this, tooth wear has a strong effect, so that apparent similarities or differences in the pattern of defects between archaeological assemblages of human remains can instead be due to differences in the preservation of teeth, the age-at-death and other factors.

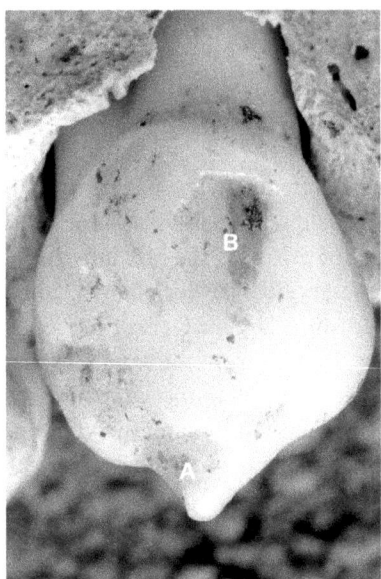

Figure 7.15 Upper left deciduous canine from the same dentition as in Figure 7.14, showing the same neonatal plane-form defect (marked 'A') around the cusp tip. The tooth also shows a large, isolated plane-form defect in the middle of the crown side (marked 'B'). This is matched on the other canines of this dentition. This is an example of an LHPC. Width of field of view 5 mm.

For all these reasons, there is currently no satisfactory method for recording enamel hypoplasia by examination with the naked eye.

Under the microscope, the full complexity of the crown surface is visible and it is clear why the appearance of a defect to the eye gives so little idea of its true nature. Furrow-form defects result to a large extent from variations in the spacing of perikymata, so it is possible to build sequences of perikymata counts within defects, between them and within the overall count of perikymata from cusp tip to cervix. These sequences can be matched between different teeth within one dentition to form an extended sequence. This makes it possible to catalogue the nature of the furrow-form defects and their position within the sequence of tooth crown formation. Standard tables of crown development can then at least give some context. The definition of normal and abnormal is again difficult, and the measurement of perikymata spacing has some potential for defining rules. Plane- and pit-form defects cannot be investigated in this way, but microscope sections can unravel the sequence of events. Although they are much more prominent features on the crown surface, they usually represent what seems to be a short disruption compared with many furrow-form defects, for which sections also provide a more detailed record.

Cross striation counts can be used to establish the age of the growth-disrupting events that caused them with some confidence, and to consider the way in which the long-period rhythm of matrix secretion brings about the change in crown surface

Summary

outline. Sections also allow the study of accentuated long-period lines, which are often associated with surface defects, although it is still not entirely clear what they represent. Where sectioning is not possible, perikymata counts can be used in a relative way to position the defects in the dental development sequence with reference to stages such as the completion of different tooth crowns. The timing and duration can be expressed as proportions of intervals between stages without knowing the actual periodicity of the perikymata.

It has been known since Bunon's time that enamel hypoplasia can be related to the medical history of a child, including infectious disease and deficiencies in nutrition. In addition, it has been known that the timing of disruption determines which teeth in the dentition are affected. Nevertheless, it has always been difficult to match an individual defect with a particular episode during development. For this reason, it is not possible to speak about one particular cause. All that can be inferred from a defect is that something disturbed enamel development. This might make it seem that enamel hypoplasia is a useful general purpose indicator of health but, as discussed in Chapter 8, it is not that simple. When seen in section together, what the crown surface defects and accentuation lines do best is to provide, for each individual, a highly detailed sequence of disruption to the normal, even progression of development. The evidence reviewed here suggests that these sequences are exquisitely sensitive to the tiniest disturbance; however, it does not yet seem possible to say what the defects represent for an individual without knowing their clinical history.

8 Health, stress and evolution: case studies in bioarchaeology and palaeoanthropology

The second main question posed in Chapter 1 asked about the impact on human health of the major social and biological transitions for which there is evidence from the fossil and archaeological record. The discussion in Chapter 7 has shown how sequences of hypoplastic defects of enamel may relate to disturbances during development and they are widely seen as one of the best indicators of health during childhood. One of the key theoretical approaches in bioarchaeology has been to interpret them as markers of physiological stress, following the ideas of Hans Selye. Health and stress are, however, complex topics which require some initial discussion before moving on to case studies which illustrate the methodological and theoretical issues.

Health, stress and prevalence

Health

Measuring health is difficult, even for living populations. What is health? According to the official definition of the WHO (World Health Organization, 2010), it is a state of complete physical, mental and social well-being, much more than just the absence of disease. As suggested in an important paper by Boorse (1977), different stakeholders in the concept of health define it in different ways. We value our own health, but individual people have their own measures of healthiness. Pain and discomfort are variably perceived as the same condition may be excruciating for some and pass unnoticed by others. A doctor might define health as an absence of diseases that they can diagnose and treat, or of conditions that lead to an impairment of normal bodily functions. Evolutionary biologists could see disease as evidence of maladaptation to the environment and health as evidence of Darwinian fitness. Physiologists might see health as a state in which the bodily functions are in equilibrium and disease as the state in which they are not. Everyone views the concept through the lens of their own particular concerns.

For those whose job it is to measure health, there is no index that can do this in a positive sense. The only possible approach is to catalogue negative impacts on health: morbidity (defined here as a higher than expected prevalence of disease or level of incapacity) and mortality, or departures from normal death rate.

Health is essentially measured through 'unhealth'. This is the approach taken in the WHO's annual reports on global health indicators (for example, World Health Organization, 2010). These statistics catalogue mortality rates and life expectancy, particularly for babies and young children, cause-specific mortality rates for mothers at childbirth, communicable and non-communicable diseases, and injury. Even with the WHO's resources, these are complex statistics to determine and must involve many assumptions.

It is not clear that bioarchaeologists can take a similar approach to understanding health in the past. In prime place amongst the difficulties is the nature of archaeological assemblages. The people whose remains are interred in a cemetery make a poor basis for understanding health in the once-living population because they are all dead and it is not possible to be more *un*healthy than that. This might make them seem a good group in which to estimate morbidity, but they are far from a simple cross-section of people in the population. They are those who were at most risk of dying – mainly the very young, the old and the sick. It thus seems unlikely that the prevalence of a disease amongst the dead would be a good representative for prevalence amongst the living. In addition, of the 18 infectious diseases listed by the WHO, only tuberculosis and leprosy leave any trace on the skeleton and, even with these, differential diagnosis poses problems.

It is difficult to make a case that the diseases most commonly seen in archaeological human remains, such as dental caries, are general indicators of morbidity in a similar way. Undoubtedly, oral health must have an impact on the quality of life and ability to eat (Sheiham, 2005) and it should not be ignored, but it is rarely a killer. It also has an aetiology which is strongly affected by carbohydrates in the diet so it would, for example, be a very poor indicator to choose for overall health changes at the origins of agriculture, because carbohydrate consumption is exactly what was transformed during the agricultural transition.

Estimating mortality is similarly difficult for bioarchaeologists. The dead people from a cemetery provide only one-half of the WHO's mortality indices in deaths per 100 000, which also require a count of people remaining alive in the population. It might instead be possible to use the proportion of children in the death assemblage, as the WHO has pointed out that this is an important health statistic. Unfortunately, a glance at the mortality tables provided online by the WHO (www.who.int/whosis/whostat/2010/en/) shows that there is something amiss with most archaeological death assemblages. Modern age-at-death graphs are U-shaped, with a peak for young children, a peak for old people and a broad valley for the older children, young adults and middle-aged people in between. Even in affluent countries with low fertility and readily available health care, the shape of the graph is similar, except that the childhood peak is lower and the old age peak higher and further along the scale.

In contrast, most archaeological age-at-death graphs have a single broad peak in the middle, with low numbers of children and few elderly people. Nowhere in today's spectrum, from countries with large families in desperate need, to countries with affluent one-child families who have never known a day's hardship, is there a graph like it. The logical conclusion is that the archaeological graph arises because the cemetery is not a random sampling of those who died. In many cultural contexts, children were rarely buried alongside adults, so they are absent from the graph. It is also difficult to estimate age-at-death in older adults from the skeleton and dentition. The variability in age indicators increases with age and most of them have a tendency to underestimate, so the age estimates accumulate somewhere in the middle.

If fossils preserved in natural accumulations or cemetery assemblages are not good representatives of the health of the living populations from which they were separated at the time of their death, the logical step is to look instead for evidence of their state of health during the years prior to their death. This is the point which engages the discussions from previous chapters in this book. For example, there is evidence (page 16) that chronic malnutrition during childhood, often coupled with disease, can result in shorter adult stature. Children's growth may cease altogether for an episode of malnutrition with or without disease, but, like most mammals, human children adapt to hard times by following them with a period of catch-up growth when conditions improve. If, however, the episodes last a long time or recur, the catch-up may not be so complete. Adult stature (or for archaeology the lengths of limb bones, which are highly correlated with stature) might therefore be a good indicator of morbidity in childhood.

There is considerable research suggesting that, within each age group of adults for one population and males and females separately, shorter people show higher mortality than those of average height (Waaler, 1984; Peck and Vågerö, 1989; Allebeck and Bergh, 1992; Song *et al.*, 2003). There are also social differences: people from less well-off socio-economic groups tend to be smaller and to live less long than people from more fortunate circumstances. This is encouraging for an archaeologist because it suggests that health during childhood has an effect on later morbidity and mortality in adulthood, which might reasonably be detectable as a simple measurement of long bone lengths. However, the mortality differences point to the selective nature of a death assemblage and show how difficult it is to reason from the archaeological collection of skeletons excavated in a cemetery to the original living population.

The defects of enamel hypoplasia may provide a more direct way to look at morbidity during childhood in the years before the individual actually died. The advantage of these defects is that they provide a schedule for each individual which shows not just the accumulated effect, but the day-to-day disruptions due to disease and undernutrition. As will be discussed below, it is difficult to assess them as a 'disease' which has a prevalence. They cannot be counted and compared in a simple way between archaeological assemblages because of the sequential way in which they occur. There are therefore difficulties for enamel hypoplasia as a health indicator of the type used by the WHO. Enamel hypoplasia is instead best at providing a detailed record of disruptions to growth at the level of the individual. For fossil and archaeological assemblages, where the number of individuals represented may be small, this is a strength rather than a disadvantage.

Stress

One particular focus for bioarchaeologists interested in health has been the *general adaptation syndrome* (GAS) proposed by Hans Selye. This had its roots originally in the ideas of Walter Cannon, a physiologist at Harvard Medical School. Cannon's most famous contributions were the *'flight or fight'* response (Cannon, 1915) and homeostasis (Cannon, 1932). *Homeostasis* is the way in which the bodily systems maintain themselves in a steady state, bathing all cells in fluids with constant temperature, oxygen tension, glucose, and so on. The flight or fight response is the way in which mammals, when threatened, in pain, aggressive or excited, prepare themselves without conscious effort for rapid action. The heart beat rises sharply, breathing quickens, the blood vessels which supply muscles widen whilst those in other regions constrict, activity in the digestive tract is inhibited – and a whole list of other physiological changes which rapidly return to normal once the threat or stimulation is resolved. The threat itself may be perceived consciously, but the body's response is controlled unconsciously through the HPA axis (involving the hypothalamus and the pituitary and adrenal glands) and the autonomic nervous system. Cannon saw flight-or-fight as a response by the body to maintain homeostasis when this was threatened. His main point was that the immediate physiological reactions were the same, whatever the nature of the threat: cold, bleeding, trauma, toxin or emotion (Cannon, 1915).

Selye was a Hungarian-born endocrinologist, trained in Prague, who then moved to Johns Hopkins University in 1931 and then to Montreal in 1936. He started a series of experiments in which laboratory animals were repeatedly exposed to such threats and recorded how the body's response changed. He described the response as 'stress' (Selye, 1976, p. 53): 'Stress is the non-specific response of the body to any demand. A stressor is an agent that produces stress at any time.' By non-specific he meant that the stress response involved a number of components which were effectively the same for any stressor. This idea has a struck a chord with anthropologists interpreting the defects of enamel hypoplasia because a similar defect might represent a whole range of growth-disrupting factors.

The stress concept is very widely accepted and can be seen in countless publications, but, for a general audience, stress is an awkward term for Selye to have used for the body's response. For most ordinary people, 'stress' implies emotional *dis*tress or, by extension, the mental state in which people are left by continuous alarms or demands. On the one hand, the word is used in the sense of a trigger or stressor, whilst on the other hand it also implies their response to such stressors, which may be both physiological and emotional. Most people would not use it to describe their physiological response to pain, cold or severe hunger. Engineers use the word 'stress' to describe the forces acting on a structure and 'strain' to describe the measureable deflection that results. This again is the other way around to Selye's usage, as the strain is the response of the structure and the stress is the trigger to this response. Anthropologists also use the engineering concept of stress in terms of the forces that act on the skeleton during walking or other activities. The bone remodels in response to these stresses, changing its shape into a form that resists them more effectively. Care is therefore needed and it is a good idea to specify Selye's meaning as 'Selyean stress'.

Selye distinguished the immediate response to a stressor from longer term responses to repeated exposure to stressors (Seyle, 1976, p. 53): 'The general adaptation syndrome (GAS) represents the chronologic development of the response to stressors when their action is prolonged. It consists of three phases: the alarm reaction, the stage of resistance and the stage of exhaustion.' The alarm reaction was the immediate flight-or-fight response as initially described by Cannon, which resolves once the stressor is removed. The stage of resistance was marked by long-term physiological and structural changes; for example, enlargement of the adrenal gland to repeated exposure to stressors over some time. Selye described situations in which this conferred resistance to the stressors. The stage of exhaustion was when this resistance broke down and the animals succumbed to a range of diseases which were seen as characteristic of this sequence of events. Selye was a prolific author and his ideas were widely accepted. Even today, these ideas remain fundamental assumptions underlying much work in medicine, biology and anthropology. However, more than 80 years after the original experiments, it is not surprising that the field has moved on.

One issue remains the way in which 'stress' as a word continues to be used with various meanings, even in endocrinology and medicine generally. It may mean the factor that stimulates the reaction, the reaction itself or the consequences of over-stimulation. Careful authors can emphasise that they are describing the body's reaction as the 'stress response' whilst continuing to use 'stressor' to describe a stimulating factor. The cumulative effect of continued activation of the stress response may then be described as 'chronic stress' or, in more detail, as the 'systemic effects of chronic hyperactivity of the stress system' (Kaltsas and Chrousos, 2007, p. 312).

It is now recognised that stressors are poorly defined. In effect, they are anything that stimulates the stress response system. Similarly, for different purposes, a definition of stress response may include a large variety of physiological and behavioural mechanisms and it is not uncommon for stress to be defined as a response that is triggered by a stressor (Romero et al., 2009). This is another of the circular arguments that bedevil discussions of health. One more area of recent debate is the extent to which the stress response is in fact non-specific (Goldstein and Kopin, 2007). It is recognised that the threat of the stressor may be perceived either consciously or unconsciously and that there is a certain degree of specificity, both in relation to this perception and to the nature of the threat. In addition, the acceptable limits for different parameters of homeostasis many vary depending on other activities which are taking place at the same time.

The measurement of stress provokes heated discussion, even in living people (Ice and James, 2007). It involves behavioural and emotional assessments, as well as measurements of hormone levels, immune system function and blood pressure. There are issues for the interpretation of all these and, in any case, for archaeological human remains or fossil hominoids, there is no possibility of measuring any of them. The bioarchaeological approach has been to consider conditions thought to be caused by chronic stress and to use them as markers. Chronic hyperactivity of stress response is known to interfere with the control of growth through suppressing the production of growth hormone in the pituitary gland and through the reduction of the response to it and other growth factors in growing tissues (Kaltsas and Chrousos, 2007). Psychosocial factors seem to cause children to be small for their age and to reach developmental landmarks at an older age than expected, but it is difficult to disentangle these effects from related factors such as deficient nutrition and poor hygiene (Tanner, 1989). Chronic stress also appears to interfere with bone turnover (Kaltsas and Chrousos, 2007) and may be one factor in osteoporosis, although the causes of this condition are complex. Similarly, there are effects on thyroid and adrenal function, gastrointestinal and reproductive function, metabolism and psychiatric conditions.

The position of the defects of enamel hypoplasia in this discussion is not clear-cut. In the first discussions of Selyean stress in bioarchaeology (Goodman *et al.*, 1988), enamel hypoplasia was singled out 'as a promising marker of previous stress' (Goodman *et al.*, 1980, p. 516). Chapter 7 reviews clinical evidence that the defects are related to particular events during childhood which disrupt the regular rhythm of enamel matrix secretion. At the head of the list are infectious diseases in which the symptoms include episodes of fever – a sustained rise in systemic temperature that in some way particularly affects the ameloblasts that are about to switch into their maturation phase. Next come episodes of severe malnutrition. In any case, deficient nutrition and infectious disease are known to reinforce one another's effects on growth (page 16), so they need to be considered together. In a Selyean sense, however, they are not themselves stressors, even though doubtless there may be times during the period of disease or malnutrition in which the stress response is triggered, consciously or unconsciously, but there is no evidence that the stress response itself is the *mechanism* by which the enamel defects are created.

In a more general way, enamel hypoplasia follows other aspects of growth in its relationship to socio-economic factors (Enwonwu, 1973), although this in itself must involve malnutrition and disease as well. Psychosocial factors are known to affect the production of growth hormone and children with hypopituitarism achieve their dental development stages later than normal children, although the skeletal development stages are affected to a greater extent (Garn *et al.*, 1965). Deficient growth hormone production has, however, not been directly linked with the defects of enamel hypoplasia. Similarly, it has not been possible to find any clinical study in which psychosocial factors alone are implicated in enamel defects, even though it seems reasonable enough to suppose that this might be the case. To summarise, it is difficult to isolate the Selyean stress response from all the other potential factors that might cause enamel hypoplasia. There is currently no experimental evidence to link defects with episodes during which known stressors were deliberately introduced to trigger a stress response while other known defect-causing factors were excluded.

'Stress' is so widely used as a term in bioarchaeological publications that care needs to be taken in discussing its use. In fact, it has moved away from the strict Selyean definition, as shown by Larsen's textbook *Bioarchaeology*, in which it is 'physiological disruption resulting from impoverished environmental circumstances' (Larsen, 1997, p. 6). The text makes it clear that these circumstances encompass a wide range of factors, including dietary deficiency, disease and social conditions. They might well also include stressors in the original Selyean sense, but these are indistinguishable from the rest. Bioarchaeologists have also expanded the term outside its meaning in ordinary conversation, where stress implies repeated psychological pressure and the reaction to it. Nor is it the same as the usage in other areas of biological anthropology that seek to measure stress in living people, relating it to their family circumstances and working situations. Many bioarchaeologists go further than Larsen and use the term 'stress' in the sense of 'stressor'. So a defect of enamel hypoplasia is seen as the record of an episode of stress. The word has ceased to have a clear meaning within bioarchaeology and means different things in the wider field of anthropology, not to mention more general readers, so it is not used in this book for discussing enamel hypoplasia.

Prevalence and hypoplasia?

Enamel hypoplasia is often reported in terms of prevalence. *Prevalence* can be defined in epidemiological terms as the number of cases of a disease counted in a particular population, divided by the total number of individuals in that population (including those both with and without the condition). It is usually expressed in terms of cases per thousand people. Applying the concept of prevalence to the defects of enamel hypoplasia is complicated because the defects relate to a sequence of events in the past. If dental development is divided into a series of stages, recording a positive diagnosis of 'hypoplasia' for an individual implies that there is a defect in at least one of the stages. Making a negative diagnosis, however, implies not only that a defect was not found in any of the stages, but also that all stages of the sequence were observable. If any stage is missing, perhaps because the appropriate teeth are not preserved or are worn, there is no knowing whether or not it had a defect in it. For example, the 1 to 2 year stage in the dental development sequence is situated on the occlusal surfaces of the permanent first molars, lower incisors and upper first incisor (Figure 4.7). These are also the first permanent teeth to erupt, so they tend to be the most heavily worn and lose the occlusal parts of their crowns at a relatively young age. It is therefore common in archaeological dentitions for that part of the sequence to be missing and therefore unobservable. To say that such a dentition is defect-free is like claiming an individual shows no evidence of injuries when a quarter of the skeleton is missing. It is therefore necessary to specify which part of the development sequence a prevalence applies to. Where this is not specified, the prevalence would be strongly affected not only by tooth wear, but by tooth loss and preservation, not to mention dental development itself where the appropriate stage has not yet been formed. For this reason, in many studies, the simple hypoplasia prevalences which represent whole dentitions cannot be taken at face value.

This is not the only complication, however. Even when a development stage is represented by part of the crown in several teeth which are preserved, these parts cannot necessarily be regarded as equal for the purposes of observing defects. The 1–2 year stage discussed above is not too problematic because it is found in parts of incisors and first molar crowns which are all characterised by widely spaced occlusal perikymata with similar effects on the width and prominence of a furrow-form defect (page 177). The following 2–3 year stage, however, is a very different matter. It is found in the mid-crown part of incisors and canines, but in the cervical part of first molar crowns. Mid-crown perikymata produce the optimum prominence for defects, which are narrower and less prominent when formed by cervical perikymata. Without careful examination under a microscope, a defect giving a positive diagnosis in the canine could be missed in the cervical region of the first molar in the same individual. Incisors and canines, with single roots, are frequently lost in archaeological assemblages where the molars are held firmly in the jaws by their multiple roots. The apparent overall prevalence for hypoplasia could therefore be greatly reduced in a museum collection of skulls which, handled over many years, had lost many of its incisors and canines. For the prevalence to have any value in comparison between assemblages, it is necessary to specify not only the part of the development sequence, but also the tooth type. Once again, there are few studies for which this is the case.

Case studies in bioarchaeology and palaeoanthropology

The defects of enamel hypoplasia are amongst the most commonly observed features of fossil and archaeological dentitions. Much work has been published in this area of research and it is not possible to present all the studies here, particularly in view of the difficulties of comparison between them. Instead, several key questions have been identified as case studies.

The harsh world of the Neanderthals?

The stocky, robust build of Neanderthals (Holliday, 1997) and their broad projecting mid-face are seen as adaptations to life in cold environments, although this has been questioned (Rae et al., 2010). Their big, wide jaws and large, heavily worn anterior teeth suggest an adaptation to the habitual application of strong forces, although, again, this is debated (Clement et al., 2011; Harvati et al., 2011). Their limb bone morphology implies heavy physical exertion (Trinkaus et al., 1999; Pearson et al., 2006). Most known Neanderthal fossils represent young adults (Trinkaus, 1995) and a large proportion of them show healed fractures with a pattern similar to that of modern rodeo riders, suggesting personal contact with similar large and hostile creatures in the hunt (Berger and Trinkaus, 1995). The life of a Neanderthal has been characterised as 'nasty, brutish and short' (Pettitt, 2000, p. 362). Was this true? It implies a difficult environment for the growth of children and a study of enamel hypoplasia seems a logical way to address the question.

The investigation of hypoplastic defects in Neanderthals has its origin in the study by Molnar and Molnar (1985) of the Krapina teeth. This assemblage of about 400 bone fragments and teeth (Schwartz and Tattersall, 2002) was found in a rock shelter near Zagreb in Croatia and comes from deposits dated to around 130 000 BP. It includes 196 teeth, both isolated and preserved in jaws, representing at least 35 'dental individuals' (Radovcic et al., 1988). It is one of the largest dental groups in the fossil hominin record. Seventy-six of the 196 teeth had only slight wear, or none, so it is clear that many young individuals are represented. The teeth are beautifully preserved and it was clear to Molnar and Molnar that many crowns were marked by defects, ranging from small disturbances to the perikymata to larger furrow- and pit-form defects. Out of 18 dental individuals examined, 13 had at least some hypoplastic teeth as defined by them, mostly in the canines, followed by premolars and incisors. This is over two-thirds of the total number of individuals and they were concerned this might be because they were looking too closely at this well-preserved collection. For comparison, they recorded dentitions from two Copper/Bronze Age sites in Hungary: Budakalasz and Tape. About half and one-third, respectively, of these dentitions showed defects, suggesting that the Krapina prevalence was indeed high. Their paper did not, however, give details of the wear state of these later sites and, if the teeth were more worn, they would be expected to preserve fewer defects on their crown surfaces.

Ogilvie and co-workers (Ogilvie et al. 1989; Ogilvie and Trinkaus, 1990) studied defects in a very large collection of dental impressions from 669 teeth, including most of the Neanderthal specimens from Europe and Asia available at the time. They included only those molars with at least some of the occlusal surface remaining, and incisors with at least 3 mm of crown left. This still represents considerable wear, but given the worn state of most Neanderthal dentitions it is easy to see why such specimens would be included to maximise the study collection. There would, however, be a reduction in the prominence of defects by more general abrasion. Varying combinations of teeth were preserved in different specimens, so they calculated a frequency of furrow- and pitted-form defects for each tooth type separately. A total of 269 of the teeth were from the Krapina assemblage and 40% of these showed defects. This was no surprise, but they also recorded defects in 35% of other European Neanderthal teeth and in 28% of Near Eastern Neanderthal teeth. Taken together, the defects were most commonly recorded in permanent canines, followed by incisors. This is a common pattern and, in many studies, probably represents the different geometry of crown growth in molars. The lower limit for identifying defects was not defined, but particular care was taken, including the use of a binocular microscope. There are rarely any doubts about prominent defects and, as they noted, the replication process would tend to make the smallest defects invisible.

Ogilvie and co-workers wished to estimate the timing of defects using the Swärdstedt table (page 179). This was not straightforward, because it uses measurements and assumes a standard height of the unworn crown based on the mean value for a study of modern humans. Neanderthal teeth are typically larger, so they had to calculate a correction factor. In addition, they had to add estimated ages for third molars, which are not included in the original table. As has been found in many studies of modern humans, this procedure suggested that the modal age for growth disruption in all teeth, except the third molars, was 4 years of age. There was a separate mode at around 11 years for the third molars, which are widely separated in their crown development schedule. It is usual to suggest, as did Ogilvie and co-workers, that the 4-year peak is related to weaning. They were criticised at the time by Neiburger (1990), as many factors such as childhood fevers might be more likely to disrupt enamel formation. It is certainly true that studies of weaning in recent hunter-gatherers suggest a younger age (page 152). Ogilvie and Trinkaus (1990) accepted that it would not be possible to identify any cause with confidence, but pointed to the lack of growth disruptions before 3 years of age, the period of greatest danger for modern children in terms of infectious disease.

Another study of the Krapina dental assemblage was carried out by Hutchinson et al. (1997). No details were given of the recording method, but they compared their results with studies of foraging and agriculturalist people from the USA (page 215) and found that the frequency of defects in Krapina teeth did not stand out as particularly high. Guatelli-Steinberg et al. (2004) compared a carefully selected group of Neanderthals with a similar group of recent Inuit dentitions. This comparison has a particular resonance because the body build, lifestyle and heavy tooth wear of the recent Inuit, when living a traditional hunting and gathering life, have often been suggested as a good model for the supposedly cold-adapted Neanderthals. They included only permanent incisors and canines in which an estimated 70% or more of the original crown height was preserved and which still showed perikymata over most of the crown side. Once more, the Krapina assemblage was used, together with a group of Neanderthals from southern France. The Inuit dentitions were a mixed collection from Point Hope in Alaska. Dental replicas made from impressions were examined under low magnification and furrow-form defects identified from perikymata spacing. Four out of ten (40%) Krapina individuals showed defects, compared with eight out of 21 (38%) Point Hope individuals. Defect duration was estimated from the number of perikymata in the occlusal wall, the total number of perikymata included in the defect and the width of the defect (which was correlated with the count of perikymata) depending on the extent of damage due to abrasion. Based on these measures, the average duration of defects in the Inuit group was greater than for the Neanderthals.

The real comparison to be made, however, is between the Neanderthals and Upper Palaeolithic modern humans. If the Neanderthals were replaced by not only morphologically different people, but also by a far more complex material culture, then it is logical to ask if this was accompanied by a better quality of life. As Skinner (1996) has pointed out, the expectation has been that Upper Palaeolithic technology provided an advantage over the Middle Palaeolithic technology associated with Neanderthals. He selected for study two groups of dentitions. The Middle

Palaeolithic group combined Neanderthals from Krapina and western Europe with early modern humans from Jebel Irhoud and Qafzeh (page 131). It included 59 dental individuals comprising 128 deciduous and 154 permanent teeth. The Upper Palaeolithic group was drawn from sites in France and included 47 individuals with 162 deciduous and 125 permanent teeth. To minimise the effects of tooth wear, the individuals chosen were young: age-at-death estimates ranged from 1 to 18 years.

One of the difficulties was that some dentitions preserved a number of teeth, whereas others did not, and there was considerable variety in which teeth were preserved. For this reason Skinner reported both the proportion of individuals affected and the teeth affected. He tabulated different teeth separately, but the comparisons were made on the basis of all teeth together. Few details were given about the criteria used for scoring defects, or about the method of estimating the ages of growth disruption, but at least all the scoring was carried out by a single researcher. Skinner found that the bulk of the deciduous teeth were not affected, with most of the defects either neonatal or of the isolated canine type (page 214), and there were more of these in the Upper than the Middle Palaeolithic group. A total of 26% of the Upper Palaeolithic permanent teeth had defects, as opposed to 18% of Middle Palaeolithic permanent teeth. This result must be affected by the teeth present because defects tend, for example, to be more prominent on incisors and canines than on molars. An examination of Skinner's data tables, however, shows that the proportions of these two groups of teeth preserved were approximately equal in the two groups, so this may not be so large a problem as it first appears. Approximately half of dental individuals showed a defect on one or more tooth in both groups. Defects were plotted on a graph as their estimated age at initiation, showing a mode between 3 and 5 years for Middle Palaeolithic contexts, with relatively low frequencies in permanent teeth for younger ages. This contrasted with a more even spread through all age groups from 1 to 6 years in the Upper Palaeolithic group. This graph must, however, be strongly affected by the location of the defects on different teeth.

So what is the answer to the question? Different studies, using a variety of approaches, have shown different things, but there does not yet seem to be any consistent evidence that Neanderthals had an unusually high frequency of defects in their tooth crowns. There are, however, significant challenges. For a hominin, the Neanderthals have an unusually large total assemblage, within which there is remarkably little morphological variation even over its large geographical range. This ought to be a good thing, but the largest collection of teeth by far is the Krapina assemblage, which contains a particularly large proportion of little-worn teeth from young individuals. They are ideal for studying enamel defects, but the heavily worn teeth of other fossils are not. Many of the Krapina teeth are unusually large for a Neanderthal, so they could only be described as typical in the sense that the assemblage is so much bigger than the others and so dominates plots of tooth size. Krapina is also very early and comes from the deposits of an interglacial period (Schwartz and Tattersall, 2002); not the best for assessing the supposed cold-adapted behaviour of Neanderthals. It is thus particularly important that

Guatelli-Steinberg and co-workers found little difference between the Krapina and other Neanderthal teeth. There are no equivalent large assemblages of teeth from Upper Palaeolithic contexts, so the key comparison group is always going to be a mixed bag drawing from different sites.

The only other large collection of hominin teeth from Europe forms part of the accumulation at the remarkable site of Sima de los Huesos, a vertical shaft in the large limestone karst system of the Sierra de Atapuerca in Spain (Bermúdez de Castro *et al.*, 2004). The date of deposition is likely to have been before 530 000 BP (Bischoff *et al.*, 2007) and the hominin remains have been assigned to *H. heidelbergensis* (Arsuaga *et al.*, 1997), although there is debate over the correct placement (Balter, 2009). The skulls are not particularly similar to others assigned to *H. heidelbergensis*, but the mandibles are fairly similar to the Mauer mandible, which is the type fossil for the species (Mounier *et al.*, 2009). The Sima teeth are very similar to those of Neanderthals (Martinón-Torres *et al.*, 2007) and, at the time of writing, there are 519 teeth (only eight deciduous) representing at least 28 individuals. Cunha *et al.* (2004) took dental impressions and made epoxy resin replicas, which they examined in the scanning electron microscope. Their definition of the minimum level of defect was that there should be a reduction in enamel thickness as well as a variation in the spacing of the perikymata. Of the 25 individuals with reasonably complete permanent dentitions included in the study, seven (28%) showed enamel defects by this definition, together with six isolated teeth. This prevalence is lower than that of Krapina. Perikymata counts were used to estimate the timing and duration of the disruption to enamel secretion. They assumed that crown formation times were similar to those of modern humans and used the Reid and Dean tables for crown surface deciles (Figures 4.6 and 4.7) to derive the ages for defects in six individuals at 2.4, 2.6, 2.8, 4.6, 5.2 and 9.8 years of age.

Do the young hominins of the South African Plio-Pleistocene cave sites represent vulnerable individuals who fell easy prey to carnivores?

The Plio-Pleistocene cave sites of Sterkfontein, Swartkrans and Kromdraai near Krugersdorp in South Africa have preserved large accumulations of mammal bones. They include not just the hominins *Australopithecus*, *Paranthropus* and *Homo*, but also the remains of monkeys, a wide variety of large and small carnivores, antelopes and gazelles, horses and zebras, elephants and a long list of small creatures. The underground caverns were filled primarily with deposits which fell down through narrow shafts from the surface, including the remains of the animals. The detailed analysis of the assemblages, bone fragmentation and tooth marks by Brain (1981) concluded that they had been accumulated at the caves by carnivores, the earlier part of the depositional sequence by big cats and hyenas and the later part by hominins. The evidence from Swartkrans was particularly strong, with marks of carnivore teeth on *Paranthropus* skull bones. Later studies have in general supported this idea (Pickering *et al.*, 2004a; 2004b; Kibii, 2007).

Defects of enamel hypoplasia were first described for Swartkrans *Paranthropus* by Robinson (1952), who found that 28% of 47 isolated teeth showed such defects, with four upper molars having marked defects. Robinson reported occasional slight defects only in Kromdraai *Paranthropus* and Sterkfontein *Australopithecus*. White (1978) undertook a larger study. He did not define his criteria for identifying defects, but, to control for attrition, he presented figures separately for all teeth at any stage of wear and a control group of less worn teeth. In total, 17% of 177 Swartkrans *Paranthropus* teeth showed defects and 8% of 108 Sterkfontein *Australopithecus* teeth. This pattern was repeated in the control group and, by individual rather than by teeth, it was 30% of Swartkrans and 12% of Sterkfontein individuals. The defects tended to be found in younger individuals, with incomplete dental development, rather than in older individuals. White therefore suggested that higher frequencies of defects might reflect reduced 'fitness' in individuals who had experienced growth disruption during infancy, so that they would more easily fall prey to carnivores. Further excavation produced more permanent teeth from the sites and Moggi-Cecchi (2000) was able to examine 259 from Swartkrans and 284 from Sterkfontein to test White's conclusions. Instead, he found that there was little difference between these two assemblages in defect frequency.

As has been explored in Chapter 5, there is now strong evidence that *Paranthropus* tooth crowns formed over a shorter period than those of *Australopithecus*, with a faster crown extension rate and fewer, more widely spaced perikymata down the crown side. The shorter total development period of a *Paranthropus* crown would give a smaller window of time within which an enamel defect might be initiated (Guatelli-Steinberg, 2004). The wider perikymata spacing might also be expected to make furrow-form defects less prominent. It is clear from this that the apparent size and position of enamel defects as seen with the naked eye or modest magnification cannot simply be compared between the two genera. Guatelli-Steinberg imaged impressions of crown surfaces in the scanning electron microscope. She set careful limits for furrow-form defects, requiring perikymata to be visible. The lower limit for the identification of a defect was that the perikyma spacing appeared wider than in adjacent parts of the crown. To estimate defect prevalence, only incisors and canines with more than 50% of their expected crown height surviving were included, and the defects had to match between teeth in one individual. Out of 12 *Australopithecus* individuals, four (33%) had defects by this definition. Out of ten *Paranthropus* individuals, two (20%) had at least partially matching defects. Looking in detail at the canines, all *Paranthropus* specimens affected by hypoplasia showed just a single defect, whereas the *Australopithecus* specimens showed multiple defects. This might support the idea that the quicker forming *Paranthropus* crowns, with fewer perikyma, included fewer defects than the slower forming *Australopithecus* crowns. The *Paranthropus* defects, however, were not significantly wider than those of *Australopithecus*. Guatelli-Steinberg suggested that this might be due to differences in crown development geometry and other developmental factors.

It thus seems that there is as yet no clear answer to the question about variation in defect frequencies amongst the Pliocene and early Pleistocene hominins. Where the tooth size, morphology and pace of development are so variable between the genera, it is difficult to make comparisons without using a microscope. A study of the Taung type specimen for *A. africanus* used a furrow-form defect (estimated to be initiated at 2.5 years of age) on its permanent first molar to help derive a new age-at-death (Lacruz *et al.*, 2005).

Do living African and Asian great apes have an unusually disrupted development?

Amongst living non-human primates, the defects of enamel hypoplasia are most commonly observed and most prominent in chimpanzees, gorillas and orangutans. They are least often reported in the prosimians (such as lemurs, galagos and bushbabies) and with increasing frequency in New World monkeys, followed by Old World monkeys, then gibbons and finally the great apes (Guatelli-Steinberg, 2001; Newell *et al.*, 2006). As discussed in Chapter 3, chimpanzee tooth crowns take a particularly long time to develop. Is there a relationship between the number of defects seen and the time involved in forming the tooth crowns – the size of the so-called 'window of opportunity' within which a given number of growth-disrupting events might be expressed as defects?

Primate teeth vary considerably in size, with chimpanzees, gorillas and orangutans being among the largest and the prosimians among the smallest. Similarly, there are variations in the packing of the perikymata, in their overall spacing and in the decrease between the occlusal and cervical parts of the crown side. Some primate genera seem to have considerably more prominent perikymata than others. It is also necessary to consider abrasion of the crown side, which can rapidly reduce the sharpness of definition of the perikymata, even when there is only a relatively small amount of crown height lost by occlusal attrition. There are undoubtedly variations in diet between primate taxa which would have a strong effect on tooth wear.

Williams (1897) was concerned to investigate the effect of defective enamel structure on the aetiology of dental caries. He wanted to establish a standard of normal enamel structure against which modern urban human teeth could be judged and hypothesised that it might be found in wild primates. On the contrary, in his microscope sections of gorilla, chimpanzee and orangutan teeth, he found abundant evidence for defective structures, as seen in human teeth. Sir Frank Colyer gathered a large odontological collection in London at the Royal College of Surgeons of England, including many primates, both from the wild and from captivity. He described a number of cases of enamel hypoplasia in apes and monkeys (Colyer, 1936; 1947). The second edition of his great book *Variations and Diseases of the Teeth of Animals* (Miles and Grigson, 1990) made the observation that the incisor crowns of baboons and orangutans often have an irregular surface with very slight transverse grooving which appears to represent exaggerated perikymata. This is an important point for the discussion that follows.

Schuman and Sognnaes (1956) took this idea further. They gained access to large collections of wild chimpanzee and rhesus macaque teeth, together with smaller numbers of teeth from gorillas, orangutans and gibbons. The chimpanzee incisors were commonly marked on their labial surfaces with furrow-form defects as were, less commonly, the molars. Many teeth showed irregularities of the perikymata and most showed accentuated developmental lines in enamel sections. A proportion of sections showed interglobular dentine, but this was not necessarily matched by hypoplastic defects. The crowns of the smaller number of gorilla, orangutan and gibbon teeth were, by contrast, relatively smooth, with uniform perikymata, but all showed some interglobular dentine even though clear defects of enamel hypoplasia were absent. The rhesus macaques showed no irregularities of any kind: no defects, no irregular perikymata, no pronounced incremental lines in enamel and no interglobular dentine.

In a large study, Guatelli-Steinberg and Skinner (2000) compared chimpanzees, gorillas and orangutans with monkeys in both West Africa and East Malaysia. Hypoplasia was assessed by simple inspection of the lower canines, in which the crown height was also measured. In Africa, the defects were most commonly recorded in chimpanzees and gorillas, with middling frequency in red colobus monkeys (*Colobus badius*) and white-collared mangabeys (*Cercocebus torquatus*), and least commonly in sooty (*C. atys*) and agile (*C. galeritus*) mangabeys. In Malaysia, the defects were most commonly seen in orangutans, with middling frequency in gibbons and macaques, and least commonly in colobine monkeys. The ape/non-ape dichotomy was therefore seen in two parts of the tropics, although the gibbons and monkeys showed substantial variation. Does this consistent difference result from features of ecology, diet and behaviour, or is it due to some feature of dental anatomy or development in the larger primates? Hannibal and Guatelli-Steinberg (2005) examined variation within genera, contrasting Bornean and Sumatran orangutans, lowland and mountain gorillas, and chimpanzees. Defects were recorded by observation with a hand lens and were more commonly recorded in orangutan teeth, followed by chimpanzee teeth, and were less common in gorilla teeth. Whereas there was little difference between Bornean and Sumatran orangutans, defects were much less frequently observed in mountain gorillas than in lowland gorillas. They suggested that one likely explanation lay in the diet of gorillas, which are primarily folivores, in contrast with orangutans and chimpanzees which are frugivores. In addition, they noted the slightly wider spacing of perikymata in gorilla tooth crowns.

Newell *et al.* (2006) devised a direct test of the relationship between the rate of development in different primate taxa and the frequency of observation of hypoplastic defects. Having previously noted a correlation in 20 species of non-human primates between the frequency of recorded defects and time of permanent first molar eruption, they examined teeth from five genera of New World monkeys: *Cebus* (capuchin); *Saimiri* (squirrel monkey); *Aotus* (night monkey); *Alouatta* (howler monkey); and *Ateles* (spider monkey). Only unworn or little-worn dentitions were used, with defects being matched between antimeres. They found that monkeys such as *Cebus*, with longer dental development schedules, did indeed have more recorded defects, apparently supporting the idea that a longer development time provides a longer span within which a growth disruption could be expressed as a defect. With scanning electron microscope observations of crown surface replicas, they found that *Cebus* had overall a closer spacing of perikymata than the other monkeys, that they were more sharply defined and there was a more pronounced change from occlusal-type to cervical-type perikymata. This might well have made defects more prominent and points to the possibility that differences in crown development morphology might well play an important role in the appearance of defects.

Guatelli-Steinberg *et al.* (2012) scored hypoplasia and measured perikymata spacing using light microscopy of the surface of replicas taken from relatively little-worn teeth from Bornean and Sumatran orangutans, chimpanzees, bonobos, lowland gorillas and mountain gorillas. They confirmed that defects were least common in mountain gorilla teeth and found that in all genera they were more common in males than in females. Canines are considerably larger in male great apes than in females and take longer to form. They took this to imply that the defects were 'accumulating' at a similar rate in the two sexes. There was, however, little evidence that lateral enamel formation time had much effect. Instead, in a small number of microscope sections, they found that defects were more commonly apparent in enamel in which there was a large angle between the brown striae of Retzius and the surface (page 97). This has the effect of making the perikymata more prominent and is a reasonable explanation for the prominence of defects.

Skinner (1986) noted that, in living great apes, the incisors and canines were often marked with repeated furrow-form defects that appeared to have a regular spacing. They were distinguished as rLEH (Guatelli-Steinberg and Skinner, 2000). Skinner and Hopwood (2004) made perikymata counts by scanning electron microscope examination of crown surface replicas to estimate the periodicity of defects. For orangutans from Borneo, they concluded there was a regular periodicity of six months and for orangutans from Sumatra, a periodicity of 12 months. They estimated the duration of the growth disruption causing the defects to be six to seven weeks in both cases. To do this, they had to make assumptions about the cross striation repeat interval (page 118). Perikymata counts were not available for chimpanzees and gorillas but, from measurements of the crown surface, they deduced that the defects were repeated at six-monthly intervals. They explained this rhythm by the regular climatic alternation between dry and rainy seasons. Further work by Skinner and Pruetz (2012) investigated this idea further by examining replicas taken from the lower canines of three chimpanzees from Fongoli in Senegal. This area is characterised by a long and very marked dry season and the canines of all three animals showed marked repeated defects. These appeared to represent approximately annual disturbances, lasting about six months each.

Macho *et al.* (1996) made microscope sections of molars from the fossil baboon *Theropithecus oswaldii*, found at Koobi Fora and Olorgesailie in East Africa, dated between about 0.7 and 1.9 million years ago. They recognised accentuated striae in the enamel (with associated pronounced lines in the dentine) and used cross striation counts to establish the interval between them. There were two cycles: a longer cycle of 132–138 days and a shorter cycle of 90–108 days. Once again, they suggested that seasonal variations in weather conditions might be the cause. Dirks *et al.* (2002) sectioned teeth from two male and two female anubis/hamadryas baboons from the Awash National Park in Ethiopia and from Uganda. They used cross striation counts to establish the crown formation timing for different teeth and also the timing of accentuated lines in the enamel. The chronology of development was established in relation to the neonatal line of the permanent first molars and in relation to the known date of death for these animals. This made it possible to determine the timing of the accentuated lines in relation to the seasons. They speculated on the possible triggers for the accentuated lines in the two females. The first lines after the neonatal line, at 54 or 59 weeks, could perhaps be matched with their mother's first mating period following their birth. After an interval there were repeated accentuated lines which might have represented menarche and the menstrual cycle. When plotted against the record of rainfall, some lines matched pronounced dry seasons, although they felt that it was difficult to tease apart the different factors.

Mystery of localised hypoplasia of the deciduous (primary) canine

A substantial number of deciduous dentitions show a remarkable hypoplastic defect of the canines. Sometimes just one, or perhaps two, three or all four are marked by a large, approximately circular, pit-form defect on the mesial part of the buccal/labial side of the crown (Figures 7.14 and 7.15). Often it is present when there are no signs of defects anywhere else on the affected tooth, or in any other tooth in the deciduous dentition. The defect, known as localised hypoplasia of the primary canine (LHPC), has been observed in humans, chimpanzees, bonobos and orangutans, but not gibbons (Lukacs, 1999; Skinner and Newell, 2003). The first substantial description was by Jørgensen (1956), who concluded that they were part of the normal morphological variation in tooth crown form. This is a reasonable conclusion if there is no evidence in the rest of the dentition for a systemic growth disruption. Skinner (1986) carried out a detailed examination of dentitions from 80 children from a variety of archaeological and modern contexts and found that, overall, 45% of individuals showed these defects and 24% of canine teeth. They ranged from large pits to tiny pinprick-sized defects. There was no significant difference in frequency between the left and right teeth, but lower canines showed the condition about twice as commonly as upper canines. Skinner's explanation was that the wall of the alveolar process holding the deciduous canine socket was particularly thin (even sometimes with an opening or fenestration), which might leave the developing tooth germ vulnerable.

Taji *et al.* (2000) studied large collections of dental casts and isolated deciduous teeth at the Dental School of the University of Adelaide in Australia. Nineteen per cent of isolated canines showed defects; 47% of the individual dentition casts showed defects and, of these, 64% had just one tooth affected, 26% two teeth, 8% three teeth and 2% all four teeth. Once again, there was no significant difference between left and right teeth, but the lower canines had substantially larger frequencies than upper canines. A few of the casts were from twins and it turned out that dizygotic twins had a higher concordance for possession of at least one canine with a defect (73%) than monozygotic twins (55%), which did not suggest a strong inherited component.

Skinner and Hung (1989) sectioned deciduous canines with LHPC defects and examined them under the light microscope. Their images make it clear that they are simply large pit-form defects, with exposure of the plane of a pronounced incremental line, showing Tomes' process pits which mark the cessation of matrix secretion by a substantial group of ameloblasts. The margins of the defects are defined by normal enamel matrix. As with other pit-form defects, their position on the crown surface gives little clue to the age at which they were initiated because they are associated with a buried incremental line. Skinner and Hung were, however, able to show that this line was positioned after the neonatal line, suggesting that the defects were initiated around the middle of the year after birth. This may provide part of the answer to the mystery. Human canine crowns form relatively late in the deciduous dentition. The incisor crowns are completed within the first 2 months or so after birth and third premolars by 3 to 4 months (Appendix A, Table 12), whereas canines continue to form into the latter part of the first year. No matching defects would be expected in the incisors or third premolars. A matching defect would, however, be expected in fourth premolars and this can be seen in the example given in Figure 7.14. This implies that LHPC is generated by disruption to development in a similar way to other defects of enamel hypoplasia. What remains unexplained is why it varies so much in appearance around the deciduous dentition.

Health costs of the adoption of maize agriculture in North America

Maize (corn) is from a family of tropical grasses and it now seems clear that, together with squash, it was domesticated and first farmed in Mexico around 9000 BP (Smith, 1997; Matsuoka et al., 2002; Ranere et al., 2009). Its arrival in the archaeological record of the eastern USA is marked by a change in the ratio between the two common isotopes of carbon, as determined by an analysis of archaeological human bone specimens. The tropical grasses to which maize belongs follows a photosynthetic pathway known as C4. Plants native to the eastern USA have a C3 pathway, which can be distinguished in the ratio of carbon isotopes passed along the food chain from the plants. On the basis of this evidence, the arrival and gradual adoption of maize can be seen, up until the point that it became a major item in the diet between AD 900 and 1000 (Schoeninger, 2009). There is an associated change in the material culture found on archaeological sites, implying a transition from migratory hunting-gathering-fishing to more settled agriculture.

Dickson Mounds is a prehistoric native American site in Illinois. It is a complex consisting of cemeteries and earth mounds, with associated settlement sites, excavated in the 1930s and 1960s (Goodman and Armelagos, 1985). The approximately 3000 burials at the site were divided into three cultural groups: Late Woodland (in this area about AD 950–1100); Mississippian Acculturated Late Woodland (MALW; about AD 1100–1200); and Mississippian (about AD 1200–1300). The Woodland cultural grouping is characterised archaeologically by evidence for the occupation of seasonal camps with a largely hunting and gathering economy. Mississippian culture represents predominantly an agriculturalist way of life, with permanently occupied settlements; some large, with organised rows of houses and a plaza (Goodman et al., 1984b). Despite this Mississippian acculturation, a study of the morphological variants of teeth at the site suggested continuity of occupation, rather than migration from the main area of the Mississippian culture further south. This made Dickson Mounds an interesting assemblage in which to study the effects on health of the establishment of permanent settlements, their increasing size, the growing reliance on agriculture for the supply of food and a rise in the exchange of goods with the main Mississippian centres. In effect, this became the laboratory in which Goodman et al. (1984b) were able to test their concept of stress (page 201). They defined ten indicators of stress:

1. Decreased pace of growth in limb bone length in children;
2. increased level of sexual dimorphism in the adult pelvis and femur;
3. increased prevalence of Harris lines in tibia X-rays;
4. increased prevalence of enamel hypoplasia;
5. increased prevalence of Wilson bands;
6. increased prevalence of porotic hyperostosis in the skull;
7. increased prevalence of bony lesions of infectious disease;
8. increased prevalence of bone lesions deriving from trauma;
9. increased prevalence of joint disease;
10. increased mortality in younger age groups.

They saw their indicators as evidence of a decreased ability of people to adapt in a biological way to the new situation in which they found themselves. In the Dickson Mounds assemblages all their indicators, except Harris lines and the degree of sexual dimorphism, increased from Late Woodland to MALW and finally Mississippian contexts. Their conclusion was that the people of Dickson Mounds suffered as a consequence of their adoption of agriculture: they paid a heavy price for 'capturing more energy through economic intensification' (Goodman et al., 1988, p. 181). This idea has become a common thread that has run through much of bioarchaeology since the 1980s (Larsen, 1995; 1997).

Goodman et al. (1980) gave a more detailed account of enamel hypoplasia in 111 Dickson Mounds permanent dentitions. The positions of defects, defined as transverse areas of depressed enamel, were measured with calipers relative to the CEJ. They used these measurements to assign each defect to a six monthly stage, based on Swärdstedt's chart (page 179) and, for each individual, counted the number of stages in which they had observed a defect. The mean of these counts was lowest in dentitions from Late Woodland contexts and highest in Mississippian, with MALW in between. The trend was the same if teeth at all stages of wear were included, or if just the less-worn teeth were included. The interval between development stages with defects was more often whole years than half-years and from this they suggested that the growth disruption might have had an annual basis. When the frequencies for defects in the different stages were plotted (Goodman *et al.*, 1984a), it was found that for Late Woodland and MALW combined the mode was between the 2 and 4 year stages. For Mississippian it was between 1.5 and 3.5 years and the peak frequency was higher. Goodman *et al.* (1984b) proposed that the timing of the peaks suggested that they were related to weaning and that perhaps the weaning diet was more deficient during Mississippian times.

Age-at-death was estimated for each individual by the usual combination of indicators such as dental eruption, tooth wear, epiphyseal fusion in the skeleton and changes at the pubic symphysis (Goodman *et al.*, 1984b). Mean age-at-death was calculated separately for three groups of individuals: no defects; one defect; two or more defects (Goodman and Armelagos, 1988). It was found that, in MALW and Mississippian assemblages, the mean age-at-death was lowest in the 2+ defects group and highest in the no defects group. Late Woodland assemblages did not show this pattern of differences. The implication was that higher frequencies of defects were associated with greater morbidity.

These were important studies, at the heart of bioarchaeology. The defects of enamel hypoplasia were always considered to have the most potential of all the 'stress indicators' and Dickson Mounds was fundamental in the design of later studies. As discussed in Chapter 7, however, there are serious difficulties with the methodology in addition to the problem of Swärdstedt's chart. For example, the apparent higher frequencies for defects in Mississippian teeth might be due to less severe tooth wear or fewer old individuals with more worn teeth, or a different combination of teeth being preserved. The lower mean age-at-death amongst individuals showing defects on their teeth would be expected, because older individuals have more worn tooth crowns on which it is more difficult to see them. The peak frequencies of defects between 2 and 4 years would be expected because this is when the mid-crown of the permanent canines and incisors is formed and defects tend to be at their most prominent when they involve mid-crown perikymata. It is also likely that occlusal attrition would remove the parts of the crown that formed at younger ages. It is therefore very difficult to judge the conclusions from the published results, which could be explained in a variety of other ways.

The published figures in these reports show that at least a proportion of Dickson Mounds teeth were substantially worn. The only way to examine development in such specimens is to cut sections and examine them under the microscope. Rose (1977) and Rose *et al.* (1978) sectioned 87 permanent lower canines from Dickson Mounds and the Gibson Mounds site, also in Illinois. The Dickson Mounds material was from Mississippian and MALW contexts, whereas the Gibson Mounds material represented Middle Woodland contexts. Rose divided the sections along the EDJ into eight six-monthly development units, from 0.5 to 4.4 years of age, estimated from Massler *et al.* (1941). Within these stages, he identified accentuated lines, or Wilson bands (page 174) within the enamel, using a strict definition. In most of the teeth, the youngest units were missing as a result of dental attrition and Wilson bands were most common in the middle two-thirds of the crown side (between 1.5 and 3.5 years of estimated age). Seven per cent of the development units in the Mississippian teeth showed Wilson bands, compared with 4% of MALW and 2% of Middle Woodland development units. There were differences in Wilson band distribution between the units. In the Middle Woodland teeth they were fairly evenly distributed between 0.5 and 4 years of age; in MALW they peaked at 2.25 years and in Middle Mississippian at 1.75 and 3.25 years. In all three assemblages, the mean estimated age-at-death was lower in those individuals with Wilson bands than in those without. On the face of it, this seems to confirm the results of the study of hypoplasia at the crown surface. It must be said, however, that Wilson bands are not well understood and that a more detailed study using cross striation counts to create a schedule of Wilson bands, less strongly accentuated lines and hypoplastic defects would have helped to clarify this issue.

Settlers and missionaries: the effect of European colonisation on the New World

After the voyages of Christopher Columbus between AD 1492 and 1504, the Americas were progressively explored and settled by the Spanish. North America was colonised by Catholic missions to California, New Mexico, Arizona and Texas, the Carolinas, Georgia and Florida. Their territory of *la Florida* was much larger than the present day state and the occupation started with the establishment of the port and garrison town of St Augustine in AD 1565, primarily to protect shipping sailing up the Bahama Channel to take treasure from the Americas home to Spain. Supply for the garrison was always a problem because there were not enough colonist farmers and the Spanish sought to assimilate the settled native American agriculturalists into the colony to supply labour and food, especially maize. The driving force of this assimilation between 1587 and 1706 were Franciscan missionaries (Worth, 2001). They established missions consisting of a church, a convent and a barracks. Their impact was profound: a spread of epidemic disease, overwork, under-nutrition, intensification of maize agriculture and raids from chiefdoms outside the missions. It is no wonder that many Native Americans fled the missions to areas outside Spanish control. The population of the Spanish colony fell to a fraction of its original level. It seems highly likely that native American children under the control of the missions would suffer frequent disruptions to their growth, but were they more frequent than in children who grew up before the arrival of the Spanish?

Hutchinson and Larsen (2001) studied enamel defects in 510 dentitions from 17 assemblages of the Guale native American group on the coast and islands of modern day Georgia, then part of *la Florida*. They were able to divide them on archaeological grounds into Early Prehistoric (400 BC–AD 1000), Late Prehistoric (AD 1000–1550) and Early Mission (1600–1680) groups. In addition, they studied 262 dentitions from another 17 assemblages of Guale, Timucua, Apalachee and

Yamasee people, from the coasts and islands of modern day northern Florida, divided into Early Prehistoric (AD 0–1000), Late Prehistoric (1200–1600), Early Mission (1600–1680) and Late Mission (1680–1700). Only permanent incisors and canines were examined. Teeth judged to have more than the incisal one-third of crown height worn away by attrition were excluded on the basis that the middle one-third was more likely to show defects (Hutchinson and Larsen, 1988, but see discussion below). The identification of defects was carried out under a stereomicroscope, but no definition was given of the level of crown surface disruption required. The widths of defects (occlusal to cervical) were measured with a micrometer.

Altogether 772 individuals were represented by the teeth and the majority of their dentitions showed at least one defect in at least one tooth. When individual teeth were considered, the proportion with observed defects varied widely between teeth, but, in general, the Early Prehistoric Georgia assemblages showed a larger proportion than the later assemblages, and the Prehistoric Florida assemblages had a larger affected proportion than the Mission assemblages. The mean widths of defects were lower for Early Prehistoric Georgia assemblages than later but, for the Florida assemblages, the Prehistoric showed higher values than the Mission period. Taken together, these results do not suggest that the Spanish missions had a strong effect on the growth of the native American children studied.

This was a careful study, but there are still methodological issues as discussed in Chapter 7. The results tables show considerable variation between groups in the proportions of different teeth preserved. Even though the most worn tooth crowns were excluded, tooth wear is likely to have had an important influence on the defects recorded. Once again, cutting sections and examining them under the microscope is the only way to avoid this problem. Simpson (1999; 2001) sectioned 143 permanent lower canines from Florida, divided into Early and Late Prehistoric, Early Contact (non-mission), and Early and Late Mission groups. Rather thick (250–350 μm) light microscope sections and polished blocks were made for examination in the scanning electron microscope. He identified pathological striae or Wilson bands using definitions from Rose (page 174) and measured their position along the EDJ from the tip of the dentine horn under the cusp. Eighty-three per cent of canines from the Mission groups showed pathological striae, as compared with 48% in the Prehistoric groups. There were similar numbers of pathological striae per tooth, with a similar mean position on the EDJ. In contrast with the study of hypoplastic defects, this suggested a substantial effect, but again counts of cross striations would have allowed more detailed individual sequences to be investigated.

The European colonisation of Australia included not only conflict with indigenous people and the spread of epidemic disease, but also the taking of land for pastoral farming and social disruption. The example given here comes from the Yuendumu settlement in Central Australia. This is in the land of the Warlpiri Aboriginal people which was initially settled by Europeans in the 1880s, but saw a large expansion of pastoral farming and mining in the 1920s. The Warlpiri lived a hunter-gatherer lifestyle, but the expansion of European settlement coincided with a period of drought which caused many to move near the farming stations and mining camps. They worked on the farms and, from the 1930s, food supplements were issued which further attracted settlement around the distribution points. In 1953 a school and clinic were added. From the start, the living conditions, nutrition, water supply and health of people in these settlements were poor. Paradoxically, they became worse during the 1950s after the Australian government formally established the settlement at Yuendumu.

Between 1950 and 1970, annual visits from the University of Adelaide Dental School formed part of a study of growth and dental health. Dental impressions were taken as part of this and dental stone casts made from them. Littleton and co-workers (Littleton and Townsend, 2005; Littleton, 2005; Floyd and Littleton, 2006) scored defects of enamel from the casts in 377 individuals, the oldest born in 1890 and the youngest in 1960. These were divided into six birth cohorts: 1890–1929; 1930–39; 1940–4; 1945–9; 1950–4; 1955–60. The defects were recorded using the DDE Index (page 177) and were only scored as positive if they could be matched between left and right sides. Crown surface development for the whole permanent dentition was divided into seven stages or units, based on the ideas of Hillson (1992c; 1996; 2000), with adjustments for the figures of Reid and Dean (2000). A was the youngest stage and G the oldest. Defects were matched between teeth to assign them to one of these units. An attempt to control the impact of tooth wear was made by excluding those dental casts in which fewer than six of the seven development units could be scored.

When the frequencies of defects per development unit were plotted for each age cohort, it was apparent that people born in the 1890–1939 cohorts had lower frequencies for all development units than the later cohorts, particularly those after 1945. This increase was particularly apparent in the younger development units A–D. The percentage of individuals in each cohort with three or more units affected by defects increased by several times from 1890 to 1950. This would fit well with the idea of worsening conditions in the area of Yuendumu during this period, but it is necessary to sound a note of caution. Littleton and co-workers were concerned about the effects of tooth wear and tested its relationship with defects. They also observed, however, that the tooth wear was greatest in the 1890–1939 cohorts. It would be expected that occlusal attrition would remove development units A and B from first molars and first incisors in older individuals, and this could well have contributed to the lower frequencies of defects in these units. In addition, abrasion of the surviving crown sides would be greater in older individuals and this might well have reduced the prominence of defects, which could have contributed to the lower frequencies in units C and D. It is very difficult to assess the likely effects of this, but the teeth of older Aboriginal Australians illustrated in, for example, Campbell (1925) are very heavily worn indeed.

Social change in the Nile valley

Ancient Egypt was the part of the Nile valley with a wide flood plain downstream from the First Cataract at Aswan, and Nubia was the part between the First and Sixth

Cataracts. The valley in Nubia was narrower and rockier, with less cultivable land. It has also been suitable for reservoir and dam projects in modern times, which have led to large international survey and excavation projects in advance of the flooding. These have yielded some of the most studied assemblages of human remains from the large cemeteries, but Egypt too has produced important assemblages, some of which date back to the work of early pioneers such as Flinders Petrie.

Egypt is divided into the Delta region, or Lower Egypt, which contains huge areas of cultivation combined with the largest part of the population, and Upper Egypt, which includes everything upstream from Cairo. One of the most intriguing social transitions in the archaeological record is the unification of the kingdoms of Upper and Lower Egypt by about 3100 BC, with the imposition of strong central control by dynasties of rulers whose names were recorded in writing for the first time. It has been suggested that this was associated with a concentration of population around ceremonial centres and an intensification of irrigation agriculture (Trigger *et al.*, 1983; Hoffman, 1984). Certainly the hugely expanded range of material culture suggests a surplus which allowed trade and the employment of large groups of artisans to supply luxurious items for the elite. Did this lead to a change in the conditions of childhood which is expressed in evidence for disruptions to dental development?

Hillson (1979) recorded defects in five developmental units of crown surface formation in permanent canines, first molars and second molars. Age limits for the developmental units had been estimated from cross striation counts in microscope sections of extracted teeth from a London dental hospital. Three Predynastic (pre-unification) cemeteries (Badari, El Amrah and Abydos) were compared with Sedment, a IX Dynasty site dated about 2000 BC) and Hawara, a Roman period site dated AD 100–200. A relatively small proportion of these developmental units had recognisable defects in the Predynastic sites, although there was some variation. Both the later sites, Sedment and Hawara, had large proportions affected, particularly of mid-crown developmental units.

Starling and Stock (2007) examined some of the same material: Badari and Naqada were both Predynastic cemeteries, whereas Tarkhan was an early Dynastic cemetery dated 3100–2686 BC. Furrow-form enamel defects were recorded using the DDE Index definition of horizontal grooves (page 177) for different tooth types separately, but without dividing into developmental units. Defects were much more commonly observed in incisors and canines than in premolars and molars. The two Predynastic sites in this study had a higher proportion of teeth affected than the Dynastic site. Different methodologies and combinations of sites produced different results and it is difficult to conclude that there is strong evidence for a contrast in defects with the unification.

A much later Christian site at Kulubnarti in Nubia was the focus of a similar approach to that described above for Dickson Mounds (van Gerven *et al.*, 1990; 1995). Kulubnarti is an island on which there were several Christian settlements that survived until the arrival of the Ottoman Turks in the 16th century. From the 14th century houses became strongly fortified, suggesting a marked social change into late Christian times. Two cemeteries were excavated. One on the island was early Christian (AD 550–750) and the other, on the main Nile bank, was mostly late Christian. Enamel hypoplasia was one of a number of stress indicators (page 216) recorded, which together suggested higher levels of stress in the early Christian assemblage than in the late Christian. The proportion of canines affected by defects was not greatly different in the two sites, but the modal age of growth disruption estimated as at Dickson Mounds from measurements and the Swärdstedt chart was 3–5 years for the early Christian and 4–6 years for the late Christian cemetery. As discussed on page 180, there are difficulties with this method, but it implies the defects were more cervically placed on the late Christian canines and more mid-crown in the early Christian. This is not a large difference and illustrates the challenge involved in making comparisons of this type.

Deciduous teeth at Isola Sacra

The Isola Sacra necropolis was used for burials of people from *Portus Romae*, one of the great ports of Rome at the mouth of the river Tiber, between the 1st and 3rd centuries AD (Rossi *et al.*, 1999). It was a city of merchants and administrators as well as workers related to shipping and the activities of the harbour. Goods were transferred to and from the ports along the Tiber to Rome. The site was originally excavated between 1925 and 1940, and then again in the 1970s.

Overall, human remains representing approximately 2000 individuals were recovered from the site. Of these, 334 were infants, children or adolescents. A total of 127 had deciduous teeth with sufficient enamel remaining after wear for them to be sectioned for examination under the light microscope (FitzGerald and Saunders, 2005; FitzGerald *et al.*, 2006). Deciduous teeth can be challenging to work with because prism cross striations and brown striae of Retzius are often difficult to discern consistently, but their advantage is that a prominent neonatal line is usually clearly visible. This means that all structures in the enamel can be related to it. Perikymata are not prominent on deciduous tooth crowns and hypoplastic defects are relatively uncommon, other than those associated with the neonatal line. In deciduous enamel sections it is the accentuated lines or Wilson bands that are most noticeable, rather than a regular sequence of brown striae. It was therefore logical to study growth disruption on the basis of accentuated lines with a chronology determined by counts of prism cross striations from the neonatal line. Their definition of Wilson bands was that they were prominent striae of Retzius which were visible for at least 75% of the enamel thickness out from the EDJ.

Of the 127 individuals studied, 40% had at least one Wilson band in one of their teeth and, of the 274 teeth sectioned, 23% had at least one Wilson band. Statistical treatment was difficult because not all individuals had the same teeth preserved and they were in varying states of development in different individuals. Post-natal enamel development for each tooth was therefore divided histologically into monthly (for some purposes two-weekly) developmental units. A fully formed deciduous canine or fourth premolar crown covered 13 post-natal months of enamel development.

Incisors included 6–7 months and third premolars 10 months. Wilson bands were identified most commonly in the middle developmental units of each tooth. The maximum frequencies were considerably higher in canines and fourth premolars than in other teeth. The reason for this was not clear; perhaps it was due to some real variation in the activity of ameloblasts at different phases in crown formation, or perhaps it was because brown striae were more difficult to see in the cuspal and cervical enamel.

It was necessary to make sure that one episode of disruption which may have initiated Wilson bands in several teeth sectioned for one individual was counted only once for that individual. Additionally, it was necessary to take account of the varying 'susceptibility' of different tooth types. For this reason, in each monthly unit, they used the highest prevalence of Wilson bands found in any of the tooth types. From low values in the month after birth, the prevalence rose to around 55% in the second month. This was maintained until the fifth to ninth months, during which it rose to 80%, and then fell to 40% in the last months of the first post-natal year. The age-at-death of the children had been estimated on the basis of their state of development in the skeleton and dentition. They found that for children in which they had not identified Wilson bands, a relatively larger number died during the first year after birth and a smaller number during the sixth to tenth years. The distributions were not significantly different and, as FitzGerald and co-workers pointed out, the great majority of the children in the study in fact survived their first year.

Detailed life histories of identifiable individuals in assemblages from Victorian London

London has two collections of closely documented human remains which have been very important in the development of methods in biological anthropology. Both include burials in church crypts from the eighteenth and nineteenth centuries and the associated records provide independent evidence of the occupants' sex, date of birth and age-at-death, as well as a variety of other information. St Bride's church in Fleet Street was one of many to be rebuilt by Sir Christopher Wren following the destruction of the Great Fire of London in AD 1666. Burials took place in the crypt of the new church until it was sealed in the 1850s. Excavation of the crypt took place after the church was badly damaged in the Second World War. In total, 227 of the individuals buried there were associated with coffin plates and parish records which gave details of their age-at-death, sex and other information (Scheuer and MacLaughlin-Black, 1994; Scheuer, 1998). Christ Church in Spitalfields, also in London, was built by Nicholas Hawksmoor in the early 1700s as part of a campaign of church building in the East End of London, where many immigrants settled. These included the Huguenots from France, who brought with them their silk weaving industry and, having arrived with nothing, became prosperous. Christ Church was their parish church and they make up a large proportion of the burials that took place in the crypt until 1867. The crypt was excavated in the 1980s and the assemblage of skeletons from it have proved an important resource for anthropology because, once again, coffin plates and parish records give biographical information for 389 individuals (Molleson et al., 1993; Reeves and Adams, 1993; Cox, 1996).

The Spitalfields collection has already been mentioned in relation to a test of cross striation periodicity (page 117). Hillson et al. (1999) further described two minor furrow-form defects in a permanent first molar and first incisor from a boy buried in the crypt at Christ Church, Spitalfields in London. He was born in May 1820 and died on 14 December 1822 aged 2 years 6 months 3 weeks. The permanent first molar and first incisor were sectioned. By counting cross striations from the neonatal line it was possible to establish that one defect was initiated at approximately 1 year 7 months of age during December 1821. The occlusal wall of the defect included two widely spaced perikymata and thus occupied 18 days, given the cross striation repeat interval of nine days. A second defect was initiated at approximately 1 year 9 months of age during January 1822 and, again, the occlusal wall occupied 18 days. In both cases, the floor of the defect contained closely spaced perikymata in comparison with the overall trend down the crown side. It was possible to resolve details of the defects and, in particular, it could be seen that there was no interruption to the regular rhythm of either the cross striations or the brown striae of Retzius. Just below the surface, there was no consistent variation in the spacing of brown striae, although those associated with the start of both defects were slightly more prominent. The increase in perikymata spacing within the occlusal wall was associated with more prism boundaries reaching the surface, showing that a wider band of ameloblasts ceased matrix secretion. The narrower perikymata in the defect floors were not associated with a decrease in the width of the band and the mechanism for that must lie in the spacing of the brown striae, just before they meet the surface. It was not possible to measure this.

Teeth from another boy buried in the Spitalfields crypt showed a complex pit-form defect (Hillson and Antoine, 2011). Parish records showed that he was born on 1 February 1778 and died on 15 May 1782, aged 3 years 4 months 15 days. Large defects occupied the cuspal third of the permanent first molar and incisor crowns (Figure 7.10), but sections showed that they were caused by a disruption affecting just one brown stria at just over 1 year of age in February 1779.

Another example from the Spitalfields crypt illustrates the potential of studies based on perikymata counts alone (King et al., 2002). The teeth in this study came from a girl born on 5 January 1762 who died on 30 January 1777, aged 15 years 25 days. Dental impressions were taken and epoxy replicas were imaged in the scanning electron microscope. Thirteen furrow-form defects were identified on the surfaces of the permanent tooth crowns, including first and second incisors, canines, third premolars and first molars. It was possible to match these in the different teeth by the counts of perikymata within and between defects. Altogether, out of 193 perikyma grooves counted from the first to be seen on the second incisor to the completion of the canine crown, 25% were involved in one or other of these defects. The teeth were not sectioned, so the cross striation repeat interval was not known.

By assuming an age of 1.1 years for the first perikyma on the first incisor it was, however, possible to estimate the approximate timing for the sequence of defects.

King et al. (2005) made a further study of ten adolescents and young adults from the St Bride's crypt and 17 similarly aged individuals from the Spitalfields crypt. Matched sequences of defects were made for each individual. Measurements of perikyma groove spacing using an engineer's measuring microscope (page 183) were used to define the trend down the crown side and furrow-form defects were defined on the basis of departures. Disruption to growth was measured in each individual by the percentage of the total count of perikyma grooves in the matched sequence taken up by defects. Individuals from St Bride's showed higher percentages than those from Christ Church, but the ranges overlapped considerably. Females showed higher values than males, but there was an *increase* amongst older individuals. As the younger individuals were almost entirely boys, the sex difference may be an expression of the age trend. The age relationship was not strong, but it was nonetheless there to see and it is opposite to the relationship reported in other studies. The estimated age at which defects were initiated was in the range 2–4 years.

Summary

Referring back to the question outlined in Chapter 1, have the defects of dental enamel suggested that the key transitions of human evolution, social and cultural development had an impact on the health and well-being of children? The published evidence reviewed here does not really provide a clear answer and it may be that the question as phrased cannot be answered. As discussed in this chapter, one reason for suggesting this is the difficulty of measuring health, particularly in an archaeological context. The position of enamel hypoplasia within this discussion is difficult to define but, as shown in Chapter 7, there is strong experimental and clinical evidence that the defects seen at the crown surface are related to disease, malnutrition, social and other factors known to affect many aspects of growth. Leaving aside the question of health, it therefore seems reasonable to ask simply whether or not the defects are seen either more or less frequently in dentitions from contexts which are placed during or after major transitions in the fossil and archaeological record.

The results of the investigations summarised in this chapter are, however, inconsistent. One example is the much-discussed Neanderthal to modern human transition. Several large studies of enamel hypoplasia have been carried out but, whereas some of these suggest that Neanderthals showed defects more commonly than early or recent *H. sapiens*, others do not. The same is true of *Australopithecus* and *Paranthropus*, between which some studies have suggested differences, whereas others have not. Part of the difficulty is that the results have been based on assessment and scoring of the defects with the naked eye or modest magnification, with all the problems outlined in Chapter 7. If the dividing line between 'defect' and 'normal' cannot be defined with absolute clarity, then different studies are not directly comparable. Size and other morphological differences between teeth, as suggested in wider studies of primates, are also likely to have an effect on the numbers of increments present and therefore the width of the window within which defects would be recordable. Tooth wear, prominent in all this material, also has a strong effect on the preservation of defects, as does the differential survival of different teeth in the dentition.

The logical alternative when teeth are worn is to examine the disruptions to enamel development seen in microscope sections as accentuated lines. Substantial numbers of sections have been cut from a number of archaeological collections. So, for example, it is possible to suggest from this evidence that the major social and cultural changes accompanying the transition from Woodland to Mississippian cultures in Illinois were associated with more frequent disruptions to dental development. In the same way, the enforced settlement of Native Americans around Spanish missions on the east coast of North America seems to have been associated with a greater proportion of teeth showing marked incremental lines in their enamel.

There have been strenuous attempts to define criteria for identifying accentuated lines or Wilson bands and distinguishing them from 'normal' lines. The real difficulty here lies in interpretation. To date, no clinical studies have shown a direct link between prominent enamel lines and disease, nutritional deficiency, or any of the other factors which have been linked with the defects seen at the crown surface. The strongest evidence is that of the study of Schwartz *et al.* (2006) of a captive gorilla where the accentuated lines match events in the daily diary account of its life, including what seem, on the face of it, to be relatively minor emotional disturbances. This encapsulates the problem, both for surface defects and incremental lines seen in section. They appear to be initiated by a wide range of factors from trivial to lethal and transient to long term. There seems to be little in the appearance of the defect itself that directly indicates the severity of disruption. In comparison with the indicators of health (or *un*health, see page 199) used for living people, the causes of the defects are just too vague to be useful in that way.

The gorilla study, however, seems to suggest the way forward. It is very difficult to standardise the recording of crown surface defects in large archaeological assemblages, including dentitions in which teeth are variably preserved and worn. The defects are just too complicated to score reliably in any simple way, so any apparent differences between contexts, sites or assemblages could be explained by a whole variety of factors not related to actual differences in growth disruption. In addition, defects which look similarly marked in different individuals do not necessarily reflect the same level of disturbance; at least, this has never been proved. Instead, this review suggests that the answer is to embrace the complexity of enamel hypoplasia, rather than trying to simplify it to produce an 'all or nothing' diagnosis in a large number of dentitions. Both defects and internal lines can be used instead to generate highly detailed development sequences for single individuals. This is, in any case, the only realistic option for much of palaeoanthropology, where most fossil finds represent single, or few, individuals. Even for a large archaeological collection,

Summary

when the most suitable individuals for examination are selected out, the numbers are considerably reduced. An approach which can yield much detailed information about a few individuals, rather than imprecise information about many, is actually an advantage in these circumstances.

A lot of work still needs to be done on the building of defect sequences before much progress can be made with enamel hypoplasia. This may seem a bleak conclusion given the large volume of published data, but it is meant instead to encourage a wide range of new approaches. The final case studies reviewed in this chapter are intended to suggest alternative methodologies. At the highest level of detail, a combination of crown surface replicas and sections provides a sequence related to chronological age in which the nature of the defects can be established in relation to the ameloblasts which were disturbed. Where teeth cannot be sectioned, either direct observation of uncoated tooth crowns in an environmental scanning electron microscope or a conventional scanning electron microscope examination of a coated epoxy replica makes it possible to establish counts of perikymata in relation to key stages of dental development, such as crown completion. Simple proportional measurements of perikymata spacing over small areas of the crown surface can be used to test whether or not a disruption can be classified as a defect, according to an agreed rule. Similarly, so long as perikymata counts are used to express the duration and separation of defects as proportions of the crown surface formation period, it is not necessary to make assumptions about the periodicity of perikymata to compare sequences between individuals. All this can be accomplished with a relatively ordinary scanning electron microscope which, as explained in Chapter 7, is still by far the best way to image the crown surface and defects.

9 Conclusions

Dental histology has never been in a better position to address questions about the pattern and rate of development in human evolution. Its basis, the circadian rhythm of short-period incremental structures in enamel and dentine, is now well established and widely accepted.

The traditional methods of sectioning and light microscopy still yield the most detailed sequences and current procedures for counting and measuring along the EDJ allow the difficulties of tooth wear and the complex weave of enamel structure to be circumvented. However, not only is sectioning destructive, but it is a very difficult technique to master and much practice is needed to obtain good results. Few people in the world can do it well. Similarly, the effective use of a polarising microscope for dental histology is not widely taught and the two most common reasons for difficulties in counting the incremental structures required are a poorly centred plane of section or an incorrectly adjusted condenser lens. Experience is also necessary for a clear understanding of what is actually being seen in the microscope image. Sectioning and light microscopy will, however, remain the main way in which data for recent humans and other primates are collected for the foreseeable future; for example, from the large collections of extracted teeth available in dental hospitals.

Alternative techniques are available for fossils or archaeological specimens where sectioning cannot be contemplated. Scanning electron microscopy or confocal optical imaging of replicas or original uncoated crown surfaces make it possible routinely to count and measure perikymata. This has the disadvantage that the periodicity of the perikymata and the duration of cuspal enamel formation are unknown. It is nevertheless possible to investigate lateral enamel development on a comparative basis, and it is also now possible to establish periodicity through confocal light microscopy without sectioning and to use synchrotron radiation micro-CT techniques to image non-destructively the whole volume of enamel. The former requires the use of a rare instrument, the portable scanning confocal optical microscope, and the latter requires a large high-energy physics facility with staff experienced in the appropriate type of imaging. Given the costs of these techniques, neither is likely to be used to image a large number of fossil specimens, but both are nonetheless yielding important results in key areas of discussion. They will continue to make it possible to carry out investigations which would otherwise be impossible.

Conclusions

The review of published studies in this book confirms that, even taking into account our large size, humans develop more slowly than any other living primate. Our sequence of dental development is remarkably long, with the apices of permanent third molar roots completed on average at 20 years of age, compared with just 12 years in chimpanzees. The position of the first molar in the human sequence is also distinctive in that it erupts at a similar age to our permanent incisors. In every other primate for which there are data, the first molar erupts considerably in advance of the incisors. The histological evidence assembled for fossil primates to date suggests consistently that the Pliocene hominins had a pattern of dental development unlike that of living humans and, although it has no exact modern analogues, more like other primates in its shorter timescale. Early Pleistocene fossils assigned to *Homo* also appear to have developed their teeth and skeleton to a faster schedule even if, once more, the development pattern seems unlike that of any living hominin. Assuming that these taxa were the makers and users of the first stone tools, from 2.6 million years ago, it appears that the associated cognitive abilities were developed over a shorter period than they are in living humans. It therefore also seems reasonable to suggest that these cognitive abilities may not have been so complex and, if this is so, it is likely that there are no good analogues in living primates for the associated social and cultural context.

The first taxa to have a slow dental development sequence similar to that of living humans were associated with the much more varied Middle Palaeolithic or Middle Stone Age tool assemblages which appeared in the archaeological record after 250 000 years ago. Fossils assigned to both *H. sapiens* and *H. neanderthalensis* have been found in association with these assemblages, although there is currently discussion about whether or not both taxa had a slow tempo of development, or perhaps some aspects of *H. neanderthalensis* development were more rapid. At the time of writing, this is the main focus of debate, particularly as it has proved possible to use synchrotron radiation micro-CT to image histological details of important fossils for which it would be impossible to cut microscope sections.

The part of the sequence where the elongation of human development is most striking is in the permanent molars. Not only is it seen in the late completion of the third molars, but also in the lack of overlap between the molars in the formation timing for their crowns. Many Pliocene hominin fossils represent relatively young individuals, so the first part of the molar series is the best known. The latter part is more difficult because there are currently few fossils with developing second and third molars. There may, of course, be exciting new finds which fill the gap, but it is also possible that new non-destructive imaging techniques could be applied to several fully formed molars from single dentitions, making it possible to establish the degree of overlap in molar crown formation. The third molar is a particularly interesting tooth in human evolution because the reduction in its size is one of the most striking trends in the genus *Homo*. It is associated with a more general reduction in the size of the whole dentition, jaws and face, but it is proportionately greater in the third molar. This tooth would therefore make a good focus for future research.

It is well established that disruption to the even course of dental development can be seen in accentuated lines within tooth sections and in defects of enamel hypoplasia at the surface. As described in this book, the basis for these features is reasonably well understood at a histological level, but their complex variation still requires a lot of research. Enough is known, however, to show that they are far more complicated than many studies have taken into account and the relationships between the causes and forms of defects are simply unknown at the present time. In spite of this, it is often assumed that the surface defects represent a single pathological condition that can be diagnosed by simple examination of tooth crowns. Large data sets have been collected, both in clinical and in archaeological contexts. It has been possible tentatively to link prominent defects with severe deprivation and malnutrition in clinical studies of large groups of people, but it has never been possible to match defects with identifiable episodes in an individual clinical history. For the accentuated lines within the enamel, there is just one example in which they match exactly the events in a detailed daily log book. These include, however, trivial events as well as serious ones, which calls into question the use of the lines as indicators of growth disruption for fossil individuals in which there is no log book check.

Close examination of crown surface defects shows clearly that those which can be seen with the naked eye are just the most visible part of a spectrum which includes microscopic perturbations. Small wonder that students worry about the definitions of 'defect' and 'normal' when they start. It is extremely difficult to standardise and this must account in a large part for the variation between studies which has been shown in this review. Added to this are the action of tooth wear in removing evidence of defects and the fact that the same type and size of defect is more prominent on some parts of a crown surface than others. Similarly, a defect that appears to the eye to occupy half of crown formation may on closer inspection represent a single day's disruption within otherwise relatively normal development.

There really is no way in which such a complicated phenomenon can be simplified enough to allow rapid recording of large assemblages of dentitions. The review presented in this book suggests instead that a detailed examination of the development sequence of individuals is likely to be more rewarding for research in the future. Studies which combine sectioning with crown surface examination are likely to provide the greatest detail and the most well-established chronology, but surface examination alone can still provide useful information. The techniques outlined here allow a relatively simple comparison by proportional measurements using commonly available equipment. Looked at in this way, the archaeological collections of the world show a fascinating variety of defect forms and chronologies. It is even possible to contemplate studies of living people with known clinical histories based on impressions of their teeth. The key to all of this is careful microscopy and detailed dental histology.

Appendix A: Tables

Table 1 Begun classification of Hominidea

Magnafamily Hominidea
  Superfamily Proconsuloidea
    Proconsul, Afropithecus, Heliopithecus
  Superfamily Hominoidea
    Family Hominidae
      Subfamily Griphopithecinae
        Griphopithecus, Equatorius, Nacholapithecus, Kenyapithecus
      Subfamily Homininae
        Tribe Dryopithecini
          Dryopithecus, Hispanopithecus, Rudapithecus, Ouranopithecus
        Tribe Hominini
          Subtribe Hominina
            Homo, Australopithecus, Paranthropus, Ardipithecus, Sahelanthropus, Orrorin, Pan
          Subtribe Gorillina
            Gorilla
      Subfamily Pongidae
        Pongo
      Tribe Sivapithecini
        Sivapithecus, Ankarapithecus, Gigantopithecus
      Tribe Lufengpithecini
        Lufengpithecus, Khoratpithecus
    Family Hylobatidae

Source: Begun (2010), Table 1, p. 69.

Note: Hylobatidae are the gibbons and siamangs. Hominidae includes the living species *Homo sapiens* (humans), *Pan troglodytes* (chimpanzee), *Pan paniscus* (bonobo), *Gorilla gorilla* (western gorilla), *Gorilla beringei* (eastern gorilla), *Pongo pygmaeus* (Bornean orangutan) and *Pongo abelii* (Sumatran orangutan).

Table 2 Harrison classification of Hominoidea

Superfamily Hominoidea
Family Hylobatidae
Family Hominidae
Stem Hominidae
Griphopithecus, Kenyapithecus, Pierolapithecus, Anoiapithecus, Dryopithecus, Oreopithecus
Subfamily Ponginae
Ankarapithecus, Sivapithecus, Lufengpithecus, Khoratpithecus, Gigantopithecus, Pongo
Stem Homininae or Hominini
Udabnopithecus, Nakalipithecus, Ouranopithecus, Samburupithecus, Chororapithecus, Sahelanthropus, Orrorin, Ardipithecus
Subfamily Homininae
Extant African great apes
Pan, Gorilla
Tribe Hominini
Australopithecus, Paranthropus, Homo

Source: Harrison (2010), pp. 532–3. For common names, see Table 1 above.

Table 3 Abbreviations for teeth used in tables

Tooth	Abbreviation for upper dentition	Abbreviation for lower dentition
Deciduous		
First incisor	UDI1	LDI1
Second incisor	UDI2	LDI2
Canine	UDC	LDC
Third premolar[1]	UDP3	LDP3
Fourth premolar[1]	UDP4	LDP4
Permanent		
First incisor	UI1	LI1
Second incisor	UI2	LI2
Canine	UC	LC
Third premolar[2]	UP3	LP3
Fourth premolar[2]	UP4	LP4
First molar	UM1	LM1
Second molar	UM2	LM2
Third molar	UM3	LM3

Notes: left and right are not distinguished.

[1] Deciduous third and fourth premolars are often named first and second molars instead.

[2] Permanent third and fourth premolars are often named first and second premolars instead. See text for further discussion.

Table 4 Gingival emergence in human deciduous teeth

Tooth	Iceland girls (mean)	Iceland girls (SD)	Iceland boys (mean)	Iceland boys (SD)	Nigerian girls (mean)	Nigerian girls (SD)	Nigerian boys (mean)	Nigerian boys (SD)	UK white girls (median)	UK white boys (median)	UK black girls (median)	UK black boys (median)
UDI1	0.8	0.2	0.8	0.2	0.7	0.3	0.7	0.2	0.8	0.8	0.8	0.7
UDI2	0.9	0.3	0.9	0.3	1.1	0.3	1.1	0.3	1.1	0.9	1.0	0.9
UDC	1.5	0.2	1.5	0.2	1.6	0.4	1.7	0.3	1.7	1.4	1.6	1.4
UDP3	1.3	0.2	1.3	0.2	1.3	0.3	1.4	0.2	1.4	1.3	1.3	1.3
UDP4	2.1	0.4	2.2	0.3	2.1	0.4	2.0	0.4	2.3	2.2	2.2	2.1
LDI1	0.6	0.2	0.7	0.3	0.9	0.3	0.9	0.2	0.8	0.7	0.7	0.7
LDI2	1.0	0.2	1.0	0.3	1.1	0.3	1.1	0.3	1.2	1.1	1.1	1.0
LDC	1.5	0.2	1.6	0.3	1.6	0.4	1.6	0.3	1.8	1.7	1.6	1.6
LDP3	1.3	0.2	1.4	0.2	1.4	0.3	1.4	0.2	1.5	1.4	1.4	1.3
LDP4	2.0	0.3	2.1	0.2	2.1	0.5	2.1	0.4	2.4	2.3	2.3	2.3

Age in years after birth. SD denotes standard deviation.

Data: Iceland (Magnusson, 1982), Table 2, p. 94, a cross-sectional study of 498 boys and 429 girls. Nigeria (Folayan et al., 2007), Table 3, p. 445, a cross-sectional study of 925 boys and 732 girls. UK (Lavelle, 1975), Table 1, p. 289, a cross-sectional study of 3600 white children and 600 black children. Mean and standard deviation, or median, were determined by probit analysis in all three studies.

Table 5 Gingival emergence in human permanent teeth taken from a study combining a population with relatively fast development rate and a population with slow rate

Tooth	African girls (mean; median for M3)	African girls (SD)	African boys (mean; median for M3)	African boys (SD)	Asian girls (mean; median for M3)	Asian girls (SD)	Asian boys (mean; median for M3)	Asian boys (SD)
UI1	6.6	0.7	6.9	0.7	7.0	0.7	7.2	0.8
UI2	7.7	0.9	8.0	0.9	8.0	0.7	8.4	0.9
UC	10.3	1.4	10.9	1.4	10.6	0.8	11.2	1.1
UP3	9.4	1.1	9.9	1.4	9.7	1.1	10.0	1.6
UP4	10.2	1.1	10.7	1.6	10.7	1.1	11.1	1.4
UM1	6.1	0.8	6.3	0.7	6.3	0.7	6.7	0.7
UM2	11.4	1.3	11.5	1.1	11.5	0.8	12.2	9
UM3	18.5	2.4	18.9	2.5	21.0	2.6	20.7	2.4
LI1	5.6	0.6	5.8	0.6	6.3	0.6	6.6	0.6
LI2	6.6	0.7	6.9	0.9	7.2	0.8	7.5	0.9
LC	9.2	1.0	10.0	1.2	9.7	0.8	10.6	1.0
LP3	9.6	1.1	10.1	1.4	9.8	1.0	10.6	1.3
LP4	10.2	1.4	10.9	1.5	10.7	1.4	11.4	1.5
LM1	5.7	0.8	6.0	0.7	6.1	0.7	6.5	0.6
LM2	11.1	1.4	11.4	1.1	11.1	0.9	11.9	1.1
LM3	17.8	2.7	18.2	2.6	20.3	2.7	20.0	2.4

Age in years after birth. SD denotes standard deviation.

Data: Hassanali and co-workers (Hassanali and Odhiambo, 1981; Hassanali, 1985) including 802 African girls, 881 African boys, 582 Asian girls and 582 Asian boys living in similar socio-economic circumstances in Nairobi, Kenya. These studies used the Kärber method for maximum likelihood estimation. Left and right are not distinguished, except in M3, where only the left side is reported.

Table 6 Eruption of deciduous teeth in chimpanzees, gorillas and orangutans

Tooth	Chimpanzee males & females (median)	Chimpanzee males & females (5th percentile)	Chimpanzee males & females (95th percentile)	Lowland gorilla males (mean)	Lowland gorilla males (SD)	Orangutan males (mean)	Orangutan males (SD)	Orangutan females (mean)	Orangutan females (SD)
UDI1	0.24	0.14	0.36	0.17	0.03	0.46	0.02	0.39	0.05
UDI2	0.30	0.12	0.49	0.20	0.02	0.74	0.17	0.66	0.17
UDC	1.02	0.65	1.33	0.81	0.07	1.11	0.16	0.99	0.23
UDP3	0.35	0.23	0.52	0.40	0.06	0.60	0.10	0.65	0.18
UDP4	0.78	0.52	1.13	0.99	0.13	0.99	0.13	0.88	0.14
LDI1	0.26	0.14	0.39	0.11	0.02	0.39	0.08	0.32	0.07
LDI2	0.37	0.21	0.47	0.24	0.04	0.58	0.09	0.47	0.07
LDC	1.12	0.69	1.49	0.78	0.09	1.10	0.18	0.98	0.16
LDP3	0.39	0.26	0.60	0.42	0.04	0.61	0.09	0.58	0.16
LDP4	0.73	0.46	1.09	0.76	0.06	0.90	0.30	0.72	0.16

Age in years after birth. Rounded to two decimal places. SD denotes standard deviation. Males and females combined. Only data from left side presented here.

Data: Chimpanzee values are for 22 male and 36 female captive animals from the Laboratory for Experimental Medicine and Surgery in Primates in Tuxedo, New York. Kuykendall et al. (1992), p. 385, Table 2. It was not possible to calculate probit statistics so the median (50th percentile) age of tooth emergence is presented, with the 5th and 95th percentiles. Gorilla values are for seven males living at several different zoos, taken from Smith et al. (1994), Appendix Table A33, computed from Keiter (1981), Table 6, p. 235. Orangutan values are for 13 males and 12 females from a variety of zoos, computed from Fooden and Izor (1983), Table IV, p. 290.

Table 7 Gingival emergence of permanent teeth in chimpanzees

Tooth	Nissen & Riesen females (median age at eruption)	Nissen & Riesen females (minimum)	Nissen & Riesen females (maximum)	Nissen & Riesen males (median age at eruption)	Nissen & Riesen males (minimum)	Nissen & Riesen males (maximum)	Kuykendall (probit median)	Kuykendall (minimum)	Kuykendall (maximum)
UI1	5.5	4.5	6.8	5.5	4.9	6.5	5.6	4.6	6.4
UI2	7.0	5.8	8.3	6.7	5.8	7.7	6.1	5.2	6.8
UC	9.2	7.6	10.1	9.2	8.0	9.8	*8.0	7.3	8.7
UP3	7.0	6.1	8.1	6.7	6.1	8.2	6.7	4.9	8.0
UP4	7.8	6.3	8.3	7.2	6.3	8.3	6.5	4.9	7.6
UM1	3.2	2.8	3.8	3.3	3.0	3.8	3.2	2.3	4.4
UM2	6.8	5.9	7.6	6.7	5.7	7.8	6.7	5.2	7.4
UM3	11.5	9.8	13.1	11.3	10.0	13.6	–	–	–
LI1	5.9	5.0	7.0	5.3	5.2	6.4	*5.5	4.8	6.3
LI2	5.9	5.0	7.3	6.4	5.6	6.9	6.0	4.9	6.4
LC	8.8	7.9	9.1	9.5	8.1	10.1	*8.0	7.3	8.7
LP3	7.3	6.3	8.1	7.5	6.3	8.3	7.4	4.9	8.0
LP4	7.4	6.1	9.1	7.4	6.3	8.3	6.6	4.9	8.0
LM1	3.1	2.7	3.8	3.3	3.0	3.6	3.1	2.1	4.0
LM2	6.3	5.9	7.3	6.7	5.6	7.0	6.6	4.8	6.8
LM3	10.8	9.0	13.1	10.4	9.0	11.1	–	–	–

Age in years after birth. Nissen and Riesen gave separate figures for males and females. Kuykendall *et al.* did not distinguish males and females and only data from the left side are presented here.

Note: Nissen and Riesen. For young animals, monthly examinations established age at eruption closely. Older animals could only be examined when anaesthetised, so a wider age interval had to be recorded and, for calculating the median, the midpoint of the interval was used. The minimum value was the youngest closely established age, or youngest upper limit for an age interval, whichever was the younger figure. Similarly, the maximum was the oldest closely established age, or oldest lower limit for an age interval.

Note: Kuykendall. Probit models were used to calculate the median values for each tooth, based on the percentages of individuals in which it had emerged in successive age groups. For some teeth, such models could not be used, because they went from 0 to 100% erupted in one age group. In such cases, the 50th percentile of known eruption ages was tabulated and they are here indicated by '*'.

Data: All are for captive animals. Nissen and Riesen (1964), Table 1A, p. 286, eight males and seven females kept in the Yerkes Laboratories of Primate Biology, Orange Park, Florida. Kuykendall *et al.* (1992), Table 5, p. 388, 22 males and 36 females kept in the colonies of the Laboratory for Experimental Medicine and Surgery in Primates, New York University Medical Center.

Table 8 Gingival emergence of permanent teeth in gorillas and orangutans

Tooth	Lowland gorilla males & females (mean)	Lowland gorilla males & females (range)	Orangutan males & females (approximate values)
UI1	6.0	5.3–6.8	6.0–7.0
UI2	6.5	5.5–7.5	6.0–8.0
UC	8.9	7.5–10.3	8.0–10.0
UP3	7.1	6.0–8.3	6.0–7.0
UP4	7.0	5.9–8.0	6.0–8.0
UM1	3.5	3.0–4.0	About 3.5
UM2	6.8	5.9–7.6	About 5.0
UM3	11.4	9.7–13.1	About 10.0
LI1	5.8	4.9–6.6	6.0–7.0
LI2	6.1	5.3–7.0	6.0–8.0
LC	7.7	6.4–9.0	8.0–10.0
LP3	7.3	6.1–8.6	6.0–7.0
LP4	7.1	5.9–8.4	6.0–8.0
LM1	3.5	3.0–4.0	About 3.5
LM2	6.6	5.7–7.5	About 5.0
LM3	10.4	8.7–12.1	About 10.0

Age in years after birth. Left and right sides are not distinguished.

Data: Gorilla values are from Willoughby (1978) and orangutan values are from Fooden and Izor (1983). Both were reported by Smith et al. (1994), Appendix Table A64.

Table 9 Gingival emergence of deciduous teeth in rhesus macaques

Tooth	Females (25th percentile)	Females (median)	Females (75th percentile)	Males (25th percentile)	Males (median)	Males (75th percentile)
UDI1	0.03	0.05	0.07	0.03	0.05	0.07
UDI2	0.08	0.10	0.12	0.08	0.10	0.13
UDC	0.17	0.19	0.22	0.16	0.18	0.21
UDP3	0.16	0.18	0.20	0.17	0.20	0.22
UDP4	0.37	0.42	0.46	0.40	0.44	0.49
LDI1	0.03	0.04	0.06	0.02	0.04	0.06
LDI2	0.05	0.06	0.08	0.04	0.06	0.08
LDC	0.17	0.20	0.23	0.16	0.19	0.22
LDP3	0.17	0.19	0.22	0.19	0.21	0.23
LDP4	0.34	0.37	0.42	0.37	0.41	0.45

Age in years after birth. Left and right not distinguished.

Data: Hurme and van Wagenen (1953), p. 299, Tables II and III. This was a longitudinal study of 53 females and 44 males kept in a colony at Yale University Medical School. Weekly dental examinations were carried out over 15 years by one person. Curves were fitted by hand to plots of age versus cumulative percentage of tooth emergence. The median and percentiles were then read from these curves.

Table 10 Gingival eruption of permanent teeth in rhesus macaques

Tooth	Females (25th percentile)	Females (median)	Females (75th percentile)	Males (median)
UI1	2.39	2.49	2.62	2.49
UI2	2.57	2.73	2.91	2.73
UC	3.27	3.46	3.70	4.04
UP3	3.21	3.37	3.56	3.55
UP4	3.39	3.65	4.00	3.65
UM1	1.35	1.44	1.53	1.49
UM2	3.17	3.33	3.51	3.25
UM3	5.81	6.23	6.74	5.4 ±
LI1	2.31	2.42	2.56	2.40
LI2	2.40	2.54	2.71	2.51
LC	2.94	3.13	3.34	3.84
LP3	3.19	3.37	3.58	3.72
LP4	3.26	3.52	3.79	3.66
LM1	1.25	1.31	1.38	1.36
LM2	2.99	3.14	3.30	3.12
LM3	5.42	5.74	6.14	5.3 ±

Age in years after birth. Left and right not distinguished.

Data: Hurme and van Wagenen (1961), p. 111, Table 5, p. 128, Table 14. This was a longitudinal study of 42 females and 42 males kept in a colony at Yale University Medical School. Monthly dental examinations were made. For females, curves were fitted by hand to plots of age versus cumulative percentage of tooth emergence. The median and percentiles were then read from these curves. Relatively few males were kept in the colony and the adult males proved difficult to examine without anaesthesia, so examinations had to be made at wider intervals. There were thus statistical difficulties with the males and only the median is presented.

Table 11 First appearance of mineralised tissue in human deciduous tooth germs

Tooth	Nomata (age at onset of mineralisation estimated from Kunitomo)	Nomata (crown-rump length at onset of mineralisation)	Nomata (age at onset of mineralisation estimated from Streeter)	Sunderland (youngest appearance of mineralised tissue in any child)	Sunderland (youngest age group with mineralised tissue in all children)	Kraus & Jordan (their estimate of age at first calcification)	Kraus & Jordan (crown-rump length at first calcification)	Kraus & Jordan (age at first calcification estimated from Streeter)
UDI1	17.0	114	16	15	19	–	–	–
UDI2	19.7	142	18	16	21	–	–	–
UDC	21.3	159	20	20	22	–	–	–
UDP3	19.3	138	18	16	19	15.5	118–136	16–17
UDP4	21.3	159	20	20	22	19	135–224	17–26
LDI1	17.7	121	16	16	19	–	–	–
LDI2	17.7	121	16	17	19	–	–	–
LDC	19.7	142	18	19	22	–	–	–
LDP3	19.7	142	18	16	19	15.5	113–146	16–18
LDP4	23.7	190	22	20	22	18	145–178	18–21

Gestational ages in weeks. Crown-rump lengths in mm.

Data: Nomata (1964), Table 12, p. 72. This was a histological study of tooth germs (microtome sectioned and stained with haematoxylin and eosin) dissected out of the jaws of 140 foetuses at the Tokyo Medical and Dental University. Both crown-rump lengths and gestational ages estimated from the tables of Kunitomo (1928) were reported. Sunderland et al. (1987), from graphs in Figures 1 to 5, pp. 168–172, combining males and females for compatibility with other studies. They did not give crown-rump lengths, but instead gave ages that they reported had been estimated by comparison with standard tables (Streeter, 1920). For this reason, Nomata's crown-rump lengths have here also been converted into ages using Streeter's Table 1, p. 153. This results in younger ages for the onset of mineralisation, more similar to those of Sunderland et al. (1987). Kraus and Jordan (1965) dissected tooth germs from aborted foetuses in the large collection at the University of Pittsburgh Cleft Palate Research Center, stained them with Alizarin Red S and dissected away the developing crowns. Figures in this table are for the mesiobuccal cusp and are taken from their Table 1 p. 53 (317 foetuses), Table 2 p. 65 (230 foetuses), Table 3 p. 79 (226 foetuses) and Table 4 p. 99 (242 foetuses). They reported that their own age estimates came from Patten (1968) and Streeter (1920), although without giving details. They did not report age estimates for their ranges of crown-rump lengths, so these are estimated here from Table 1 of Streeter (1920).

Table 12 Crown and root completion in human deciduous teeth: development state at birth

Tooth	Lunt and Law, extent of crown formation at birth	Crown completion (mean)	Crown completion (SD)	Apex closed (mean)	Apex closed (SD)
UDI1	5/6 complete	0.12	0.24	2.26	0.15
UDI2	2/3 complete	0.28	0.24	2.58	0.49
UDC	1/3 complete	0.83	0.26	3.33	0.13
UDP3	Cusps united; occlusal completely calcified plus 1/2 to 3/4 crown height	0.35	0.11	2.87	0.53
UDP4	Cusps united; occlusal incompletely calcified; calcified tissue covers 1/5 to 1/4 crown height	0.78	0.26	3.92	0.60
LDI1	3/5 complete	0.10	0.20	1.98	0.11
LDI2	3/5 complete	0.32	0.07	2.39	0.40
LDC	1/3 complete	0.81	0.12	3.51	0.35
LDP3	Cusps united; occlusal completely calcified	0.48	0.18	2.91	0.35
LDP4	Cusps united; occlusal incompletely calcified	0.92	0.26	3.54	0.74

Ages in years after birth (chronological age). SD denotes standard deviation.

Data: Crown formation at birth from Lunt and Law (1974), Table 4, p. 605. Crown and apex completion from Liversidge and Molleson (2004), Table 2, p. 311, using their stage D for crown complete and their H2 (apical foramen closed to <1 mm) as the apex closed stage. Their material comprised: children's remains from the crypt of Christ Church, Spitalfields in London (53 with independently known age and 68 with age estimated using regression equations from developmental tooth heights – see Table 13 below); radiographs of 61 child patients attending the Royal London Hospital; Medieval children's remains from Newark Bay, Orkney (59 with unknown age-at-death) and Whithorn, Scotland (74 also of unknown age).

Table 13 Developmental tooth height regression on age for human children

Tooth	A	B
DI1	−0.653	0.144
DI2	−0.581	0.153
DC	−0.656	0.210
DP3	−0.814	0.222
DP4	−0.904	0.292

Tooth	b0	b1	b2	b3	b4	b5	Minimum tooth height (mm)	Maximum tooth height (mm)
I1	1.0627	−0.5654	0.1518	−0.00765	0.00012	–	1.6	22.7
UI2	−0.4486	0.6520	−0.0080	–	–	–	2.5	21.9
LI2	1.61016	−0.8697	0.2249	−0.01285	0.000233	–	2.5	21.9
C	0.0644	0.2530	−0.0061	0.00952	−0.000724	0.0000147	1.3	24.9
P3	1.6140	0.5355	–	–	–	–	1.0	21.3
P4	2.2326	0.5604	–	–	–	–	1.3	21.3
M1	0.1258	−0.1992	0.1297	−0.00832	0.00017	–	1.4	20.4
M2	0.1198	1.6049	−0.1141	0.00341	–	–	2.4	21.0
M3	8.1775	0.6666	–	–	–	–	5.1	20.8

Note: males and females are combined; upper and lower are combined with the exception of UI2 and LI2. These regressions are only appropriate where the crown or root is still forming; for the permanent teeth they are also appropriate only between the given minimum and maximum values.

For deciduous teeth, age in years after birth = $a + bx$.

For permanent teeth, age in years after birth = $b0 + b1x + b2x^2 + b3x^3 + b4x^4 + b5x^5$, where x = developmental tooth height in mm.

Data: Both deciduous and permanent equations are based on studies of children's remains from the crypt of Christ Church, Spitalfields in London. Deciduous equations are from Table 7, p. 432 in Liversidge et al. (1998), using 64 children of independently known age-at-death between birth and 5.4 years. Permanent equations are from Liversidge and Molleson (1999), using 76 children of independently known age-at-death between 1 and 19 years.

Table 14 Initiation times for human permanent first molars

Crown-rump length (mm)	Percentage of individuals showing M1 calcification	Christensen & Kraus age estimate from Patten	Age estimate from Streeter
0–275	0 ($n = 30$)	0–28	0–31
275–287	30 ($n = 9$)	28	31–32
288–303	60 ($n = 12$)	28–32	32–34
304–332	73 ($n = 11$)	32–36	34–37
333–390	100 ($n = 12$)	36+	37+

Gestational ages in weeks.

Data: Christensen and Kraus (1965), read from Figure 2, p. 1340. Their age estimates were based on tables in Patten (1968). For consistency with Appendix A, Table 11 above, the crown-rump lengths are also used to derive age estimates from Table 1 of Streeter (1920).

Table 15 Development of human permanent incisor and canine tooth germs

Tooth	3 days	2 months	2 months 17 days	3 months	7 months 16 days	8 months	8–9 months	11 months	1 year 9 months	2 years
UI1	Bell	Bell	Bell	Bell	+	+	+	+	+	+
UI2	–	–	Cap	Bud	Bud	Bell	Bell	?	+	+
UC	Bud	Bud	Bell	Bell	Bell	+	Bell	?	+	+
LI1	Cap	Bell	Bell	Bell	+	+	+	+	+	+
LI2	Cap	Cap	Bell	Bell	+	+	+	+	+	+
LC	Cap	Cap	Bell	Bell	+	+	+	+	+	+

Age in years after birth.

Data from Ooë (1981), list on p. 146. 'Bud', 'bell' or 'cap' refer to the form of the tooth germ and imply that dentine and enamel were not being deposited. '+' shows that the teeth were at the bell stage and dentine and enamel *were* being deposited. Based on demineralised, stained histological preparations of jaws from 16 Japanese infants of known age-at-death.

Table 16 Human permanent crown initiation

Tooth	MFH girls		MFH boys		Haaviko girls		Haaviko boys		Liversidge girls		Liversidge boys	
	Mean	SD	Mean	SD	Median	IQR	Median	IQR	Mean	SE	Mean	SE
UM2	–	–	–	–	3.8	1.6	3.7	1.4	–	–	–	–
UM3	–	–	–	–	9.4	4.0	9.0	4.1	–	–	–	–
LC	0.5	0.1	0.5	0.1	–	–	–	–	–	–	–	–
LP3	1.7	0.2	1.8	0.2	–	–	–	–	–	–	–	–
LP4	2.9	0.4	3.0	0.4	–	–	–	–	–	–	–	–
LM1	0.1	0.1	0.0	0.1	–	–	–	–	–	–	–	–
LM2	3.5	0.4	3.7	0.4	3.9	1.9	3.9	2.1	3.4	0.2	3.5	0.2
LM3	9.6	1.0	9.2	1.0	9.6	4.3	9.8	6.2	–	–	–	–

Age in years after birth. SD denotes standard deviation, IQR interquartile range and SE standard error of mean.

Data: MFH = Moorrees, Fanning and Hunt as scaled by Harris and Buck (2002), Table 1, p. 16 and Table 2, p. 18. Includes longitudinal data from the Fels growth study, 136 boys and 110 girls X-rayed six-monthly, and 58 boys and 41 girls from the Harvard study. Haavikko (1970), Table 4, pp. 126–127, including radiographs of 615 boys and 437 girls taken at the Institute of Dentistry, University of Helsinki. Liversidge *et al.* (2006), Table IX, p. 465, combining data from 4480 girls and 4522 boys in separate studies from Australia, Belgium, Finland, France, Korea and Quebec. Probit methods were used for all statistics.

Table 17 Human permanent crown completion

Tooth	MFH girls		MFH boys		Demirjian & Levesque girls	Haaviko girls		Demirjian & Levesque boys	Haaviko boys		Liversidge girls		Liversidge boys	
	Mean	SD	Mean	SD	Median	Median	IQR	Median	Median	IQR	Mean	SE	Mean	SE
UI1	4.9	0.5	5.3	0.6	–	3.3	–	–	3.3	–	–	–	–	–
UI2	5.7	0.6	5.9	0.6	–	4.4	2.3	–	4.6	1.7	–	–	–	–
UC	–	–	–	–	–	4.5	2.1	–	4.6	1.3	–	–	–	–
UP3	–	–	–	–	–	6.3	1.2	–	6.8	2.0	–	–	–	–
UP4	–	–	–	–	–	6.6	1.9	–	7.1	2.0	–	–	–	–
UM1	–	–	–	–	–	3.5	0.8	–	3.6	1.5	–	–	–	–
UM2	–	–	–	–	–	6.9	2.2	–	7.3	1.4	–	–	–	–
UM3	–	–	–	–	–	12.8	4.7	–	13.2	5.6	–	–	–	–
LI1	–	–	–	–	–	–	–	–	–	–	–	–	–	–
LI2	–	–	–	–	–	–	–	–	3.3	1.4	–	–	–	–
LC	3.9	0.5	4.0	0.5	2.9	4.1	1.6	3.3	4.3	1.6	–	–	–	–
LP3	5.0	0.6	5.2	0.6	4.2	5.4	1.4	4.5	5.9	1.5	5.3	0.1	5.5	0.2
LP4	6.2	0.7	6.2	0.7	5.6	6.4	1.5	5.9	7.0	2.5	6.1	0.2	6.4	0.1
LM1	2.2	0.3	2.1	0.3	–	3.5	0.8	–	3.5	0.9	–	–	–	–
LM2	6.2	0.7	6.5	0.7	5.9	7.0	1.7	6.3	7.4	1.5	6.3	0.1	6.7	0.1
LM3	12.3	1.3	12.0	1.2	–	13.3	4.3	–	13.7	5.0	–	–	–	–

Age in years after birth. SD denotes standard deviation, IQR interquartile range and SE standard error of mean.

Data: MFH = Moorrees, Fanning and Hunt as scaled by Harris and Buck (2002), Table 1, p. 16 and Table 2, p. 18. Includes longitudinal data from the Fels growth study, 136 boys and 110 girls X-rayed six-monthly, and 58 boys and 41 girls from the Harvard study. Haavikko (1970), Table 4, pp. 126–127, including radiographs of 615 boys and 437 girls taken at the Institute of Dentistry, University of Helsinki. Liversidge et al. (2006), Table IX, p. 465, combining data from 4480 girls and 4522 boys in separate studies from Australia, Belgium, Finland, France, Korea and Quebec. Demirjian and Levesque (1980), Table III, p. 1117, comprised cross-sectional data from 2705 girls and 2732 boys in Montréal in Canada. Probit methods were used for all statistics.

Table 18 Human permanent root apex closure

Tooth	MFH girls		MFH boys		Demirjian & Levesque girls	Haaviko girls		Demirjian & Levesque boys	Haaviko boys		Liversidge girls		Liversidge boys	
	Mean	SD	Mean	SD	Median	Median	IQR	Median	Median	IQR	Mean	SE	Mean	SE
UI1	–	–	–	–	–	9.3	0.9	–	9.8	2.6	–	–	–	–
UI2	–	–	–	–	–	9.6	0.9	–	10.8	1.9	–	–	–	–
UC	–	–	–	–	–	12.7	3.2	–	13.6	3.7	–	–	–	–
UP3	–	–	–	–	–	12.6	2.1	–	13.3	3.0	–	–	–	–
UP4	–	–	–	–	–	13.4	2.5	–	14.0	4.0	–	–	–	–
UM1	–	–	–	–	–	9.2	1.7	–	9.8	2.9	–	–	–	–
UM2	–	–	–	–	–	15.1	1.7	–	16.2	3.4	–	–	–	–
UM3	–	–	–	–	–	19.6	2.9	–	19.5	3.2	–	–	–	–
LI1	7.7	0.8	8.1	0.9	8.1	8.0	2.1	8.5	8.0	2.7	7.9	0.1	8.3	0.1
LI2	8.5	0.9	9.3	1.0	9.2	9.0	1.3	9.6	9.6	2.1	8.8	0.1	9.3	0.1
LC	11.3	1.2	13.0	1.4	12.2	11.5	2.5	13.4	13.2	1.9	11.7	0.1	13.2	0.1
LP3	12.1	1.3	13.3	1.4	12.7	12.1	1.9	13.4	12.8	2.7	12.2	0.1	13.0	0.1
LP4	13.6	1.4	14.2	1.5	13.6	12.8	2.8	14.2	13.8	3.9	13.5	0.1	14.1	0.1
LM1	8.0	0.9	8.5	0.9	9.5	9.2	1.4	10.2	9.8	3.0	9.3	0.1	10.0	0.1
LM2	13.8	1.4	14.2	1.5	14.9	14.7	1.9	15.3	15.7	3.5	14.8	0.2	15.3	0.2
LM3	20.1	2.0	19.2	2.0	–	20.8	3.4	–	20.4	3.3	–	–	–	–

Age in years after birth. SD denotes standard deviation, IQR interquartile range and SE standard error of mean.

Data: MFH = Moorrees, Fanning and Hunt as scaled by Harris and Buck (2002), Table 1, p. 16 and Table 2, p. 18. Includes longitudinal data from the Fels growth study, 136 boys and 110 girls X-rayed six-monthly, and 58 boys and 41 girls from the Harvard study. Haavikko (1970), Table 4, pp. 126–127, including radiographs of 615 boys and 437 girls taken at the Institute of Dentistry, University of Helsinki. Liversidge *et al.* (2006), Table IX, p. 465, combining data from 4480 girls and 4522 boys in separate studies from Australia, Belgium, Finland, France, Korea and Quebec. Demirjian and Levesque (1980), Table III, p. 1117, comprised cross-sectional data from 2705 girls and 2732 boys in Montréal in Canada. Probit methods were used for all statistics.

Table 19 Chimpanzee permanent crown initiation

	Kuykendall midpoint ages at attainment (years)			Anemone ages at attainment (years)		
	Males	Females	Combined sex	25th percentile	50th percentile	75th percentile
LI1	(0.42)	0.48	0.42	–	–	–
LI2	(0.42)	0.48	0.42	–	–	–
LC	(0.42)	0.47	0.42	–	–	–
LP3	1.40	1.31	1.34	–	–	–
LP4	(1.52)	1.31	1.37	–	–	–
LM1	(0.13)	–	(0.13)	–	–	–
LM2	–	1.00	1.29	1.11	1.40	1.69
LM3	3.53	3.20	3.41	3.58	3.93	4.27

Age in years after birth.

Data: Kuykendall (1996), Table 3, pp. 140–143. Midpoint age at attainment is the average age between the oldest animal which has initiated crown development and the youngest which has not. This study was based on 118 captive animals from the Laboratory for Experimental Medicine and Surgery in Primates in Tuxedo, New York, and the Yerkes Regional Primate Research Center, Emory University, Atlanta, Georgia. Anemone (2002), Table 12.2, p. 265, including 49 chimpanzees (male and female) from the Yerkes Center and the Southwest Foundation for Biomedical Research, San Antonio, Texas. Probit analysis was used to determine the median and quartiles. Figures in parentheses denote single individuals.

Table 20 Chimpanzee permanent crown completion

	Kuykendall midpoint ages at attainment (years)			Anemone ages at attainment (years)		
	Males	Females	Combined sex	25th percentile	50th percentile	75th percentile
LI1	2.94	3.13	3.13	–	–	–
LI2	2.94	3.13	3.13	–	–	–
LC	6.71	5.44	6.32	–	–	–
LP3	3.87	4.00	3.87	–	–	–
LP4	4.58	4.00	4.58	–	–	–
LM1	1.59	1.34	1.40	1.48	1.73	1.98
LM2	4.20	(4.23)	4.20	3.29	3.73	4.17
LM3	6.87	6.88	6.88	6.02	6.53	7.04

Age in years after birth.

Data: Kuykendall (1996), Table 3, pp. 140–143. Midpoint age at attainment is the average age between the oldest animal which has completed crown development and the youngest which has not. This study was based on 118 captive animals from the Laboratory for Experimental Medicine and Surgery in Primates in Tuxedo, New York, and the Yerkes Regional Primate Research Center, Emory University, Atlanta, Georgia. Anemone (2002), Table 12.2, p. 265, including 49 chimpanzees (male and female) from the Yerkes Center and the Southwest Foundation for Biomedical Research, San Antonio, Texas. Probit analysis was used to determine the median and quartiles. Figure in parenthesis denotes a single individual.

Table 21 Chimpanzee permanent root apex closed

	Kuykendall midpoint ages at attainment (years)			Anemone ages at attainment (years)		
	Males	Females	Combined sex	25th percentile	50th percentile	75th percentile
LI1	8.94	8.61	8.64	–	–	–
LI2	9.35	8.79	9.24	–	–	–
LC	–	–	–	–	–	–
LP3	9.84	8.61	8.77	–	–	–
LP4	10.43	8.61	9.06	–	–	–
LM1	7.29	7.02	7.29	5.57	6.34	7.10
LM2	(10.74)	–	10.74	7.92	8.75	9.58
LM3	–	–	–	11.32	12.31	13.31

Age in years after birth.

Data: Kuykendall (1996), Table 3, pp. 140–143. Midpoint age at attainment is the average age between the oldest animal which has completed root apex closure and the youngest which has not. This study was based on 118 captive animals from the Laboratory for Experimental Medicine and Surgery in Primates (LEMSIP) in Tuxedo, New York, and the Yerkes Regional Primate Research Center, Emory University, Atlanta, Georgia. Anemone (2002), Table 12.2 p. 265, including 49 chimpanzees (male and female) from the Yerkes Center and the Southwest Foundation for Biomedical Research, San Antonio, Texas. Probit analysis was used to determine the median and quartiles. Figure in parenthesis denotes a single individual.

Table 22 Orangutan permanent tooth development in relation to four stages of first molar development

Tooth				
UI1	Crown ½ developed	Crown ¾ developed	Crown ¾ developed to root length ≥ crown height	Start of root development to apex complete
UC	Crown ¼ developed	Crown ¼ developed	Crown ¼ developed to crown ½ developed	Crown ¾ developed to start of root development
UP4	Crown ¼ developed	Crown ¼ developed	Crown ¾ developed to start of root formation	Crown ¾ developed to start of root development
UM1	Crown ½ developed	Crown complete	Root length ≥ crown height	Root apex complete
UM2	Tooth crypt present to crown ¼ developed	Initial calcification to crown ¼ developed	Crown ½ developed to crown fully formed	Root length ≥ crown height to apex complete
UM3	–	–	Tooth crypt present to initial calcification	Crown ¼ developed to root apex partially open
LI1	Crown ¼ developed	Crown ¾ developed	Crown ½ developed to start of root formation	Root length ≥ crown height to apex complete
LC	Crown ¼ developed	Crown ¼ to ½ developed	Crown ¼ developed to crown ¾ developed	Crown ¾ developed to root apex partially open
LP4	Crown ¼ developed	Crown ¼ developed	Crown ½ developed to root length ≥ crown height	Start of root development to apex complete
LM1	Crown ½ developed	Crown complete	Root length ≥ crown height	Root apex complete
LM2	Tooth crypt present to crown ¼ developed	Crown ¼ developed	Crown ¼ developed to root length ≥ crown height	Tooth crypt present to initial calcification
LM3	–	–	Tooth crypt present to initial calcification	Crown fully formed to root apex partially open

Data: Re-tabulated from Winkler *et al.* (1996), Figure 1, p. 209 and Table 3, p. 210.

Table 23 Lower permanent tooth development in the pig-tailed macaque (*Macaca nemestrina*)

Tooth	Pig-tailed macaque (male)			Pig-tailed macaque (female)		
	Initial calcification (mean)	Crown completion (mean)	Apical closure (mean)	Initial calcification (mean)	Crown completion (mean)	Apical closure (mean)
LP3	0.97	2.20	4.21	0.96	2.08	4.03
LP4	1.09	2.17	4.21	0.99	2.04	4.03
LM1	Present by 3 to 6 months	0.84	2.56	Present by 3 to 6 months	0.75	2.42
LM2	1.12	2.15	4.32	0.98	2.02	4.07
LM3	3.01	4.16	6.66	2.77	3.79	6.62

Age in years after birth.

Data: Swindler *et al.* (1982), Table 1, p. 49. 40 *Macaca nemestrina* kept at the Regional Primate Research Center Field Station at Medical Lake, Washington. The means are for the age at which the development stages were first observable.

Table 24 Lower permanent tooth development in the yellow baboon (*Papio cynocephalus*)

Tooth	Yellow baboon (male)			Yellow baboon (female)		
	Initial calcification (mean)	Crown completion (mean)	Apical closure (mean)	Initial calcification (mean)	Crown completion (mean)	Apical closure (mean)
LI	0.57	1.97	3.93	0.60	1.98	4.07
LC	0.86	3.08	6.48	0.91	1.98	6.31
LP3	1.21	2.25	4.69	1.18	2.08	4.62
LP4	1.24	2.22	4.90	1.21	1.94	4.63
LM1	Before birth	0.82	2.42	Before birth	0.66	2.42
LM2	1.28	2.25	4.57	1.13	1.98	4.17
LM3	2.93	4.05	6.48	2.74	4.06	–

Age in years after birth. LI1 and LI2 not distinguished.

Data: Swindler and Meekins (1991), Tables 1–7, pp. 573–575. Radiographs of 20 *Papio cynocephalus* kept at the Regional Primate Research Center Field Station at Medical Lake, Washington. The means are for the age at which the development stages were first observable.

Table 25 Rate of enamel and dentine apposition

Tissue	Mean rate	Standard deviation	Range
Dentine: deciduous upper first incisor	3.77	0.67	3.16–4.42
Enamel: deciduous fourth premolar	3.92	0.26	3.60–4.30
Enamel: permanent teeth	2.71	0.53	2.36–2.95

Rate in micrometres per 24 hours.

Data: The rates were obtained using sodium fluoride labelling in the teeth of a child and the values are taken from the text of Schour and Poncher (1937).

Table 26 Regression formulae for the relationship between cuspal enamel thickness and enamel formation time

For modern humans $y = 8.7 + 0.37x - 0.00005x^2$ $R^2 = 0.97$, standard error $= 0.1$
For African apes $y = 8.68 + 0.25x - 0.00003x^2$ $R^2 = 0.97$, standard error $= 0.01$
For australopithecines $y = 6.64 + 0.21x - 0.00001x^2$ $R^2 = 0.97$, standard error $= 0.01$
For early *Homo* $y = 3.76 + 0.26x = 0.00002x^2$ $R^2 = 0.99$, standard error $= 0.01$
For *Homo ergaster/erectus* $y = 3.7 + 0.27x - 0.00003x^2$ $R^2 = 0.99$, standard error $= 0.01$

y is cuspal enamel formation time in days and x is enamel thickness under the cusp tip in micrometres.

From Dean et al. (2001) Figure 1, p. 629.

Table 27 Ages for initiation of permanent crowns in chimpanzees, gorillas, orangutans and humans, estimated from enamel histology

	Chimpanzee Reid et al.*	Chimpanzee Smith et al.*	Gorilla Schwartz et al.*	Gorilla Beynon et al.*	Orangutan Beynon et al.*	Human forensic case*	Human T49 Picardie*	North European humans (mean)	South African humans (mean)
UI1	0.3, 0.2	–	0.2	0.6	0.2	0.3	0.4	0.4	0.4
UI2	1.3, 0.9	–	0.5	0.9	1.4	0.7	1.1	1.1	0.1
UC	0.4, 0.9	–	0.3	0.4	0.4	0.4	0.8	0.8	0.8
UP3	1.2, 1.3	–	1.1	1.3	1.7	1.7	1.9	1.8	1.8
UP4	1.7, 1.2	–	0.7	1.5	1.8	2.4	2.7	2.6	2.6
UM1	−0.2, −0.2	−0.1	−0.1	−0.1	<0.0	0.0	−0.1	0.0	0.0
UM2	1.4	–	1.4	2.2	1.9	2.9	2.8	3.0	3.0
UM3	3.8	–	–	4.9	–	–	7.7	8.0	8.0
LI1	0.2, 0.5, 0.3, 0.2	–	0.2	0.9	0.4	0.3	0.3	0.3	0.3
LI2	0.2, 0.7, 0.7, 0.2	–	0.2	0.9	1.1	0.7	0.4	0.4	0.4
LC	0.4, 0.6, 0.5, 0.4	–	0.3	0.8	0.9	0.4	0.6	0.6	0.6
LP3	1.1, 1.4, 1.2, 1.1	–	0.7	1.3	1.2	1.7	1.9	1.8	1.8
LP4	1.4, 1.8, 2.0, 1.4	–	1.1	1.6	1.8	2.7	2.7	2.6	2.6
LM1	−0.1, 0.2, −0.2	−0.2 to −0.1	−0.2	−0.1	<0.0	0.0	−0.1	–	–
LM2	1.8, 1.7, 2.0	1.3 to 2.1	1.1	2.6	2.3	–	2.9	3.0	3.0
LM3	3.6, 3.6	–	–	4.9	–	6.4	7.8	8.0	8.0

Age in years after birth (negative figures denote before birth).

* Used microscope sections for all teeth from single individuals, including a permanent first molar in which the neonatal line was identifiable.

Data. Estimates based on counts of prism cross striations. Chimpanzee, Reid et al. (2007b) is from Table 10, p. 213. Gorilla, Schwartz et al. (2006) is from Table II, p. 1213. Gorilla and Orangutan, Beynon et al. (1991a) are scaled from Figures 7 and 9, pp. 198–200. Human forensic case was a male of unknown age, but possible West African origin, reported in Dean et al. (1993a), Table 3, p. 259. Human T49 is a single individual from the Medieval monastery site of Saint-Nicholas d'Acy, Picardie in France (Reid et al., 1998a) from Table 3, p. 470. North European and South African humans are from Reid and Dean (2006), Tables 4 and 5, pp. 333–335 and Holt et al. (2012), Figure 1, p. 6.

Table 28 Ages for completion of permanent crowns in chimpanzees, gorillas, orangutans and humans, estimated from enamel histology

	Chimpanzee Reid et al.*	Chimpanzee Smith et al.*	Gorilla Schwartz et al.*	Gorilla Beynon et al.*	Orangutan Beynon et al.*	Human forensic case*	Human T49 Picardie*	Human Picardie mean	North European humans (mean)	South African humans (mean)	Human Spitalfields range (number of teeth)
UI1	4.3	–	2.9	4.7	5.0	3.5	4.2	4.4	5.0	4.2	3.3–4.5 (5)
UI2	5.7	–	–	4.9	6.6	4.4	4.5	4.7	5.1	4.8	–
UC	6.9	–	–	–	8.0	4.8	5.2	5.2	5.3	4.8	–
UP3	5.6	–	–	–	6.4	4.5	5.4	5.4	4.7	5.5	–
UP4	5.2	–	–	5.0	6.0	5.5	5.7	5.6	5.4	5.8	–
UM1	2.3	2.1–2.3	2.5	2.9	2.6	2.4	2.6	2.8	3.0	3.0	–
UM2	–	–	–	5.2	5.3	6.1	6.0	6.1	6.4	6.3	–
UM3	–	–	–	–	–	–	11.0	11.0	11.4	11.3	–
LI1	0.5, 5.8, 5.3	–	2.7	4.5	5.1	3.3	3.6	3.8	3.8	3.4	3.1–4.5 (6)
LI2	5.2, 5.8	–	2.7	5.1	6.6	4.4	4.5	4.6	4.2	3.8	4.2–5.4 (4)
LC	5.5+	–	–	–	9.4	4.8	6.2	6.0	6.2	5.2	–
LP3	5.7, 5.5	–	–	–	8.1	4.5	5.4	5.7	5.2	4.6	–
LP4	5.6, 5.7, 5.6	–	–	–	6.1	5.8	6.1	6.1	5.7	5.4	–
LM1	2.4, 3.1, 2.6	2.0–2.6	2.3	2.7	2.9	2.7	3.0	3.3	3.3	3.1	2.3–4.3 (18)
LM2	5.5, 4.5	–	–	–	5.1	–	6.1	6.1	6.2	6.2	–
LM3	7.0	–	–	–	–	9.6	10.9	10.9	11.3	11.2	–

Age in years after birth.

* Used microscope sections for all teeth from single individuals, including a permanent first molar in which the neonatal line was identifiable.

Data: As in Table 27, except that the means for Picardie are from Reid et al. (1998a), Table 5, p. 474, based on four individuals from the site. North European and South African humans are from Reid and Dean (2006), Figures 3 and 4, pp. 343–4 and from Holt et al. (2012), Figure 1, p. 6. Unlike the other figures, the Spitalfields study was based on direct observation of the whole tooth and a range is presented for the known age-at-death children from the crypt of Christ Church, Spitalfields in London (Liversidge, 2000), Table 3, p. 715.

Table 29 Ages for initiation of permanent crowns in Neanderthal, early modern human, *Proconsul* and *Papio*, estimated from enamel histology

	Scladina Neanderthal	Irhoud early modern human	*Proconsul heseloni*	*Papio anubis*
UI1	–	–	–	–
UI2	0.6	–	–	–
UC	0.3	–	–	–
UP3	–	–	–	–
UP4	–	–	–	–
UM1	0.0	–	–	–
UM2	3.0	–	–	–
UM3	5.9	–	–	–
LI1	–	–	–	0.6, 0.6
LI2	–	5.1	–	0.7, 0.7
LC	0.3	6.7	–	0.7, 0.7
LP3	1.7	4.0	–	1.1, 1.1
LP4	–	–	–	1.5, 1.5
LM1	–	–	−0.1	−0.1, −0.1
LM2	–	–	–	1.4, 1.4
LM3	–	–	–	3.8, 3.7

Age in years after birth (negative figures denote before birth).

Data: Scladina Neanderthal, Smith *et al.* (2007d), supporting information Table 5. Based on microscope section of permanent upper first molar in which the neonatal line was found, and synchrotron radiation micro-CT scans of other teeth. Jebel Irhoud, Smith *et al.* (2007c), Table 1, p. 6131. *Proconsul heseloni*, Beynon *et al.* (1998), caption of Figure 19, p. 200 which describes neonatal line in permanent first molar. *Papio anubis*, Dirks *et al.* (2002), Table 2, p. 244.

Table 30 Ages for completion of permanent crowns in Neanderthal, early modern human, *Proconsul* and *Papio*, estimated from enamel histology

	Scladina Neanderthal	Irhoud early modern human	*Proconsul heseloni*	*Papio anubis*
UI1	3.45	–	–	–
UI2	3.87	–	–	–
UC	3.84	–	–	–
UP3	–	–	–	–
UP4	–	–	–	–
UM1	2.35	–	–	–
UM2	5.62	–	–	–
UM3	–	–	–	–
LI1	3.41	–	–	3.0, 3.29
LI2	3.38	5.53	–	3.19, 3.33
LC	3.89	7.23	–	3.98, 3.9
LP3	4.41	3.91	–	3.1, 3.21
LP4	–	–	–	3.17, 3.29
LM1	–	–	1.12	1.35, 1.38
LM2	–	–	–	2.88, 3.12
LM3	–	–	–	5.71, 5.15

Age in years after birth.

Data: Scladina Neanderthal, Smith *et al.* (2007d), supporting information Table 5. Based on microscope section of permanent upper first molar in which the neonatal line was found, and synchrotron radiation micro-CT scans of other teeth. Jebel Irhoud, Smith *et al.* (2007c), Table 1, p. 6131. Assumed crown initiation age from Picardie T49 (see Table 27), combined with synchrotron radiation micro-CT imaging. *Proconsul heseloni*, Beynon *et al.* (1998), caption of Figure 19, p. 200 which describes neonatal line in permanent first molar and Table 6, p. 193. *Papio anubis*, Dirks *et al.* (2002), Table 2, p. 244.

Table 31 Permanent crown formation times for living and fossil members of genus *Homo*

	Modern humans, South Africa (mean)	Modern humans, South Africa (range)	Modern humans, North Europe (mean)	Modern humans, North Europe (range)	Jebel Irhoud	Neanderthals (mean)	Neanderthals (range)	Early Homo KNM-WT 15000	Early Homo KNM-ER 820	Early Homo KNM-ER 1590	Early Homo KNM-ER 808	Early Homo SK 27	Early Homo Sangiran 4
UI1	3.6	3.5–3.7	4.3	4.1–4.5	–	3.3	3.2–3.5	2.6–3.2	–	3.1–3.7	–	–	–
UI2	3.6	3.5–3.8	3.9	3.7–4.3	–	3.1	3.0–3.2	2.7–3.2	–	–	2.7–3.2	–	–
UC	3.9	3.9–4.0	4.4	4.1–4.8	–	3.4	–	2.9–3.5	–	3.5–4.2	–	4.1–4.9	3.8–4.5
UP3	3.0	3.0–3.1	3.9	3.6–4.0	–	3.2	–	–	–	–	–	–	–
UP4	3.0	2.6–3.3	3.4	3.0–3.7	–	3.1	–	–	–	–	–	–	–
UM1	2.7	2.3–2.9	2.8	2.6–2.9	–	2.3	2.2–2.3	–	–	–	–	–	–
UM2	3.1	2.8–3.4	3.2	2.8–3.6	–	2.9	2.6–3.2	–	–	–	–	–	–
UM3	3.0	2.7–3.1	3.1	3.0–3.4	–	2.7	–	–	–	–	–	–	–
LI1	–	–	–	–	–	–	–	2.6–3.1	–	–	–	–	–
LI2	–	–	–	–	5.1	–	–	2.5–3.0	2.9–3.5	–	–	–	–
LC	4.7	–	5.5	5.0–5.8	6.7	3.6	–	3.1–3.7	–	–	–	–	–
LP3	3.2	3.1–3.2	4.0	3.7–4.4	–	2.7	–	–	–	–	–	–	–
LP4	3.2	2.9–3.5	3.4	2.8–3.9	–	3.0	–	–	–	–	–	–	–
LM1	3.0	–	3.0	2.9–3.1	4.0	2.6	2.3–2.9	–	–	–	–	–	–
LM2	3.1	3.0–3.2	3.0	2.8–3.3	–	2.3	2.2–2.4	–	–	–	–	–	–
LM3	–	–	–	–	–	–	–	–	–	–	–	–	–

Estimated time from crown initiation to completion in years, rounded to one decimal place. Where molar cusps are distinguished, the mesiobuccal cusp is presented: paracone (cusp 2) or protoconid (cusp 1). No distinction is made between males and females.

Data: Modern humans and Neanderthals from Smith *et al.* (2010b) SI Table 6. Neanderthals comprise: Engis 2 from Belgium, Gibraltar 2, La Quina H18 from France, Krapina Maxillae B and C from Croatia, Obi-Rakhmat 1 from Uzbekistan, Scladina 1 from Belgium, Le Moustier 1 from France. Jebel Irhoud is an early anatomically modern *H. sapiens* fossil from North Africa (Smith *et al.*, 2007c), Table 2, p. 6132. Early *Homo* are from Dean *et al.* (2001) Table 1, p. 628. Those with prefix KNM are from East Africa, SK from South Africa, and Sangiran from Indonesia. Modern human estimates were determined from examination of microscope sections. Estimates for Jebel Irhoud and Neanderthals were derived from micro-CT examination. Estimates for early *Homo*: cuspal enamel formation was estimated using regression formulae given here in Appendix A, Table 26 and imbricational enamel formation was estimated from perikymata counts, with the minimum assuming a cross striation periodicity of six days and the maximum 10 days (Dean *et al.*, 2001).

Table 32 Permanent crown formation times for living chimpanzees, gorillas and orangutans

	Chimpanzee HT 43/87	Chimpanzee HT 88/89	Chimpanzee HT 28/90	Chimpanzee HT 59/89	Chimpanzee Smith et al. (mean)	Chimpanzee Smith et al. (range)	Chimpanzee Dean (mean)	Chimpanzee Schwartz & Dean (range)	Gorilla Schwartz et al.	Gorilla Schwartz & Dean (range)	Orangutan Schwartz & Dean (range)
UI1	–	5.7	–	–	–	–	–	–	2.9	–	–
UI2	–	4.5	–	–	–	–	–	–	–	–	–
UC	–	7.1	–	–	–	–	–	–	–	–	–
UP3	–	4.3	–	–	–	–	–	–	–	–	–
UP4	–	3.6	–	–	–	–	–	–	–	–	–
UM1	–	2.6	–	–	2.5	2.1–2.9	–	–	2.5	–	–
UM2	–	3.2	–	–	2.9	–	–	–	–	–	–
UM3	–	–	–	–	2.5	2.2–2.7	–	–	–	–	–
LI1	4.4	4.9	5.0	–	–	–	–	–	2.7	–	–
LI2	4.5	5.6	5.7	–	–	–	–	–	2.7	–	–
LC	6.2	–	–	–	–	–	–	5.3–7.6	–	4.6–9.8	4.6–9.6
LP3	4.8	4.7	4.9	–	–	–	–	–	–	–	–
LP4	4.1	3.8	4.1	–	–	–	–	–	–	–	–
LM1	2.5	3.0	2.8	–	2.2	2.0–2.6	2.3	–	2.3	–	–
LM2	3.1	3.5	3.6	3.8	2.7	2.2–3.0	2.4	–	–	–	–
LM3	4.0	3.6	–	–	2.7	–	2.7	–	–	–	–

Estimated time from crown initiation to completion in years, rounded to one decimal place. Where molar cusps are distinguished, the mesiobuccal cusp is presented: paracone (cusp 2) or protoconid (cusp 1). No distinction is made between males and females.

Data: Chimpanzee: HT individuals from Reid et al. (1998b), Table 4, pp. 438–9, using mesiobuccal cusps and largest value when left and right reported; Smith et al. (2007b), Table 6, p. 209; Dean (2010), personal communication, data from Figure 2a, p. 3399. Gorilla: Schwartz et al. (2006), Table II, p. 1213; Lower canine for chimpanzees, gorillas and orangutans from Schwartz and Dean (2001), Table 4, p. 275.

Table 33 Crown formation times for fossil *Australopithecus* and *Paranthropus*

	Australopithecus Dean et al.	*Australopithecus afarensis* AL 333–52 and AL 366-1	*Australopithecus africanus* Stw 284 and Stw 402	*Paranthropus* Dean et al.	*Paranthropus* KNM-ER 733D	*Paranthropus robustus* SK63	*Paranthropus robustus* SKX 21841	*Paranthropus robustus* KB 5223
UI1	3.2–3.8	–	–	2.3–2.7	–	–	–	–
UI2	2.8–3.4	–	–	2.3–2.8	–	–	–	–
UC	3.3–3.9	–	–	3.1–3.7	–	–	–	–
UP3	–	–	–	–	–	–	–	–
UP4	–	–	–	–	2.4	–	–	–
UM1	–	–	2.7	–	–	–	–	–
UM2	–	–	3.0–3.2	–	–	–	–	–
UM3	–	–	–	–	–	–	2.7	–
LI1	2.9–3.6	–	–	2.1–2.5	–	2.4–2.7	–	–
LI2	3.0–3.7	–	–	2.3–2.8	–	2.6–2.9	–	–
LC	3.7–4.5	–	–	2.9–3.4	–	3.2–3.4	–	–
LP3	–	–	–	–	–	–	–	–
LP4	–	–	–	–	–	–	–	–
LM1	–	2.2	–	–	–	2.4	–	2.2–2.7
LM2	–	–	–	–	–	–	–	–
LM3	–	2.3–2.4	–	–	–	–	–	–

Estimated time from crown initiation to completion in years, rounded to one decimal place. Where molar cusps are distinguished, the mesiobuccal cusp is presented: paracone (cusp 2) or protoconid (cusp 1).

Data *Australopithecus* and *Paranthropus* Dean et al. (2001) ranges are from Table 1, p. 628. Cuspal enamel formation was estimated using regression formulae given here in Appendix A, Table 26. Imbricational enamel formation was estimated from perikymata counts, with the minimum assuming a cross striation periodicity of six days and the maximum 10 days. For *A. afarensis*, cross striation periodicity was established by confocal light microscopy and applied to perikymata counts (Lacruz and Ramirez-Rozzi, 2010, Table 2, p. 203), together with an estimate for cuspal enamel formation based on the enamel thickness multiplied by a correction factor. For *A. africanus* Stw 284 and 402, and *P. robustus* SKX 21841, the same approach was taken for isolated tooth fossils, respectively, from Sterkfontein and Swartkrans, South Africa (Lacruz et al. 2006, pp. 584–585). A similar approach was also used for *P. robustus* KB 5223 from Kromdraai, South Africa (Lacruz, 2007, p. 180). *Paranthropus* KNM-ER 733D was fractured and polarising microscopy was used to count cross striations in cuspal enamel and determine their periodicity which, together with perikymata counts, yielded the estimated crown formation time (Beynon and Dean, 1987, p. 775). Estimates for *P. robustus* SK63 from Swartkrans, South Africa are from Dean et al. (1993b), Tables 1–3 and text, pp. 413–15.

Table 34 Crown formation times for fossil *Paranthropus boisei/aethiopicus*

	Mean formation time assuming six cross striation periodicity	Mean formation time assuming seven cross striation periodicity	Mean formation time assuming ten cross striation periodicity
Upper and lower premolars	2.2	2.5	3.6
Upper and lower molars	2.2	2.6	3.7

Estimated time from crown initiation to completion in years, rounded to one decimal place.

Data: Sixty-six fractured premolars and molars from the Omo Group, Ethiopia. Optical microscopy was used to count brown striae of Retzius in the fractured enamel and perikymata at the crown surface. Cross striation periodicity could not be determined, so formation time was estimated on the basis of three different assumptions. Figures are from Ramirez-Rozzi (1995), Table III, p. 225.

Table 35 Permanent crown formation times for fossil *Proconsul*

	Proconsul nyanzae maximum mean estimate	*Proconsul heseloni* maximum mean estimate
DP3	–	0.3
DP4	–	0.5
I1	2.5	2.4
I2	2.5	2.4
C	3.0–4.7	2.5
P3	–	1.8
P4	–	1.7
M1	2.0	1.2
M2	2.3	1.4
M3	–	1.7

Estimated time from crown initiation to completion in years, rounded to one decimal place. Where molar cusps are distinguished, the mesiobuccal cusp is presented: paracone (cusp 2) or protoconid (cusp 1).

Data from Beynon *et al.* (1998), Table 6, p. 193. The teeth were isolated specimens from Rusinga Island, Kenya. All were sectioned, so estimates could be based on direct counts and measurements of prism cross striations.

Table 36 Permanent crown formation times for fossil *Afropithecus, Dryopithecus, Sivapithecus, Lufengpithecus* and *Gigantopithecus*

	Afropithecus turkanensis, range for KNM-WK 17024 & 24300	*Dryopithecus laietanus*, IPS 1781, 1794 & 1782	*Sivapithecus parvada*, maximum value for GSP 47585	*Sivapithecus indicus*, maximum value for NHM M13365	*Lufengpithecus hudienensis*, PDYV 30	*Gigantopithecus blacki*
UI1	–	–	–	–	–	–
UI2	–	–	–	–	–	–
UC	–	–	–	–	–	–
UP3	–	–	–	–	–	–
UP4	–	–	–	–	–	–
UM1	–	1.7–2.0	–	2.3	–	–
UM2	–	2.2	–	–	–	–
UM3	–	–	–	–	–	–
LI1	–	–	–	–	–	–
LI2	–	–	–	–	–	–
LC	–	–	–	–	–	–
LP3	–	–	–	–	–	–
LP4	–	–	–	–	–	–
LM1	–	–	2.4	–	2.0–2.2	–
LM2	2.3–3.0	2.2	–	–	–	–
LM3	–	–	–	–	–	3.5–4.1

Estimated time from crown initiation to completion in years, rounded to one decimal place. Where molar cusps are distinguished, the mesiobuccal cusp is presented: paracone (cusp 2) or protoconid (cusp 1).

Data: All are based on counts and measurements of incremental structures in conventional light microscopy of sections. *Afropithecus* estimates are the minimum and maximum values for two isolated specimens in Table 6, p. 298 of Smith *et al.* (2003). *Dryopithecus laietanus* are the values for three isolated specimens in Table 1, p. 131 of Kelley *et al.* (2001). *Sivapithecus* are the maximum mesial crown formation values for two isolated specimens in Table 1, p. 64 of Mahoney *et al.* (2007). *Lufengpithecus* is a single specimen, from Table 2, p. 196 of Schwartz *et al.* (2003). *Gigantopithecus* was a single specimen, discussed in text on p. 384 of Dean and Schrenk (2003).

Table 37 Tezonteopan growth study, with revised age estimates for three zones of upper first incisor crown development

Zone	Goodman *et al.* age estimate (years)	Revised age estimate (years)	Supplemented group – proportion of first incisors with LEH in this zone (%)	Non-supplemented group – proportion of first incisors with LEH in this zone (%)
Incisal one-third	Birth–1.5	1–2	14.6	40.0
Central one-third	1.5–3	2–3	33.3	38.1
Cervical one-third	3.0–4.5	3.0–4.5	4.8	28.6

Sources: Goodman *et al.* (1991) Table 3, with revised age estimates derived from Reid and Dean (2006) Figure 3.

Appendix B: Technical information

Laboratory techniques

Handling and cleaning loose archaeological or fossil teeth

Primate teeth are small, smooth, rounded and difficult to grip. Fossils can be delicate, so pad the table with a foam sheet in case they fall. A small tray of dry sand holds them gently in position under a camera or microscope; black sand from aquarium suppliers provides a good visual contrast. It is safest to use fingers for handling, but fine forceps can be useful, especially the tweezers used in electronics, which have plastic tips that don't scratch. Good light is essential: from a window behind the bench, a bright desk lamp or, in the field, a head torch.

Labelling is vital, but teeth should never be marked directly. It is safest to keep each tooth separate in a small Ziploc polythene bag containing its label (a Tyvek label written with permanent black marker pen). Perforate the bag ten times to provide ventilation, but keep it zipped up tight or the tooth will become separated from its label. Keep a photographic catalogue of the teeth with their labels in view to sort out mix-ups. The tooth bags can be kept in a lidded polythene box, with ventilation holes drilled just below where the lid clips on. Thin sheets of polythene foam (Jiffy Foam or similar) help to pad them. This system works well for all teeth and bones, with foam sheets cut to fit inside the bags. Clear bags show the contents on one side and the labels on the other side of the foam, so it is not necessary to open the bag to check what is inside. The less the specimens are handled, the better. Avoid centrally heated or air-conditioned stores, which are too dry and can cause dentine to crack over time.

For similar reasons, avoid washing archaeological or fossil teeth with water because the dentine may crack on drying. The effect is minimised by slow drying, but loss of water continues in storage and problems often arise years later. A lot of cleaning can be done while keeping the specimen dry. Thicker dirt can be removed gently with a wooden cocktail stick. It may then be enough to dry-brush the specimen using short, gentle strokes with an artist's hogs-bristle brush to flick the dirt and dust away. This takes a while, but keep at it, with quick brush strokes so the dirt rises up in a cloud. Scrubbing with a wet toothbrush is *absolutely forbidden* as it is very destructive to softened archaeological dentine. If wetting is required to soften the dirt, it is best to use acetone (highly flammable) on a cotton bud. Avoid scrubbing the surface; roll the bud along to pick up dirt. In general, it is better to leave the tooth slightly dirty, but whole and unscratched, rather than sparkling clean. Getting the enamel surface really clean for moulding (see next section) takes time. Dentine and bone can never be completely cleaned and, in fact, the dirt is part of what is holding them together, so cleaning simply produces an artificially smooth surface.

Making tooth replicas for microscopy

The initial mould is made using dental impression material: the 'fine body' or 'wash' rather than the putty (Hillson, 1992b). Most of the modern silicones will do, but Coltène President or Streuers RepliSet consistently produce the best resolution. These materials have a finite shelf life and the setting time varies with temperature, so test each batch before using on important fossils. The tooth has to be clean (see previous section). Once thoroughly mixed, the material is rapidly spread over the surface to be cast. It is better to cast just one side of the tooth at a time as the mould releases more easily. Once the material has set (usually six minutes), the edge of the mould is released gently with a cocktail stick. Then the mould can be pulled away, but great care has to be taken because it is possible to pull away the enamel if it is cracked. The mould needs to be left to cure for a while and then it can be kept in a labelled Ziploc plastic bag.

The replica is made simply by pouring an epoxy resin into the mould. Several types can be used, but it needs to be free flowing and fluid before setting. Bubbles may form and these need to be released by gently poking them with a cocktail stick. Once set, the replica can be sputter-coated with gold for the scanning electron microscope (this also helps when examining in a light microscope). The detail resolved by these replicas is typically 0.5 µm, but there is inevitably some distortion of the overall shape and some rounding of sharp edges.

For making the moulds, isolated teeth can be held in a slit cut into foam sheet, with the surface to be cast uppermost (for example, the crown corner between the buccal and occlusal sides, which gives good coverage but is easy to release). The impression material is draped like a blanket over it. Teeth held in the jaw require great care because the impression material can become wedged in the spaces between them. Thin strips of aluminium foil can be gently pushed into the spaces beforehand and then removed with tweezers later. It may be very difficult to release the impression if the buccal, occlusal and lingual crown sides are all covered with the silicone material. This is why it is better just to drape a blanket of impression material over the buccal and occlusal surfaces.

Making microscope sections

Cutting sections is an art which requires much practice. Beginners should start with freshly extracted teeth that are not important specimens. Absolute beginners are best taking a whole premolar (about the right size and commonly extracted by orthodontists) in the fingers and polishing one side on medium-fine, wet-and-dry abrasive paper lubricated with water until the centre line of the tooth is reached.

Turn it over and polish the other side until it is reduced to a thin translucent slice. The evenness of the section thickness is adjusted by varying finger pressure in different places. It can then be polished with finer grades of abrasive slurry, dispersed in water on a glass plate, to remove the scratches. To check it, the section can be temporarily mounted on a glass slide in a drop of water (with a cover slip if desired) and examined under a light microscope. The thickness can be judged by the interference colours seen in a polarising microscope; these should be greys or strong yellows. This simple approach does not produce good sections, but is excellent initial practice. It teaches the importance of the section plane, the evenness of section thickness and the removal of scratches.

For research purposes, the section must be of even thickness (plane-parallel), of the required thickness (usually around 100 μm, but sometimes thinner) and free of scratches. This is achieved by cutting and polishing machines, but sectioning still requires a great deal of practice before good results are achieved. The first requirement is a saw. This usually has a slowly rotating metal disc blade coated with industrial diamond and the tooth is held against it in a jig that allows two plane-parallel cuts to be made. This takes out a slice, in most cases around 500 μm thick.

It is absolutely crucial to centre the section on the tip of the dentine horn underlying the cusp of the tooth crown which is the focus of interest. This will be the central mamelon of incisors and the main point of canines, with the section plane running from labial to lingual and down the central axis of the crown and root. For molars there are usually two section planes: one including both the mesiobuccal and the mesiolingual cusps, and the other including both the distobuccal and distolingual cusps. Again, the section plane runs from buccal to lingual and is parallel to the vertical axis of the crown. The two cuts are carefully positioned on either side of this critical plane and it is worth marking up the surface of the tooth before placing it in the machine, where it is fixed to a counterweighted arm assembly which holds it gently against the outer rim of the cutting disc. Dental sticky wax makes a useful temporary adhesive for this purpose. When the saw cuts out the slice of tooth, a certain amount of tissue is lost in the cut (the width of cut or *kerf*). The thinnest cuts are made by a special saw with a sheet-like annular blade which is stretched in a frame and cuts on its inside.

The slice of tooth is then polished (lapped) down to the required thickness using a precision lapping machine. Most have a rotating glass plate on which abrasive slurry is spread (usually aluminium oxide in distilled water). Traditionally, polishing starts with coarser grades of abrasive and progresses to finer and finer grades. It can, however, be difficult to remove the scratches and experience has shown that it is best to use a fine (3 μm grit) abrasive throughout. The crucial part of the machine is a heavy jig which holds the glass slide on which the tooth slice is mounted during lapping. The jig holds the slide in position by vacuum and keeps the slice against the glass plate so that it is polished down in a plane parallel to the slide surface. It also has a micrometer adjustment which stops polishing at the required thickness, although it is necessary to keep checking by lifting off the jig, looking at the slice, and making an independent measurement using a hand-held thickness gauge.

Both sides of the tooth slice need to be polished. First, the side which will eventually be cemented permanently to the glass slide is polished. The slice is fixed temporarily to a glass slide using an adhesive such as dental sticky wax (sticking down, of course, the side of the slice which is *not* currently being polished). It is then polished down to the level of the critical plane. After releasing the slice from the temporary slide mounting and cleaning carefully (see below), it is turned over and cemented permanently to a clean glass slide using a resin designed specifically for the purpose. This bond is important. It needs to adhere well to both the slide and tooth slice because it will be subjected to large forces during polishing and it provides the main support for the section as it becomes thinner and more delicate. The adhesive also needs to make a parallel layer between the slide and section, or else the section will be polished thinner at one end than the other. The best results are achieved by using a spring-loaded bonding jig while the adhesive sets.

Once mounted again into the lapping machine jig, the section is polished down to its final thickness. As well as checking with a thickness gauge, it is a good idea to check the section in a polarising microscope. The interference colours change as the thickness changes. Once finished, it is common to mount sections permanently by cementing on a glass cover slip to protect the section. If, however, it might be necessary to reduce the thickness further, a good slide box will protect the section and the cover slip can be mounted temporarily using a drop of water. In fact, water is optically a much better mounting medium than normal Canada balsam or equivalent mounting medium.

One of the main requirements is to keep the glass slides and tooth slices scrupulously clean between each stage of the work. This is to stop the surface being scratched by particles left over from previous stages and to present clean, dry surfaces, which are essential for the adhesive to bond correctly. Cleaning is normally achieved in an ultrasonic bath, with much rinsing and then careful drying in a vacuum desiccator.

Freshly extracted teeth are strong enough to be sectioned directly without special treatment. Similarly, many fossilised teeth may be strong enough. A little support can be provided by coating the outside with a layer of cyanoacrylate ('super glue'). Many archaeological specimens, however, are too delicate to survive sectioning without substantial support. They are impregnated before cutting with resin which hardens into a block. It is important to use the hardest resin possible, so that the section polishes evenly, and at present this is methylmethacrylate. This is an irritant which requires a fume cupboard and is difficult to use. There are several commercially available alternatives. It is best to test them on less precious material before using them routinely, because not all work well with teeth. To provide adequate support, they need to have a low enough viscosity to soak right into all the microscopic spaces inside the dental tissues before they are polymerised to make a solid block.

The procedures described here are those developed at the UCL Institute of Archaeology, University College London, by Dr Daniel Antoine. The main companies that market sectioning equipment and materials are: Logitech www.logitech.uk.com; Beuhler www.buehler.com; and Struers www.struers.com.

Microscopy

Low power stereomicroscope and macrophotography

Teeth are small and even basic examination requires at least modest magnification. The simplest instrument to use is a binocular microscope or stereomicroscope. Many biological and anthropological laboratories have one of these as standard equipment. No great magnification is required (10 or 20 times is perfectly adequate), but the lighting is very important. Perikymata are such low relief structures that the direction of illumination has to be just right. It is necessary to use a very bright light, arranged so it crosses the surface of the crown or replica obliquely. This requires experiment and there are many different options, including fibre optic lamps and arrays of light-emitting diodes.

The advantage of the stereomicroscope is that, because both eyes are used, it is possible to gain a good impression of depth and height on the strongly curved surface of the crown. It is, however, important to change the focus continually because not all of the curved surface can be seen in focus at the same time. Some stereomicroscopes are fitted with a camera, although this rarely provides a good image of features such as perikymata because of the small depth of focus. It is usually better to use a single-lens reflex camera with its own lenses. Most macro lenses can produce an image which is the same size as the object being photographed. For teeth, it is usually necessary to do better than this. The simplest and cheapest way to achieve magnification is with extension tubes and a lens-reversing attachment. Using these, it is possible to take images which are twice or four times life size. The exposure times are long and it is therefore necessary to have a good tripod for the camera.

Scanning electron microscope

The best instrument for imaging the crown surface is the scanning electron microscope. Relatively low magnifications are used (typically 20 to 50 times only), but the great advantage is the huge depth of focus, which makes it possible to see the full curve of the tooth crown side in sharp focus. The scanning electron microscope is not affected by light reflections from the glossy enamel surface, which appears smooth and matte in the images. It was previously necessary to coat the tooth surface with a conducting layer, such as gold, and to place the tooth under high vacuum in the specimen chamber of the microscope. Delicate fossils could be cracked by drying under vacuum, so care needs to be taken, but the gold can be cleaned off by simply wiping or by immersing in mercury (hazardous material). Many modern microscopes, however, can be operated at near atmospheric pressure and do not require the specimen to be coated. This makes examination quick and simple. Most of the scanning electron microscope images in this book were of uncoated specimens, untreated except for careful cleaning, imaged using a Hitachi S3700 instrument operated at 20 kV and 50 kPa with an environmental secondary electron detector (ESED). Another option, particularly useful in the case of fossils and museum specimens, is to take an impression of the crown surface and to use this to cast a resin replica sputter-coated with gold (see earlier section). Some resolution is inevitably lost, but most of the structures on the crown surface are still large enough to be observed clearly.

In the conventional mode of operation of the scanning electron microscope (SE or ET mode), the specimen is scanned with a finely focussed primary electron beam, which interacts with the surface of the specimen to release secondary electrons that are attracted to a detector on the side of the chamber (the Everhart Thornley detector). At any point in the scan, the yield of secondary electrons is mainly related to the tilt of the specimen surface relative to the primary electron beam and the positioning of the detector. More secondary electrons are detected from surfaces facing towards the detector, which therefore appear brighter in the image. Conversely, surfaces facing away are darker. The effect is similar to a grassy field lit by brightly shining early morning sun, in which the strongly oblique illumination outlines the tiniest bumps with shadow.

Two images in this book were made using the back-scattered electron (BSE) mode. Some electrons from the primary beam do not penetrate far into the specimen surface and are instead scattered out again. The yield of these back-scattered electrons is measured by an additional detector. Once again, the signal is greatly affected by the tilt of the surface, but it is also affected by the elemental composition and density of the specimen. The microscope and BSE detector can be operated in a way that emphasises the composition part of the signal, so that the brighter parts of the images represent more heavily mineralised features and the darker parts represent less heavily mineralised features.

Conventional transmitted light microscopy

The traditional tool for examining sections of enamel and dentine is a compound microscope set up for observation in transmitted light. The glass slide is illuminated from below through the condenser lens, the image is magnified by an objective lens and then spread across the retina by the eyepiece lens. The tooth section, on its glass slide, is placed on the microscope stage between the condenser and objective lenses and there is usually a mechanism for moving it around in small increments. No great magnification is required for the structures described in this book. Much observation is carried out with a 10× objective lens and a 10× eyepiece lens, giving a total magnification of 100×. A lower power objective (for example, 4×) can be very useful in seeing or photographing a wider area of the section, although, paradoxically, they are more expensive. Occasionally a higher power is useful, with a 30× objective lens and 10× eyepiece lens giving 300× magnification. Higher power objective lenses tend to come closer to the section surface than lower power lenses. After all the care taken in making the section, it is very important to check when changing magnification that the objective lens will not hit the surface.

Most microscopes have a binocular viewing head which has a separate adjustment to allow for the difference in focus between the eyes. It is important to adjust this properly as it makes a big difference to what can be seen. A digital camera usually records the image through a third eyepiece lens (a trinocular head). Most purpose-designed digital microscope cameras are operated through computer software which allows the image to be manipulated in a large variety of ways. One particularly useful function is the ability to stitch neighbouring fields of view together to give a larger image of the whole tooth section. Where the microscope focus is controlled by software, it is also possible to take a stack of images of the same field of view at a slightly different focus and then to combine them to create a higher contrast image, or a three-dimensional model of structures in the section. All the light microscope images in this book are, however, simple digital images.

Although an ordinary compound microscope provides satisfactory images for most purposes, the orientation of the tiny crystallites in the mineral component and the collagen fibres in dentine is studied by a polarising, or petrological, microscope (Schmidt and Keil, 1971). This has one filter below the condenser lens (the polariser) and one above the objective lens (the analyser). Each of these polarises the light beam by constraining the waveforms of which it is composed to only one plane. They are rotated until they are at right angles to one another so that, without a specimen on the microscope stage, they cancel one another out and the field of view is completely dark.

The mineral component and collagen fibres both possess an optical property known as birefringence, which separates and polarises the light beam passing through the section into two components which are resolved by the analyser filter to create a range of colours in the image. These are known as interference colours and they change in hue and brightness with the thickness of the section, the birefringence of the materials in it and the orientation of crystallites, collagen fibres and structures. With a section thickness of 100 μm the colours for enamel and dentine are various shades of grey and yellow. Thicker sections produce brighter yellows, so a polarising microscope provides a good check during sectioning, as discussed earlier.

A proper polarising microscope has a rotating stage calibrated in degrees. If the section is rotated relative to the polariser and analyser, the colours change in brightness (the phenomenon of extinction) and may be used to measure the orientation of the mineral components and collagen (Schmidt and Keil, 1971). Extinction is related to the optic axis which runs along the length of the crystallites in the mineral component and the collagen fibres. As the stage is rotated, the crystallite or fibre will appear dark when the optic axis is perpendicular to either the polariser or the analyser, and is brightest when it is at 45 degrees. This is why the radial crystallisation of the calcospherites is visible as a Maltese cross pattern: at any one point of stage rotation, some crystallites will be perpendicular to the analyser and some perpendicular to the polariser, thus producing the dark arms of the cross. The incremental structures described in Chapter 4 can all be seen clearly without polarising microscopy, but it is required to see some variation in the brown striae of Retzius, which relates to variation in the state of mineralisation.

The key for all observations in transmitted light microscopy is to arrange a contrast in refractive index between the different components of the section. In the case of teeth, the commercial cover slip mounting medium has a refractive index which does not depart greatly from the mineral component of enamel and dentine. For this reason it is a good idea not to mount the slide permanently with a cover slip, but to store it dry and use water (which differs more in refractive index) as a temporary mountant for a cover slip whilst under the microscope. The other crucial step for the strong contrast of incremental structures is to centre and focus the condenser lens carefully and to close the condenser iris diaphragm down to a small aperture. Neglecting this is one reason why some beginners have difficulties in seeing the structures. Another way to increase the contrast is to use dark field illumination (Boyde, 1989), in which only oblique rays of light enter the slide. As its name implies, the background of the image is dark, but light scattered or reflected from the discontinuities within the enamel or dentine shows them up brightly. It is usually arranged simply by placing an opaque disc, known as a patch spot, on the glass filter carrier below the condenser lens. The key is to get the diameter of the spot right, which is achieved by trial and error.

Ideally, any microscope section should be thinner than the structures which are being observed which, in the case of enamel, is around 6 μm. Sections almost as thin as this can be polished, but it is all too easy to damage the tiny wafer of enamel and dentine that remains in the final stages. In practice, most work is based on sections 80–100 μm thick, although some workers polish them down to 40 μm. A thicker section is generally preferred for brown striae of Retzius because their visibility depends on the cumulative effect of light scattering through a thickness of enamel. The plane of focus moves up and down within the tooth section as the focus of the microscope is adjusted. Those features that lie absolutely along this plane are sharply in focus and those which are above or below it become progressively less sharp the further away they are from the plane. The term *depth of field* defines the zone within which features are acceptably in focus (the human eye can accommodate a certain degree of lack of focus). It varies with the magnification of the objective lens and its quality, so a standard 10× lens would have a depth of field of perhaps 9 μm and a 40× lens a depth of field of 1 μm. This might seem satisfactory, but, although only those features within this narrow zone are sharply in focus, out of focus elements from a thicker zone of the section also contribute to the image. With some structures there are considerable possibilities for confusion and care needs to be taken.

Confocal light microscopy

A confocal microscope scans a narrow beam of bright light from a point source across a thin plane of focus within the section through the same objective lens with which the image is observed. A small aperture admits only the light reflected from that plane, so that all reflections from structures not in the plane are eliminated. A sensitive detector records the reflected light and computer software builds the image. In the imaging of dental specimens, the depth of field may be as small as 1 μm. By adjusting the focus through the instrument software, it is possible to image

Technical information

different planes within the section and to build up a three-dimensional model of the structures inside. It is also possible to focus through the tooth crown surface, or through an already broken surface without the destructive sectioning of fossil specimens.

Confocal microscopes are becoming more common in university laboratories. Most are laser-scanning confocal microscopes, which use vibrating mirrors to scan the laser beam across the specimen. Another design is the tandem scanning microscope, which uses a spinning disc pierced with holes to scan the beam. Lacruz and co-workers (Lacruz and Bromage, 2006; Lacruz *et al.*, 2006; 2008) carried a portable version (portable scanning confocal optical microscope) into museum collections. With this, it was possible to take the plane of focus up to 50 µm inwards from the surface of fossil teeth without sectioning.

Since the 1980s there have been large advances in three-dimensional surface scanning technologies at a microscopic scale. The most practical option in the case of tooth surfaces is a confocal microscope. When focussed on the surface, there is a sharp, bright line which marks the point at which the very thin horizontal plane of focus meets the surface. The course of the line depends on the topography of the surface. The depth of the plane of focus is controlled by software, so that a stack of images (a z-series) is recorded, from the lowest to the highest point in the field of view. This image stack is converted by the software into a three-dimensional model of the specimen surface. The model can be rendered in software, with various directions of virtual lighting, points of view and other effects, to produce images that can be seen on the computer screen (Bocaege *et al.*, 2010). It is also possible to take measurements of features in these images so, for example, the spacing of the perikymata can be measured. The Alicona InfiniteFocus, designed for measuring the form and roughness of surfaces, is an optical system of this type (www.alicona.co.uk). Nikon has a useful website which explains many of these new types of microscopy (www.microscopyu.com/).

Micro-CT

Over the ten years before this book was written, there were enormous developments in the field of X-ray imaging. This has made it increasingly possible to examine the internal structure of fossil teeth non-destructively (Smith and Tafforeau, 2008; Smith, 2008). Medical computed tomography (CT) machines were introduced in the 1970s, but have never required the kind of resolution needed for the detailed structure of teeth. A modern medical CT machine has a minimum voxel size of a few hundred micrometres. A voxel is cube-shaped, the three-dimensional equivalent of a square pixel in a more familiar two-dimensional digital image. Computed tomography software uses the scan data to produce a stack of digital images, one on top of another. In effect, they are like closely spaced sections through the specimen and the software can generate them for any plane or orientation of section. In each section image, the brightness or colour of each voxel represents the absorption of X-rays by the material present at that point in the specimen. Enamel, for example, is much more heavily mineralised than dentine, so it absorbs more X-rays as they pass through and the two tissues can be distinguished by their strong contrast.

Micro-CT machines were developed in the 1980s (Elliott and Dover, 1982). Some commercial machines have voxel sizes of the order of a few micrometres, although most are currently operated at between 15 and 30 μm. These make it possible to see details such as the thickness of the enamel layer, the form of the enamel-dentine junction and the shape of the pulp chamber. This can be important information for understanding the development of teeth but, to see fine histological detail equivalent to an actual polished thin section and the conventional light microscope, it is necessary for the voxels to be smaller than 1 μm across. In addition, the contrasts in X-ray absorption between the different structures of enamel are not great enough to be visible in conventional micro-CT. For that, it is necessary to use an altogether higher order of X-ray source: a large synchrotron.

At the time of writing, most dental studies have been based on the European Synchrotron Radiation Facility at Grenoble (Tafforeau *et al.*, 2006). The synchrotron beam has a much higher intensity than that provided by a conventional X-ray tube; it can be filtered to be monochromatic and it is much more coherent. This facilitates voxel sizes down to fractions of a micrometre, greatly reduces 'artefacts' in the image (false features created by the imaging technique) and makes possible the use of a long propagation distance (the distance between the specimen and detector), which allows phase-contrast imaging to be used.

All these characteristics have proved particularly useful in fossils that show taphonomic alteration, showing details that more conventional CT cannot. For tooth fossils, it is possible to image non-destructively brown striae of Retzius and prism cross striations, using a voxel size as small as 0.7 μm. A direct comparison has been made between light microscope images of a polished thin section and a virtual section of similar thickness produced from synchrotron radiation micro-CT scan data taken from the tooth before sectioning (Tafforeau and Smith, 2008). In the published images, the structures look remarkably similar in both contrast and spacing.

Options for measurement

It is possible to take measurements directly with a basic light microscope using a device called a *graticule*. This is a glass disc with a scale printed on it, which is mounted in the eyepiece lens. The scale is calibrated by viewing a scale slide and calculating the conversion factor. A scale slide can similarly be used to calibrate image analysis software such as ImageJ (freeware, available from http://rsbweb.nih.gov/ij/index.html). These methods can be used to measure the spacing of perikyma grooves accurately in tooth sections where the brown striae of Retzius emerge at the crown surface. It is more difficult to measure spacings in a stereomicroscope image or in a photo taken using a camera with a macro lens, partly because the magnification is not high enough and partly because the curvature of the crown means that only a small area can be focussed at once.

One alternative is to use an engineer's measuring microscope, in which the stage is fitted with two micrometers to record horizontal coordinates (X and Y) and the focussing mechanism is similarly fitted with a micrometer to record vertical (Z) coordinates. With this, it is possible to record a profile down the crown (Hillson and Jones, 1989; King et al., 2005), keeping the Y coordinate constant while logging changes in X and Z. From these, it is a simple calculation to determine the direct distance between perikyma grooves. The precision is limited by the depth of focus of the objective lens but, with practice, it is a relatively rapid and direct method for taking measurements.

Software is available for building three-dimensional models of surfaces from scanning electron microscope images. For example, Alicona Mex uses images recorded at several different tilts of the specimen (www.alicona.co.uk). It is then possible to measure the widths, spacings and heights of features, including a profile of the surface in which the perikymata can be recognised and their spacing measured. Another possibility is to use an optical system such as the Alicona InfiniteFocus (Bocaege et al., 2010), which records a stack of images and different focus planes from which it builds a similar three-dimensional model.

At the time of writing, these were fairly specialist and expensive facilities, not widely available in university laboratories. As an alternative, it is possible to take simple measurements from a single straightforward scanning electron microscope image using basic image analysis software such as ImageJ, providing it is over a limited area of the crown surface and the measurements are not calibrated for scale. The measurements are therefore usually in counts of pixels and they are considered relative to one another in the same image. To see features clearly, the crown surface is never horizontal, but needs to be tilted towards the detector. This means that a horizontal scale of measurement will yield an incorrect result. Over a wider field of view the curve of the crown means that the angle of tilt varies, so the error changes. This is why the measurements need to be confined within a limited area. So long as everything is sharply in focus, the error should not be too large as a proportion of the relative measurements between perikymata in the same field of view. Graphs of relative spacing made in this way (Figures 7.11 and 7.12) have been tested against absolute spacing measurements of the same crown using the Alicona InfiniteFocus and the pattern of changes matches closely.

An optical system such as the Alicona InfiniteFocus (Bocaege et al., 2010), requires no coating for an original tooth. Gold coating is required, however, if epoxy replicas are imaged in this instrument. The microscope software creates three-dimensional models of the surface and generates profiles through it, in which the perikyma grooves are readily identifiable and spacings can be measured with resolutions better than 1 µm. Another possibility is to take stereoscopic pairs of images in the scanning electron microscope, using adjustments of eucentric tilt, and then to apply similar Alicona software. The advantage of this approach is a wider field of view and the ready availability of scanning electron microscopes.

The specimens could also be imaged in an ordinary microscope under reflected light. The depth of focus is relatively small, but imaging software on modern instruments allows the compilation of a three-dimensional model from multiple images at different focus settings, which gets around this problem. The software again makes it possible to generate profiles.

References

Aiello, L. C., Montgomery, C. and Dean, C. (1991). The natural history of deciduous tooth attrition in hominoids. *Journal of Human Evolution*, 21, 397–412.

Aiello, L. C. and Wells, J. C. K. (2002). Energetics and the evolution of the genus *Homo*. *Annual Review of Anthropology*, 31, 323–38.

Aine, L., Maki, M., Collin, P. and Keyrilainen, O. (1990). Dental enamel defects in celiac-disease. *Journal of Oral Pathology & Medicine*, 19, 241–5.

Allebeck, P. and Bergh, C. (1992). Height, body mass index and mortality: do social factors explain the association? *Public Health*, 106, 375–82.

AlQahtani, S. J., Hector, M. P. and Liversidge, H. M. (2010). Brief communication: the London atlas of human tooth development and eruption. *American Journal of Physical Anthropology*, 142, 481–90.

Altman, D. G. and Chitty, L. S. (1994). Charts of fetal size: 1. Methodology. *BJOG: An International Journal of Obstetrics & Gynaecology*, 101, 29–34.

Altman, D. G. and Chitty, L. S. (1997). New charts for ultrasound dating of pregnancy. *Ultrasound in Obstetrics and Gynecology*, 10, 174–91.

Altmann, J. and Alberts, S. C. (2005). Growth rates in a wild primate population: ecological influences and maternal effects. *Behavioral Ecology and Sociobiology*, 57, 490–501.

Anderson, D. L., Thompson, G. W. and Popovitch, F. (1976). Age of attainment of mineralisation stages of the permanent dentition. *Journal of Forensic Sciences*, 21, 191–200.

Andresen, V. (1898). Die Querstreifung des Dentins. *Deutsche Monatsschrift für Zahnheilkunde*, 16, 386–9.

Andrews, P. (1996). Palaeoecology and hominoid palaeoenvironments. *Biological Reviews*, 71, 257–300.

Anemone, R. L. (2002). Dental development and life history in hominid evolution. In *Human Evolution Through Developmental Change*, eds. N. Minugh-Purvis and K. J. McNamara. Baltimore & London: Johns Hopkins University Press, pp. 249–80.

Antoine, D. M., Hillson, S. and Dean, M. C. (2009). The developmental clock of dental enamel: a test for the periodicity of prism cross-striations in modern humans and an evaluation of the most likely sources of error in histological studies of this kind. *Journal of Anatomy*, 214, 45–55.

Appleton, J. (1991). The effect of lead acetate on dentine formation in the rat. *Archives of Oral Biology*, 36, 377–82.

Arey, L. B. (1974). *Developmental Anatomy*, revised 7th edn. Philadelphia, London: W. B. Saunders.

Arsuaga, J. L., Martínez, I., Gracia, A. and Lorenzo, C. (1997). The Sima de los Huesos crania (Sierra de Atapuerca, Spain). A comparative study. *Journal of Human Evolution*, 33, 219–81.

Asper, H. (1916). Über die "Braune Retzius'sche Parallelstreifung" im Schmelz der Menschlichen Zähne. *Schweizerische Vierteljahrsschrift für Zahnheilkunde*, 26, 275–314.

Atsalis, S. and Margulis, S. W. (2006). Sexual and hormonal cycles in geriatric *Gorilla gorilla*. *International Journal of Primatology*, 27, 1663–87.

Bäckman, B. (1989). *Amelogenesis imperfecta. An Epidemiologic, Genetic, Morphologic and Clinical Study*. Umeå: Departments of Pedodontics and Oral Pathology. University of Umeå.

Balter, M. (2009). New work may complicate history of Neandertals and *H. sapiens*. *Science*, 326, 224.

Balter, V., Blichert-Toft, J., Braga, J., Telouk, P., Thackeray, F. and Albarède, F. (2008). U-Pb dating of fossil enamel from the Swartkrans Pleistocene hominid site, South Africa. *Earth and Planetary Science Letters*, 267, 236–46.

Barrickman, N. L., Bastian, M. L., Isler, K. and van Schaik, C. P. (2008). Life history costs and benefits of encephalization: a comparative test using data from long-term studies of primates in the wild. *Journal of Human Evolution*, 54, 568–90.

Begun, D. R. (2007). Fossil record of Miocene hominoids. In *Primate Evolution and Human Origins. Vol. 2 of Handbook of Palaeoanthropology*, ed. W. Henke and I. Tattersall. Berlin: Springer, pp. 921–77.

Begun, D. R. (2010). Miocene hominids and the origins of the African apes and humans. *Annual Review of Anthropology*, 39, 67–84.

Bentley, G. R., Goldberg, T. and Jasienska, G. (1993). The fertility of agricultural and non-agricultural traditional societies. *Population Studies*, 47, 269–81.

Berger, T. D. and Trinkaus, E. (1995). Patterns of trauma among the Neandertals. *Journal of Archaeological Science*, 22, 841–52.

Bermúdez de Castro, J. M., Martinón-Torres, M., Carbonell, E. *et al.* (2004). The Atapuerca sites and their contribution to the knowledge of human evolution in Europe. *Evolutionary Anthropology: Issues, News, and Reviews*, 13, 25–41.

Bermúdez de Castro, J. M., Martinón-Torres, M., Prado, L. *et al.* (2010). New immature hominin fossil from European Lower Pleistocene shows the earliest evidence of a modern human dental development pattern. *Proceedings of the National Academy of Sciences*, 107, 11739–44.

Berten, J. (1895). Hypoplasie des Schmelzes (Congenitale Schmelzdefecte; Erosionen). *Deutsche Monatsschrift für Zahnheilkunde*, 13, 425–39.

Beynon, A. D. and Dean, M. C. (1987). Crown-formation time of a fossil hominid premolar tooth. *Archives of Oral Biology*, 32, 773–80.

Beynon, A. D., Dean, M. C., Leakey, M. G., Reid, D. J. and Walker, A. (1998). Comparative dental development and microstructure of *Proconsul* teeth from Rusinga Island, Kenya. *Journal of Human Evolution*, 35, 163–209.

Beynon, A. D., Dean, M. C. and Reid, D. J. (1991a). Histological study on the chronology of the developing dentition in gorilla and orangutan. *American Journal of Physical Anthropology*, 86, 189–203.

Beynon, A. D., Dean, M. C. and Reid, D. J. (1991b). On thick and thin enamel in Hominoids. *American Journal of Physical Anthropology*, 86, 295–309.

Bhat, M. N. and Nelson, K. B. (1989). Developmental enamel defects in primary teeth in children with cerebral palsy, mental retardation, or hearing defects: a review. *Advances in Dental Research*, 3, 132–42.

Bhutani, V. K. (1997). Extrauterine adaptations in the newborn. *Seminars in Perinatology*, 2, 1–12.

Biggerstaff, R. H. (1967). Time trimmers for the Taungs child, or how old is *Australopithecus africanus*? *American Anthropologist*, 69, 217–20.
Bischoff, J. L., Williams, R. W., Rosenbauer, R. J. et al. (2007). High-resolution U-series dates from the Sima de los Huesos hominids yields $600^{+\infty}_{-66}$ kyrs: implications for the evolution of the early Neanderthal lineage. *Journal of Archaeological Science*, 34, 763–70.
Black, G. V. and McKay, F. S. (1916). Mottled teeth: endemic developmental imperfection of teeth heretofore unknown in literature of dentistry. *Dental Cosmos*, 58, 129–56.
Blakey, M. L. and Armelagos, G. J. (1985). Deciduous enamel defects in prehistoric Americans from Dickson Mounds: prenatal and postnatal stress. *American Journal of Physical Anthropology*, 66, 371–80.
Blankenship, J. A., Mincer, H. H., Anderson, K. M., Woods, M. A. and Burton, E. L. (2007). Third molar development in the estimation of chronologic age in American blacks as compared with whites. *Journal of Forensic Sciences*, 52, 428–33.
Boas, F. (1935). The tempo of growth of fraternities. *Proceedings of the National Academy of Sciences of the United States of America*, 21, 413–18.
Bocaege, E., Humphrey, L. T. and Hillson, S. (2010). Technical note: a new three-dimensional technique for high resolution quantitative recording of perikymata. *American Journal of Physical Anthropology*, 141, 498–503.
Boesch, C. and Boesch-Achermann, H. (2000). *The Chimpanzees of the Taï Forest: Behavioural Ecology and Evolution*. Oxford University Press.
Bogin, B. (1999). *Patterns of Human Growth*. 2nd edn. Cambridge Studies in Biological Anthropology, 23. Cambridge University Press.
Bogin, B. (2003). The human pattern of growth and development in a paleontological perspective. In *Patterns of Growth and Development in the Genus Homo*, ed. J. L. Thompson, G. E. Krovitz and A. J. Nelson. Cambridge University Press, pp. 15–44.
Bogin, B. (2006). Modern human life history: the evolution of human childhood and fertility. In *The Evolution of Human Life History*, ed. K. Hawkes and R. R. Paine. Santa Fe: School of American Research Press, pp. 197–230.
Boorse, C. (1977). Health as a theoretical concept. *Philosophy of Science*, 44, 542–73.
Boyde, A. (1970). The surface of the enamel in human hypoplastic teeth. *Archives of Oral Biology*, 15, 897–8.
Boyde, A. (1989). Enamel. In *Teeth*, ed. B. K. B. Berkovitz, A. Boyde, R. M. Frank et al. New York, Berlin & Heidelberg: Springer Verlag, pp. 309–473.
Boyde, A. (1990). Developmental interpretations of dental microstructure. In *Primate Life History and Evolution*, ed. C. J. DeRousseau. New York: Wiley-Liss, pp. 229–67.
Brain, C. K. (1981). *The Hunters or the Hunted?* University of Chicago Press.
Brescia, N. J. (1961). *Applied Dental Anatomy*. St Louis: C.V. Mosby.
Brody, S. (1945). *Bioenergetics and Growth: With Special Reference to the Efficiency Complex in Domestic Animals*. New York: Reinhold.
Bromage, T. G. (1991). Enamel incremental periodicity in the pig-tailed macaque: a polychrome fluorescent labeling study of dental hard tissues. *American Journal of Physical Anthropology*, 86, 205–14.
Bromage, T. G. and Dean, M. C. (1985). Re-evaluation of the age at death of immature fossil hominids. *Nature*, 317, 525–7.
Bryant, T. (1884). *A Manual for the Practice of Surgery*, 4th edn. London: J & A Churchill.
Buffon, G. L. L. de (1777). *Histoire naturelle, générale et particulière. Supplement IV. Servant de suite à l'histoire naturelle de l'homme*. Paris: Imprimerie Royale.

Bunon, R. (1743). *Essay sur les maladies des dents, ou l'on propose les moyens de leur procurer une bonne conformation dès la plus tendre enfance, & d'en assurer la conservation pendant tout le cours de la vie*. Paris: Briasson.

Bunon, R. (1746). *Expériences et demonstrations faites à l'hôpital de la Salpêtriere, et à S. Côme en présence de l'Académie Royale de Chirurgie*. Paris: Briasson.

Butler, P. M. (1967). The prenatal development of the human first upper permanent molar. *Archives of Oral Biology*, 12, 551–63.

Campbell, T. D. (1925). *Dentition and Palate of the Australian Aboriginal*. Publications under the Keith Sheridan Foundation. Adelaide: University of Adelaide.

Cannon, W. B. (1915). *Bodily Changes in Pain, Hunger, Fear, and Rage; an Account of Recent Researches into the Function of Emotional Excitement*, 1st edn. New York & London: D. Appleton.

Cannon, W. B. (1932). *The Wisdom of the Body*, 1st edn. New York: W.W. Norton.

Charnov, E. L. and Berrigan, D. (1993). Why do female primates have such long lifespans and so few babies? Or life in the slow lane. *Evolutionary Anthropology: Issues, News, and Reviews*, 1, 191–4.

Chitty, L. S. and Altman, D. G. (2002). Charts of fetal size: limb bones. *BJOG: An International Journal of Obstetrics & Gynaecology*, 109, 919–29.

Chitty, L. S., Altman, D. G., Henderson, A. and Campbell, S. (1994a). Charts of fetal size: 2. Head measurements. *BJOG: An International Journal of Obstetrics & Gynaecology*, 101, 35–43.

Chitty, L. S., Altman, D. G., Henderson, A. and Campbell, S. (1994b). Charts of fetal size: 3. Abdominal measurements. *BJOG: An International Journal of Obstetrics & Gynaecology*, 101, 125–31.

Chitty, L. S., Altman, D. G., Henderson, A. and Campbell, S. (1994c). Charts of fetal size: 4. Femur length. *BJOG: An International Journal of Obstetrics & Gynaecology*, 101, 132–5.

Chitty, L. S., Campbell, S. and Altman, D. G. (1993). Measurement of the fetal mandible: feasibility and construction of a centile chart. *Prenatal Diagnosis*, 13, 749–56.

Christensen, G. J. and Kraus, B. S. (1965). Initial calcification of the human permanent first molar. *Journal of Dental Research*, 44, 1338–42.

Clement, A. F., Hillson, S. W. and Aiello, L. C. (2011). Tooth wear, Neanderthal facial morphology and the anterior dental loading hypothesis. *Journal of Human Evolution*, 62, 367–76.

Cohen, A. A. (2004). Female post-reproductive lifespan: a general mammalian trait. *Biological Reviews*, 79, 733–50.

Colyer, J. F. (1936). *Variations and Diseases of the Teeth of Animals*. London: John Bale & Danielsson.

Colyer, J. F. (1947). Dental disease in animals. *British Dental Journal*, 82, 2–10, 31–5.

Combs, G. F. (1992). *The Vitamins. Fundamental Aspects in Nutrition and Health*. San Diego: Academic Press.

Commission on Oral Health (1982). An epidemiological index of developmental defects of dental enamel (DDE Index). *International Dental Journal*, 32, 159–67.

Conroy, G. C. and Mahoney, C. J. (1991). Mixed longitudinal study of dental emergence in the chimpanzee, *Pan troglodytes* (primates, pongidae). *American Journal of Physical Anthropology*, 86, 243–54.

Cox, M. (1996). *Life and Death in Spitalfields: 1700–1850*. York: Council for British Archaeology.

Cucina, A., Vargiu, R., Mancinelli, D. et al. (2006). The necropolis of Vallerano (Rome, 2nd to 3rd century AD): an anthropological perspective on the ancient Romans in the Suburbium. *International Journal of Osteoarchaeology*, 16, 104–17.

Cunha, E., Rozzi, F. R., De Castro, J. M. B., Martinón-Torres, M., Wasterlain, S. N. and Sarmiento, S. (2004). Enamel hypoplasias and physiological stress in the Sima de los Huesos Middle Pleistocene hominins. *American Journal of Physical Anthropology*, 125, 220–31.

Czermák, J. (1850). Beiträge zur mikroskopischen Anatomie der menschlichen Zähn. *Zeitschrift für wissenschaftliche Zoologie*, 2, 295–322.

Dahlberg, A. A. and Menegaz-Bock, R. M. (1958). Emergence of the permanent teeth in Pima Indian children. *Journal of Dental Research*, 37, 1123–40.

Dart, R. A. (1925). *Australopithecus africanus*: the man-ape of South Africa. *Nature*, 115, 195–9.

Deacon, T. W. (1992). Primate brains and senses. In *The Cambridge Encyclopedia of Human Evolution*, ed. S. Jones, R. Martin and D. Pilbeam, Cambridge University Press, pp. 109–14.

Dean, M. C. (1993). Daily rates of dentine formation in macaque tooth roots. *International Journal of Osteoarchaeology*, 3, 199–207.

Dean, M. C. (1995). The nature and periodicity of incremental lines in primate dentine and their relationship to periradicular bands in OH 16 (*Homo habilis*). In *Structure, Function and Evolution of Teeth. Dental Morphology Meeting, Florence, September 1992*, ed. J. Moggi-Cecchi. Florence: International Institute for the Study of Man, pp. 239–65.

Dean, M. C. (1998a). A comparative study of cross striation spacings in cuspal enamel and of four methods of estimating the time taken to grow molar cuspal enamel in *Pan*, *Pongo* and *Homo*. *Journal of Human Evolution*, 35, 449–62.

Dean, M. C. (1998b). Comparative observations on the spacing of short-period (von Ebner's) lines in dentine. *Archives of Oral Biology*, 43, 1009–21.

Dean, M. C. (1999). Hominoid tooth growth: using incremental lines in dentine as markers of growth in modern human and fossil primate teeth. In *Human Growth in the Past: Studies from Bones and Teeth*, ed. R. Hoppa and C. M. FitzGerald. Cambridge University Press, pp. 111–27.

Dean, M. C. (2007). A radiographic and histological study of modern human lower first permanent molar root growth during the supraosseous eruptive phase. *Journal of Human Evolution*, 53, 635–46.

Dean, M. C. (2009). Extension rates and growth in tooth height of modern human and fossil hominin canines and molars. In *Frontiers of Oral Biology: Interdisciplinary Dental Morphology*, ed. T. Koppe, G. Meyer and G. R. Alt. Basel: Karger, pp. 68–73.

Dean, M. C. (2010). Retrieving chronological age from dental remains of early fossil hominins to reconstruct human growth in the past. *Philosophical Transactions of the Royal Society B: Biological Sciences*, 365, 3397–410.

Dean, M. C. (2012). A histological method that can be used to estimate the time taken to form the crown of a permanent tooth. In *Forensic Microscopy for Skeletal Tissues, Methods and Protocols*, ed. L. S. Bell. New York: Humana Press, pp. 89–100.

Dean, M. C., Beynon, A., Reid, D. J. and Whittaker, D. (1993a). A longitudinal study of tooth growth in a single individual based on long and short period incremental markings in dentine and enamel. *International Journal of Osteoarchaeology*, 3, 249–64.

Dean, M. C., Beynon, A. D., Thackeray, J. F. and Macho, G. A. (1993b). Histological reconstruction of dental development and age at death of a juvenile *Paranthropus robustus* specimen, SK 63, from Swartkrans, South Africa. *American Journal of Physical Anthropology*, 91, 401–20.

Dean, M. C., Leakey, M. G., Reid, D. J. et al. (2001). Growth processes in teeth distinguish modern humans from *Homo erectus* and earlier hominins. *Nature*, 414, 628–31.

Dean, M. C. and Lucas, V. S. (2009). Dental and skeletal growth in early fossil hominins. *Annals of Human Biology*, 36, 545–61.

Dean, M. C. and Reid, D. J. (2001). Perikymata spacing and distribution on hominid anterior teeth. *American Journal of Physical Anthropology*, 116, 209–15.

Dean, M. C. and Scandrett, A. E. (1995). Rates of dentine mineralization in permanent human teeth. *International Journal of Osteoarchaeology*, 5, 349–58.

Dean, M. C. and Scandrett, A. E. (1996). The relation between long-period incremental markings in dentine and daily cross-striations in enamel in human teeth. *Archives of Oral Biology*, 41, 233–41.

Dean, M. C. and Schrenk, F. (2003). Enamel thickness and development in a third permanent molar of *Gigantopithecus blacki*. *Journal of Human Evolution*, 45, 381–8.

Dean, M. C. and Smith, B. H. (2009). Growth and development of the Nariokotome youth, KNM-WT 15000. In *The First Humans: Origin and Early Evolution of the Genus Homo*, ed. F. E. Grine and J. G. Fleagle. London: Springer, pp. 101–20.

Dean, M. C. and Vesey, P. (2008). Preliminary observations on increasing root length during the eruptive phase of tooth development in modern humans and great apes. *Journal of Human Evolution*, 54, 258–71.

Dean, M. C. and Wood, B. A. (1981). Developing pongid dentition and its use for ageing crania in comparative cross-sectional growth studies. *Folia Primatologia*, 36, 111–27.

Deaner, R. O., Barton, R. A. and van Schaik, C. (2003). Primate brains and life histories: renewing the connection. In *Primate Life Histories and Socioecology*, ed. P. M. Kappeler and M. E. Pereira. University of Chicago Press, pp. 233–65.

Demirjian, A. and Goldstein, H. (1976). New systems for dental maturity based on seven and four teeth. *Annals of Human Biology*, 3, 411–21.

Demirjian, A., Goldstein, H. and Tanner, J. M. (1973). A new system of dental age assessment. *Human Biology*, 45, 211–27.

Demirjian, A. and Levesque, G. Y. (1980). Sexual differences in dental development and prediction of emergence. *Journal of Dental Research*, 59, 1110–22.

Deutsch, D., Tam, O. and Stack, M. V. (1985). Postnatal changes in size, morphology and weight of developing postnatal deciduous anterior teeth. *Growth*, 49, 202–17.

Dirks, W., Reid, D. J., Jolly, C. J., Phillips-Conroy, J. E. and Brett, F. L. (2002). Out of the mouths of baboons: stress, life history, and dental development in the Awash National Park Hybrid Zone, Ethiopia. *American Journal of Physical Anthropology*, 118, 239–52.

Domínguez-Rodrigo, M., Rayne Pickering, T., Semaw, S. and Rogers, M. J. (2005). Cutmarked bones from Pliocene archaeological sites at Gona, Afar, Ethiopia: implications for the function of the world's oldest stone tools. *Journal of Human Evolution*, 48, 109–21.

Edmund, A. G. (1960). Tooth replacement phenomena in the lower vertebrates. *Royal Ontario Museum Life Sciences Division Contributions*, 52, 1–190.

Eli, I., Sarnat, H. and Talmi, E. (1989). Effect of the birth process on the neonatal line in primary tooth enamel. *Pediatric Dentistry*, 11, 220–3.

Eliot, M. M., Souther, S. P., Anderson, B. G. and Arnim, S. S. (1934). A study of the teeth of a group of school children previously examined for rickets. *American Journal of Diseases of Children*, 48, 713–29.

Elliott, J. C. and Dover, S. D. (1982). X-ray microtomography. *Journal of Microscopy*, 126, 211–13.

Emery Thompson, M., Jones, J. H., Pusey, A. E. *et al.* (2007). Aging and fertility patterns in wild chimpanzees provide insights into the evolution of menopause. *Current Biology*, 17, 2150–6.

Engle, W. A. (2004). Age terminology during the perinatal period. *Pediatrics*, 114, 1362–4.

Engle, W. A. (2006). A recommendation for the definition of "late preterm" (near-term) and the birth weight-gestational age classification system. *Seminars in Perinatology*, 30, 2–7.

Ensor, B. E. and Irish, J. D. (1995). Hypoplastic area method for analysing dental enamel hypoplasia. *American Journal of Physical Anthropology*, 98, 507–18.

Ensor, B. E. and Irish, J. D. (1997). Reply to Blakey and Armelagos with additional remarks on the hypoplastic area method. *American Journal of Physical Anthropology*, 102, 296–9.

Enwonwu, C. O. (1973). Influence of socio-economic conditions on dental development in Nigerian children. *Archives of Oral Biology*, 18, 95–107.

Erben, R. G. (2003). Bone-labeling techniques. In *Handbook of Histology Methods for Bone and Cartilage*, ed. Y. H. An and K. L. Martin. Totowa, NJ: Humana Press, pp. 99–117.

Eveleth, P. B. and Tanner, J. M. (1990). *Worldwide Variation in Human Growth*, 2nd edn. Cambridge University Press.

Fauchard, P. (1728). *Le chirurgien dentiste, ou traité des dents. Ou l'on enseigne les moyens de les entretenir propres & saines, de les embellir, d'en réparer la perte & de remedier à leurs maladies, à celles des geneives & aux accidens qui peuvent survenir aux autres parties voisines des dents. Avec des observations & des reflexions sur plusieurs cas singuliers*, 1st edn. Paris: Jean Mariette.

Fejerskov, O., Larsen, M. J., Richards, A. and Baelum, V. (1994). Dental tissues effects of fluoride. *Advances in Dental Research*, 8, 15–31.

Finney, D. J. (1971). *Probit Analysis*, 3rd edn. Cambridge University Press.

Finney, D. J. (1978). *Statistical Method in Biological Assay*. New York: Macmillan.

FitzGerald, C. and Rose, J. (2000). Reading between the lines: dental development and sub-adult age assessment using the microstructural growth markers of teeth. In *Biological Anthropology of the Human Skeleton*, ed. M. A. Katzenberg and S. R. Saunders. New York: Wiley, pp. 163–86.

FitzGerald, C., Saunders, S., Bondioli, L. and Macchiarelli, R. (2006). Health of infants in an Imperial Roman skeletal sample: perspective from dental microstructure. *American Journal of Physical Anthropology*, 130, 179–89.

FitzGerald, C. M. (1998). Do enamel microstructures have regular time dependency? Conclusions from the literature and a large-scale study. *Journal of Human Evolution*, 35, 371–86.

FitzGerald, C. M. and Saunders, S. R. (2005). Test of histological methods of determining chronology of accentuated striae in deciduous teeth. *American Journal of Physical Anthropology*, 127, 277–90.

Floyd, B. and Littleton, J. (2006). Linear enamel hypoplasia and growth in an Australian Aboriginal community: not so small, but not so healthy either. *Annals of Human Biology*, 33, 424–43.

Folayan, M., Owotade, F., Adejuyigbe, E., Sen, S., Lawal, B. and Ndukwe, K. (2007). The timing of eruption of the primary dentition in Nigerian children. *American Journal of Physical Anthropology*, 134, 443–8.

Fooden, J. and Izor, R. J. (1983). Growth curves, dental emergence norms, and supplementary morphological observations in known-age captive orangutans. *American Journal of Primatology*, 5, 285–301.

Fournier, J. A. (1881). *Syphilis and Marriage*, translated edn. New York: D. Appleton.

Fournier, J. A. (1884). Syphilitic Teeth. *Dental Cosmos*, 26, 12–25, 141–55.

Fournier, J. A. (1907). *The Treatment and Prophylaxis of Syphilis*, English translation edn. New York: Rebman.

Garn, S. M., Lewis, A. B. and Blizzard, R. M. (1965). Endocrine factors in dental development. *Journal of Dental Research*, 44, 243–8.

Garn, S. M., Lewis, A. B. and Kerewsky, S. (1965). Genetic, nutritional, and maturational correlates of dental development. *Journal of Dental Research*, 44, 228–42.

Garn, S. M., Lewis, A. B., Koski, K. and Polachek, D. L. (1958). The sex difference in tooth calcification. *Journal of Dental Research*, 37, 561–7.

Gates, R. E. (1966). Computation of the median age of eruption of permanent teeth using probit analysis and an electronic computer. *Journal of Dental Research*, 45, 1024–8.

Gingerich, P. D., Smith, B. H. and Rosenberg, K. (1982). Allometric scaling in the dentition of primates and prediction of body weight from tooth size in fossils. *American Journal of Physical Anthropology*, 58, 81–100.

Gleiser, I. and Hunt, E. E. (1955). The permanent mandibular first molar: its calcification, eruption and decay. *American Journal of Physical Anthropology*, 13, 253–84.

Godfrey, L. R., Samonds, K. E., Jungers, W. L. and Sutherland, M. R. (2003). Dental development and primate life histories. In *Primate Life Histories and Socioecology*, ed. P. M. Kappeler and M. E. Pereira. University of Chicago Press, pp. 177–203.

Goldstein, D. S. and Kopin, I. J. (2007). Evolution of concepts of stress. *Stress: the International Journal on the Biology of Stress*, 10, 109–20.

Goodall, J. (1986). *The Chimpanzees of Gombe: Patterns of Behavior*. Cambridge, MA: Belknap Press of Harvard University Press.

Goodman, A. H. and Armelagos, G. J. (1985). Disease and death at Dr Dickson's Mounds. *Natural History*, 9(85), 12–19.

Goodman, A. H. and Armelagos, G. J. (1988). Childhood stress and decreased longevity in a prehistoric population. *American Anthropologist*, 90, 936–44.

Goodman, A. H., Armelagos, G. J. and Rose, J. C. (1980). Enamel hypoplasias as indicators of stress in three prehistoric populations from Illinois. *Human Biology*, 52, 515–28.

Goodman, A. H., Armelagos, G. J. and Rose, J. C. (1984a). The chronological distribution of enamel hypoplasias from prehistoric Dickson Mounds populations. *American Journal of Physical Anthropology*, 65, 259–66.

Goodman, A. H., Lallo, J., Armelagos, G. J. and Rose, J. C. (1984b). Health changes at Dickson Mounds, Illinois (A.D. 950–1300). In *Palaeopathology at the Origins of Agriculture*, ed. M. N. Cohen and G. J. Armelagos. New York: Academic Press, pp. 271–306.

Goodman, A. H., Martinez, C. and Chavez, A. (1991). Nutritional supplementation and the development of linear enamel hypoplasias in children from Tezonteopan, Mexico. *American Journal of Clinical Nutrition*, 53, 773–81.

Goodman, A. H. and Rose, J. C. (1990). Assessment of systemic physiological perturbations from dental enamel hypoplasias and associated histological structures. *YearBook of Physical Anthropology*, 33, 59–110.

Goodman, A. H. and Song, R.-J. (1999). Sources of variation in estimated ages at formation of linear enamel hypoplasias. In *Human Growth in the Past: Studies from Bones and Teeth*, ed. R. Hoppa and C. M. FitzGerald. Cambridge University Press, pp. 210–40.

Goodman, A. H., Thomas, R. B., Swedlund, A. C. and Armelagos, G. J. (1988). Biocultural perspectives of stress in prehistoric, historical and contemporary population research. *Yearbook of Physical Anthropology*, 31, 169–202.

Gradstein, F. M., Ogg, J. G. and Smith, A. G. (2005). *A Geologic Time Scale 2004*. Cambridge University Press.

Grether, W. F. and Yerkes, R. M. (1940). Weight norms and relations for chimpanzee. *American Journal of Physical Anthropology*, 27, 182–97.

Grün, R., Stringer, C., McDermott, F. *et al.* (2005). U-series and ESR analyses of bones and teeth relating to the human burials from Skhul. *Journal of Human Evolution*, 49, 316–34.

Guatelli-Steinberg, D. (2001). What can developmental defects of enamel reveal about physiological stress in non-human primates? *Evolutionary Anthropology*, 10, 138–51.

Guatelli-Steinberg, D. (2004). Analysis and significance of linear enamel hypoplasia in Plio Pleistocene hominins. *American Journal of Physical Anthropology*, 123, 199–215.

Guatelli-Steinberg, D. (2009). Recent studies of dental development in Neandertals: implications for Neandertal life histories. *Evolutionary Anthropology*, 18, 9–20.

Guatelli-Steinberg, D., Ferrell, R. J. and Spence, J. (2012). Linear enamel hypoplasia as an indicator of physiological stress in great apes: reviewing the evidence in light of enamel growth variation. *American Journal of Physical Anthropology*, 148, 191–204.

Guatelli-Steinberg, D., Larsen, C. S. and Hutchinson, D. L. (2004). Prevalence and the duration of linear enamel hypoplasia: a comparative study of Neandertals and Inuit foragers. *Journal of Human Evolution*, 47, 65–84.

Guatelli-Steinberg, D. and Reid, D. J. (2008). What molars contribute to an emerging understanding of lateral enamel formation in Neandertals vs. modern humans. *Journal of Human Evolution*, 54, 236–50.

Guatelli-Steinberg, D., Reid, D. J. and Bishop, T. A. (2007). Did the lateral enamel of Neandertal anterior teeth grow differently from that of modern humans? *Journal of Human Evolution*, 52, 72–84.

Guatelli-Steinberg, D., Reid, D. J., Bishop, T. A. and Larsen, C. S. (2005). Anterior tooth growth periods in Neandertals were comparable to those of modern humans. *Proceedings of the National Academy of Sciences USA*, 102, 14197–202.

Guatelli-Steinberg, D., Reid, D. J., Bishop, T. A. and Larsen, C. S. (2007a). Imbricational enamel formation in Neandertals and recent modern humans. In *Dental Perspectives on Human Evolution: State of the Art Research in Dental Paleoanthropology*, ed. S. E. Bailey and J. J. Hublin. Dordrecht: Springer, pp. 211–30.

Guatelli-Steinberg, D., Reid, D. J., Bishop, T. A. and Larsen, C. S. (2007b). Not so fast: a reply to Ramirez Rozzi and Sardi (2007). *Journal of Human Evolution*, 53, 114–18.

Guatelli-Steinberg, D. and Skinner, M. (2000). Prevalence and etiology of linear enamel hypoplasia in monkeys and apes from Asia and Africa. *Folia Primatologica*, 71, 115–32.

Gurven, M. and Kaplan, H. (2007). Longevity among hunter gatherers: a cross cultural examination. *Population and Development Review*, 33, 321–65.

Gustafson, A. G. (1955). The similarity between contralateral pairs of teeth. *Odontologisk Tidskrift*, 63, 245–8.

Gustafson, G. and Gustafson, A. G. (1967). Microanatomy and histochemistry of enamel. In *Structural and Chemical Organization of Teeth*, ed. A. E. W. Miles. London: Academic Press, pp. 135–62.

Gysi, A. (1931). Metabolism in adult enamel. *Dental Digest*, 37, 661–8.

Haavikko, K. (1970). The formation and the alveolar and clinical eruption of the permanent teeth. *Proceedings of the Finnish Dental Society*, 66, 101–70.

Hamada, Y. and Udono, T. (2002). Longitudinal analysis of length growth in the chimpanzee (*Pan troglodytes*). *American Journal of Physical Anthropology*, 118, 268–84.

Hamada, Y., Udono, T., Teramoto, M. and Hayasaka, I. (2004). Body, head and facial growth: comparison between macaques (*Macaca fuscata*) and chimpanzee (*Pan troglodytes*) based on somatometry. *Annals of Anatomy*, 186, 451–61.

Hamilton, M. A., Russo, R. C. and Thurston, R. V. (1977). Trimmed Spearman-Karber method for estimating median lethal concentrations in toxicity bioassays. *Environmental Science & Technology*, 11, 714–19.

Hannibal, D. L. and Guatelli-Steinberg, D. (2005). Linear enamel hypoplasia in the great apes: analysis by genus and locality. *American Journal of Physical Anthropology*, 127, 13–25.

Harcourt, J. K., Johnson, N. W. and Storey, E. (1962). *In vivo* incorporation of tetracycline in the teeth of man. *Archives of Oral Biology*, 7, 431–7.

Harris, E. F. (2007). Mineralization of the mandibular third molar: a study of American blacks and whites. *American Journal of Physical Anthropology*, 132, 98–109.

Harris, E. F. and Buck, A. L. (2002). Tooth mineralization: a technical note on the Moorrees-Fanning-Hunt standards. *Dental Anthropology*, 16, 15–21.

Harrison, T. (2010). Apes among the tangled branches of human origins. *Science*, 327, 532.

Harvati, K., Singh, N. and Lòpez, E. N. (2011). A three-dimensional look at the Neanderthal mandible. In *Continuity and Discontinuity in the Peopling of Europe*, ed. S. Condemi and G.-C. Weniger. Dordrecht: Springer, pp. 179–92.

Harvey, P. H. and Clutton-Brock, T. H. (1985). Life history variation in primates. *Evolution*, 39, 559–81.

Hassanali, J. (1985). The third permanent molar eruption in Kenyan Africans and Asians. *Annals of Human Biology*, 12, 517–23.

Hassanali, J. and Odhiambo, J. W. (1981). Ages of eruption of the permanent teeth in Kenyan African and Asian children. *Annals of Human Biology*, 8, 425–34.

Hawkes, K., O'Connell, J. F. and Blurton Jones, N. G. (2003). Human life histories: primate trade-offs, grandmothering sociology, and the fossil record. In *Primate Life Histories and Socioecology*, ed. P. M. Kappeler and M. E. Pereira. University of Chicago Press, pp. 204–27.

Hill, K., Boesch, C., Goodall, J., Pusey, A., Williams, J. and Wrangham, R. (2001). Mortality rates among wild chimpanzees. *Journal of Human Evolution*, 40, 437–50.

Hillson, S. W. (1979). Diet and dental disease. *World Archaeology*, 11, 147–62.

Hillson, S. W. (1992a). Dental enamel growth, perikymata and hypoplasia in ancient tooth crowns. *Journal of the Royal Society of Medicine*, 85, 460–6.

Hillson, S. W. (1992b). Impression and replica methods for studying hypoplasia and perikymata on human tooth crown surfaces from archaeological sites. *International Journal of Osteoarchaeology*, 2, 65–78.

Hillson, S. W. (1992c). Studies of growth in dental tissues. In *Culture, Ecology & Dental Anthropology*, ed. J. R. Lukacs. Delhi: Kamla-Raj Enterprises, pp. 7–23.

Hillson, S. W. (1996). *Dental Anthropology*. Cambridge University Press.

Hillson, S. W. (2000). Dental pathology. In *Biological Anthropology of the Human Skeleton*, ed. M. A. Katzenberg and S. R. Saunders. New York: Wiley, pp. 249–86.

Hillson, S. W. (2005). *Teeth*. 2nd edn. Cambridge Manuals in Archaeology. Cambridge University Press.

Hillson, S. W. and Antoine, D. M. (2011). The mechanisms that produce the defects of enamel hypoplasia. *American Journal of Physical Anthropology*, Supplement: Program of the 80th Annual Meeting of the American Association of Physical Anthropologists, 163.

Hillson, S. W., Antoine, D. M. and Dean, M. C. (1999). A detailed developmental study of the defects of dental enamel in a group of post-Medieval children from London. In *Dental Morphology '98*, ed. J. T. Mayhall and T. Heikinnen. Oulu University Press, pp. 102–11.

Hillson, S. W. and Bond, S. (1997). Relationship of enamel hypoplasia to the pattern of tooth crown growth: a discussion. *American Journal of Physical Anthropology*, 104, 89–104.

Hillson, S. W., Grigson, C. and Bond, S. (1998). The dental defects of congenital syphilis. *American Journal of Physical Anthropology*, 107, 25–40.

Hillson, S. W. and Jones, B. K. (1989). Instruments for measuring surface profiles: an application in the study of ancient human tooth crown surfaces. *Journal of Archaeological Science*, 16, 95–105.

Hirota, F. (1982). Prism arrangement in human cusp enamel deduced by X-ray diffraction. *Archives of Oral Biology*, 27, 931–7.

Hodges, D. C. and Wilkinson, R. G. (1990). Effect of tooth size on the ageing and chronological distribution of enamel hypoplastic defects. *American Journal of Human Biology*, 2, 553–60.

Hoffman, M. A. (1984). *Egypt Before the Pharaohs*. London: ARK Paperbacks.

Holliday, T. W. (1997). Body proportions in Late Pleistocene Europe and modern human origins. *Journal of Human Evolution*, 32, 423–47.

Holt, S. A., Reid, D. J. and Guatelli-Steinberg, D. (2012). Brief communication: Premolar enamel formation: completion of figures for aging LEH defects in permanent dentition. *Dental Anthropology*, 25, 4–7.

Hsu, S. C. and Levine, M. A. (2004). Perinatal calcium metabolism: physiology and pathophysiology. *Seminars in Neonatology*, 9, 23–36.

Hu, H., Shih, R., Rothenberg, S. and Schwartz, B. S. (2007). The epidemiology of lead toxicity in adults: measuring dose and consideration of other methodologic issues. *Environmental Health Perspectives*, 115, 455–62.

Humphrey, L. T. (2010). Weaning behaviour in human evolution. *Seminars in Cell & Developmental Biology*, 21, 453–61.

Hurme, V. O. (1948). Standards of variation in the eruption of the first six permanent teeth. *Child Development*, 19, 213–31.

Hurme, V. O. (1960). Estimation of monkey age by dental formula. *Annals of the New York Academy of Sciences*, 85, 795–9.

Hurme, V. O. and van Wagenen, G. (1953). Basic data on the emergence of deciduous teeth in the monkey (*Macaca mulatta*). *Proceedings of the American Philosophical Society*, 97, 291–315.

Hurme, V. O. and Van Wagenen, G. (1956). Emergence of permanent first molars in the monkey (*Macaca mulatta*). Association with other growth phenomena. *The Yale Journal of Biology and Medicine*, 28, 538–67.

Hurme, V. O. and van Wagenen, G. (1961). Basic data on the emergence of permanent teeth in the rhesus monkey "(*Macaca mulatta*)". *Proceedings of the American Philosophical Society*, 105, 105–40.

Hutchinson, D. L. and Larsen, C. S. (1988). Determination of stress episode duration from linear enamel hypoplasias: a case study from St Catherine's Island, Georgia. *Human Biology*, 60, 93–110.

Hutchinson, D. L. and Larsen, C. S. (2001). Enamel hypoplasia and stress in la Florida. In *Bioarchaeology of Spanish Florida. The Impact of Colonialism*, ed. C. S. Larsen. Gainsville: University Press of Florida, pp. 181–206.

Hutchinson, D. L., Larsen, C. S. and Choi, I. (1997). Stressed to the max? Physiological perturbation in the Krapina Neandertals. *Current Anthropology*, 38, 904–14.

Hutchinson, J. (1857). On the influence of hereditary syphilis on the teeth. *Transactions of the Odontological Society of Great Britain*, 2, 95–106.

Hutchinson, J. (1858). Report on the effects of infantile syphilis in marring the development of teeth. *Transactions of the Pathological Society of London*, 9, 449–56.

Hutchinson, J. (1887). *Syphilis*. London: Cassell.

Ice, G. H. and James, G. D. (2007). *Measuring Stress in Humans. A Practical Guide for the Field*. Cambridge Studies in Biological and Evolutionary Anthropology. Cambridge University Press.

Infante, P. F. and Gillespie, G. M. (1974). An epidemiologic study of linear enamel hypoplasia of deciduous anterior teeth in Guatemalan children. *Archives of Oral Biology*, 19, 1055–61.

Iuliano Burns, S., Mirwald, R. L. and Bailey, D. A. (2001). Timing and magnitude of peak height velocity and peak tissue velocities for early, average, and late maturing boys and girls. *American Journal of Human Biology*, 13, 1–8.

Jernvall, J. and Thesleff, I. (2012). Tooth shape formation and tooth renewal: evolving with the same signals. *Development*, 139, 3487–97.

Johnsen, D., Krejci, C., Hack, M. and Fanaroff, A. (1984). Distribution of enamel defects and the association with respiratory distress in very low birthweight infants. *Journal of Dental Research*, 63, 59–64.

Jones, S. J. and Boyde, A. (1984). Ultrastructure of dentin and dentinogenesis. In *Dentin and Dentinogenesis*, ed. A. Linde. Boca Raton: CRC Press, pp. 81–134.

Jørgensen, K. D. (1956). The deciduous dentition. A descriptive and comparative anatomical study. *Acta Odontologica Scandinavica*, 14 (Supplement 20), 1–202.

Judge, D. S. and Carey, J. R. (2000). Postreproductive life predicted by primate patterns. *The Journals of Gerontology Series A: Biological Sciences and Medical Sciences*, 55, B201–9.

Kahumbu, P. and Eley, R. M. (1991). Teeth emergence in wild olive baboons in Kenya and formulation of a dental schedule for aging wild baboon populations. *American Journal of Primatology*, 23, 1–9.

Kaltsas, G. A. and Chrousos, G. P. (2007). The neuroendocrinology of stress. In *The Handbook of Psychophysiology*, ed. J. Cacioppo, L. G. Tassinary and G. G. Berntson. Cambridge University Press, pp. 303–18.

Kappeler, P. M., Pereira, M. E. and van Schaik, C. P. (2003). Primate life histories and socioecology. In *Primate Life Histories and Socioecology*, ed. P. M. Kappeler and M. E. Pereira. University of Chicago Press, pp. 1–20.

Karnosh, L. J. (1926). Histopathology of syphilitic hypoplasia of the teeth. *Archives of Dermatology & Syphilology*, 13, 25–42.

Kawasaki, K. and Fearnhead, R. W. (1975). On the relationship between tetracycline and the incremental lines in dentine. *Journal of Anatomy*, 119, 49–59.

Kawasaki, K., Tanaka, S. and Isikawa, T. (1979). On the daily incremental lines in human dentine. *Archives of Oral Biology*, 24, 939–43.

Keiter, M. D. (1981). Hand-rearing and development of a lowland gorilla at Woodland Park Zoo, Seattle. *International Zoo Yearbook*, 21, 229–35.

Kelley, J. (1997). Paleobiological and phylogenetic significance of life history in Miocene hominoids. In *Function, Phylogeny and Fossils: Miocene Hominoid Evolution and Adaptations*, ed. D. R. Begun, C. V. Ward and M. D. Rose. New York: Plenum Press, pp. 173–208.

Kelley, J. (2002). Life history evolution in Miocene and extant apes. In *Human Evolution Through Developmental Change*, ed. N. Minugh-Purvis and K. J. McNamara. Baltimore: Johns Hopkins University Press, pp. 223–48.

Kelley, J. (2004). Life history and cognitive evolution in the apes. In *The Evolution of Thought: Evolutionary Origins of Great Ape Intelligence*, ed. A. E. Russon and D. R. Begun. Cambridge University Press, pp. 280–97.

Kelley, J., Dean, M. C. and Reid, D. J. (2001). Molar growth in the late Miocene hominoid *Drypopithecus laietanus*. In *Dental Morphology 1998, Proceedings of the 11th International Symposium on Dental Morphology, Oulu, Finland*, ed. J. T. Mayhall and T. Heikinnen. Oulu University Press, pp. 123–34.

Kelley, J., Dean, M. C. and Ross, S. (2009). Root growth during molar eruption in extant great apes. In *Frontiers of Oral Biology: Interdisciplinary Dental Morphology*, ed. T. Koppe, G. Meyer and G. R. Alt. Basel: Karger, pp. 128–33.

Kelley, J. and Schwartz, G. T. (2010). Dental development and life history in living African and Asian apes. *Proceedings of the National Academy of Sciences*, 107, 1035–40.

Kelley, J. and Smith, T. M. (2003). Age at first molar emergence in early Miocene *Afropithecus turkanensis* and life-history evolution in the Hominoidea. *Journal of Human Evolution*, 44, 307–29.

Key, C. A. (2000). The evolution of human life history. *World Archaeology*, 31, 329–50.

Kibii, J. M. (2007). Taxonomy, taphonomy and palaeoenvironment of hominid and non-hominid primates from the Jacovec Cavern, Sterkfontein. *The South African Archaeological Bulletin*, 62, 90–7.

King, T., Hillson, S. and Humphrey, L. T. (2002). A detailed study of enamel hypoplasia in a post-Medieval adolescent of known age and sex. *Archives of Oral Biology*, 47, 29–39.

King, T., Humphrey, L. T. and Hillson, S. W. (2005). Linear enamel hypoplasias as indicators of systemic physiological stress: evidence from two known age-at-death and sex populations from postmedieval London. *American Journal of Physical Anthropology*, 128, 547–59.

Klein, H. (1945). Etiology of enamel hypoplasia in rickets as determined by studies on rats and swine. *Journal of the American Dental Association*, 18, 866–84.

Klein, R. G. (1999). *The Human Career. Human Biological and Cultural Origins*. University of Chicago Press.

Kraemer, H. C., Horvat, J. R., Doering, C. and McGinnis, P. R. (1982). Male chimpanzee development focusing on adolescence: integration of behavioral with physiological changes. *Primates*, 23, 393–405.

Kraus, B. S. and Jordan, R. E. (1965). *The Human Dentition before Birth*. Philadelphia: Lea & Febiger.

Kremenak, N. W. and Squier, C. A. (1997). Pioneers in oral biology: the migrations of Gottlieb, Kronfeld, Orban, Weinmann, and Sicher from Vienna to America. *Critical Reviews in Oral Biology & Medicine*, 8, 108–28.

Kreshover, S. J. (1944). The pathogenesis of enamel hypoplasia: an experimental study. *Journal of Dental Research*, 23, 231–8.

Kreshover, S. J. (1960a). Metabolic disturbances in tooth formation. *Annals of the New York Academy of Sciences*, 85, 161–7.

Kreshover, S. J. (1960b). Prenatal factors in oral pathologic conditions. *Oral Surgery, Oral Medicine, Oral Pathology*, 13, 569–77.

Kreshover, S. J. and Clough, O. W. (1953). Prenatal influences on tooth development II. Artificially induced fever in rats. *Journal of Dental Research*, 32, 565–72.

Kreshover, S. J., Clough, O. W. and Hancock, J. A. (1954). Vaccinia infection in pregnant rabbits and its effect on maternal and fetal dental tissues. *Journal of the American Dental Association*, 49, 549–62.

Krogman, W. M. (1970). Bertram Shirley Kraus. 1913–1970. A biographical sketch. *American Journal of Physical Anthropology*, 33, 3–7.

Kronfeld, R. (1935). Development and calcification of the human deciduous dentition. *The Bur*, 15, 18–25.

Kronfeld, R. and Schour, I. (1939). Neonatal dental hypoplasia. *Journal of the American Dental Association*, 26, 18–32.

Kunitomo, K. (1928). Age determination by body length and weight in Japanese fetuses (in Japanese). *Nippon Gakuzyutu Kyôkai*, 4, 670–4.

Kuykendall, K. (2001). On radiographic and histological methods for assessing dental development in chimpanzees: comments on Beynon *et al.* (1998) and Reid *et al.* (1998). *Journal of Human Evolution*, 40, 67–76.

Kuykendall, K. L. (1996). Dental development in chimpanzees (*Pan troglodytes*): the timing of tooth calcification stages. *American Journal of Physical Anthropology*, 99, 135–58.

Kuykendall, K. L. and Conroy, G. C. (1996). Permanent tooth calcification in chimpanzees (*Pan troglodytes*): patterns and polymorphisms. *American Journal of Physical Anthropology*, 99, 159–74.

Kuykendall, K. L., Mahoney, C. J. and Conroy, G. C. (1992). Probit and survival analysis of tooth emergence ages in a mixed longitudinal sample of chimpanzees (*Pan troglodytes*). *American Journal of Physical Anthropology*, 89, 379–99.

Lacruz, R. S. (2007). Enamel microstructure of the hominid KB 5223 from Kromdraai, South Africa. *American Journal of Physical Anthropology*, 132, 175–82.

Lacruz, R. S. and Bromage, T. G. (2006). Appositional enamel growth in molars of South African fossil hominids. *Journal of Anatomy*, 209, 13–20.

Lacruz, R. S., Dean, M. C., Ramirez-Rozzi, F. and Bromage, T. G. (2008). Megadontia, striae periodicity and patterns of enamel secretion in Plio-Pleistocene fossil hominins. *Journal of Anatomy*, 213, 148–58.

Lacruz, R. S. and Ramirez-Rozzi, F. V. (2010). Molar crown development in *Australopithecus afarensis*. *Journal of Human Evolution*, 58, 201–6.

Lacruz, R. S., Rozzi, F. R. and Bromage, T. G. (2005). Dental enamel hypoplasia, age at death, and weaning in the Taung child. *South African Journal of Science*, 101, 567–9.

Lacruz, R. S., Rozzi, F. R. and Bromage, T. G. (2006). Variation in enamel development of South African fossil hominids. *Journal of Human Evolution*, 51, 580–90.

Larsen, C. S. (1995). Biological changes in human populations with agriculture. *Annual Review of Anthropology*, 24, 185–213.

Larsen, C. S. (1997). *Bioarchaeology*. Cambridge Studies in Biological Anthropology. Cambridge University Press.

Lavelle, C. L. B. (1975). A note on the variation in the timing of deciduous tooth eruption. *Journal of Dentistry*, 3, 267–70.

Lawn, J. E., Cousens, S. and Zupan, J. (2005). 4 million neonatal deaths: when? where? why? *The Lancet*, 365, 891–900.

Leigh, S. R. (1996). Evolution of human growth spurts. *American Journal of Physical Anthropology*, 101, 455–74.

Leigh, S. R. (2001). Evolution of human growth. *Evolutionary Anthropology: Issues, News, and Reviews*, 10, 223–36.

Leigh, S. R. (2004). Brain growth, life history, and cognition in primate and human evolution. *American Journal of Primatology*, 62, 139–64.

Lewis, A. B. and Garn, S. M. (1960). The relationship between tooth formation and other maturational factors. *Angle Orthodontist*, 30, 70–7.

Lieberman, D. E., Pilbeam, D. R. and Wrangham, R. W. (2008). The transition from Australopithecus to Homo. In *Transitions in Prehistory: Essays in Honor of Ofer Bar-Yosef*, ed. J. J. Shea and D. E. Lieberman. Oxford: Oxbow Press, pp. 1–22.

Lindemann, G. (1958). Forekomsten af emaljehypoplasi hos børn, som har lidt af mave – tramsygdomme. *Odontologisk Tidskrift*, 66, 101–26.

Littleton, J. (2005). Invisible impacts but long term consequences: hypoplasia and contact in central Australia. *American Journal of Physical Anthropology*, 126, 295–304.

Littleton, J. and Townsend, G. C. (2005). Linear enamel hypoplasia and historical change in a central Australian community. *Australian Dental Journal*, 50, 101–7.

Liversidge, H. M. (1994). Accuracy of age estimation from developing teeth of a population of known age (0 to 5.4 years). *International Journal of Osteoarchaeology*, 4, 37–46.

Liversidge, H. M. (2000). Crown formation times of human permanent anterior teeth. *Archives of Oral Biology*, 45, 713–21.

Liversidge, H. M. (2003). Variation in modern human dental development. In *Patterns of Growth and Development in the Genus Homo*, ed. J. L. Thompson, G. E. Krovitz and A. J. Nelson. Cambridge University Press, pp. 73–113.

Liversidge, H. M. (2008). Timing of human mandibular third molar formation. *Annals of Human Biology*, 35, 294–321.

Liversidge, H. M. (2010). Interpreting group differences using Demirjian's dental maturity method. *Forensic Science International*, 201, 95–101.

Liversidge, H. M., Herdeg, B. and Rösing, F. W. (1998). Dental age estimation of non-adults. A review of methods and principles. In: Alt, K. W., Rösing, F. W. and Teschler-Nicola, M. (eds) *Dental Anthropology. Fundamentals, Limits and Prospects*. Vienna: Springer, pp. 419–42.

Liversidge, H. M., Chaillet, N., Mornstad, H. *et al.* (2006). Timing of Demirjian's tooth formation stages. *Annals of Human Biology*, 33, 454–70.

Liversidge, H. M., Dean, M. C. and Molleson, T. I. (1993). Increasing human tooth length between birth and 5.4 years. *American Journal of Physical Anthropology*, 90, 307–13.

Liversidge, H. M. and Molleson, T. I. (1999). Developing permanent tooth length as an estimate of age. *Journal of Forensic Sciences*, 44, 917–20.

Liversidge, H. M. and Molleson, T. I. (2004). Variation in crown and root formation and eruption of human deciduous teeth. *American Journal of Physical Anthropology*, 123, 172–80.

Liversidge, H. M. and Speechly, T. (2001). Growth of permanent mandibular teeth of British children aged 4 to 9 years. *Annals of Human Biology*, 28, 256–62.

Logan, W. H. G. and Kronfeld, R. (1933). Development of the human jaws and surrounding structures from birth to the age of fifteen years. *Journal of the American Dental Association*, 20, 379–427.

Loughna, P., Chitty, L. S., Evans, T. and Chudleigh, T. (2009). Fetal size and dating: charts recommended for clinical obstetric practice. *Ultrasound*, 17, 160–6.

Lukacs, J. R. (1991). Localized enamel hypoplasia of human deciduous canine teeth: prevalence and pattern of expression in rural Pakistan. *Human Biology*, 63, 513–22.

Lukacs, J. R. (1999). Enamel hypoplasia in deciduous teeth of great apes: do differences in defect prevalence imply differential levels of physiological stress? *American Journal of Physical Anthropology*, 110, 351–63.

Lunt, R. C. and Law, D. B. (1974). A review of the chronology of calcification of deciduous teeth. *Journal of the American Dental Association*, 89, 599–606.

Macchiarelli, R., Bondioli, L., Debénath, A. *et al.* (2006). How Neanderthal molar teeth grew. *Nature*, 444, 748–51.

Macho, G. A., Reid, C., Leakey, M. G., Jablonski, N. G. and Beynon, A. D. (1996). Climatic effects on dental development of *Theropithecus oswaldi* from Koobi Fora and Olorgesailie. *Journal of Human Evolution*, 30, 57–70.

Magnusson, T. E. (1982). Emergence of primary teeth and onset of dental stages in Icelandic children. *Community Dentistry and Oral Epidemiology*, 10, 91–7.

Mahoney, P. (2008). Intraspecific variation in M1 enamel development in modern humans: implications for human evolution. *Journal of Human Evolution*, 55, 131–47.

Mahoney, P., Smith, T. M., Schwartz, G. T., Dean, C. and Kelley, J. (2007). Molar crown formation in the Late Miocene Asian hominoids, *Sivapithecus parvada* and *Sivapithecus indicus*. *Journal of Human Evolution*, 53, 61–8.

Mann, A. E. (1975). *Some Paleodemographic Aspects of the South African Australopithecines*. University of Pennsylvania Publications in Anthropology No. 1. Philadelphia: University of Pennsylvania.

Marshall, W. A. and Tanner, J. M. (1970). Variations in the pattern of pubertal changes in boys. *Archives of Disease in Childhood*, 45, 13–23.

Martin, R. D. (1981). Relative brain size and basal metabolic rate in terrestrial vertebrates. *Nature*, 293, 57–60.

Martin, R. D. (2007). The evolution of human reproduction: a primatological perspective. *American Journal of Physical Anthropology*, 134, 59–84.

Martinón-Torres, M., Bermúdez de Castro, J. M., Gómez-Robles, A. *et al.* (2007). Dental evidence on the hominin dispersals during the Pleistocene. *Proceedings of the National Academy of Sciences USA*, 104, 13279–82.

Massler, M. and Schour, I. (1944). *Atlas of the Mouth and Adjacent Parts in Health and Disease*, 1st edn. Chicago: American Dental Association.

Massler, M., Schour, I. and Poncher, H. (1941). Developmental pattern of the child as reflected in the calcification pattern of the teeth. *American Journal of Diseases of Children*, 62, 33–67.

Matsuoka, Y., Vigouroux, Y., Goodman, M. M. and Sanchez, G. (2002). A single domestication for maize shown by multilocus microsatellite genotyping. *Proceedings of the National Academy of Sciences of the United States of America*, 99, 6080–4.

Matsuzawa, T., Sakura, O., Kimura, T., Hamada, Y. and Sugiyama, Y. (1990). Case report on the death of a wild chimpanzee (*Pan troglodytes verus*). *Primates*, 31, 635–41.

Mellanby, E. (1919). An experimental investigation on rickets. Two lectures delivered at the Royal College of Surgeons of England. *The Lancet*, 193, 407–12.

Mellanby, E. (1934). *Nutrition and Disease. The Interaction of Clinical and Experimental Work*. Edinburgh & London: Oliver & Boyd.

Mellanby, M. (1918). An experimental study of the influence of diet on teeth formation. *The Lancet*, 192, 766–70.

Mellanby, M. (1929). *Diet and Teeth: an Experimental Study. Part I. Dental Structure in Dogs*. Medical Research Council, Special Report Series, No. 140. London: His Majesty's Stationery Office.

Mellars, P. (2004). Neanderthals and the modern human colonization of Europe. *Nature*, 432, 461–5.

Miani, A. and Miani, C. (1971). Circadian advancement rhythm of the calcification front in dog dentin. *Minerva Stomatologica*, 20, 169–78.

Miles, A. E. W. and Grigson, C. (1990). *Colyer's Variations and Diseases of the Teeth of Animals*, revised edn. Cambridge University Press.

Mincer, H. H., Harris, E. F. and Berryman, H. E. (1993). The A.B.F.O. study of third molar development and its use as an estimator of chronological age. *Journal of Forensic Sciences*, 38, 379–90.

Moggi-Cecchi, J. (2000). Enamel hypoplasia in South African early hominids: a reappraisal. *American Journal of Physical Anthropology*, Supplement 30, 230–1.

Molleson, T. I., Cox, M., Waldron, H. A. and Whittaker, D. K. (1993). *The Spitalfields Project. Volume 2 – the Anthropology. The Middling Sort.* CBA Research Report 86. York: Council for British Archaeology.

Molnar, S. and Molnar, I. M. (1985). The incidence of enamel hypoplasia among the Krapina Neandertals. *American Anthropologist*, 87, 536–49.

Molnar, S., Przybeck, T. R., Gantt, D. G., Elizondo, R. S. and Wilkerson, J. E. (1981). Dentin apposition rates as markers of primate growth. *American Journal of Physical Anthropology*, 55, 443–50.

Moon, H. (1877). On irregular and defective tooth development. *Transactions of the Odontological Society of Great Britain*, 9, 223–43.

Moorrees, C. F. A., Fanning, E. A. and Hunt, E. E. (1963a). Age variation of formation stages for ten permanent teeth. *Journal of Dental Research*, 42, 1490–502.

Moorrees, C. F. A., Fanning, E. A. and Hunt, E. E. (1963b). Formation and resorption of three deciduous teeth in children. *American Journal of Physical Anthropology*, 21, 205–13.

Mounier, A., Marchal, F. and Condemi, S. (2009). Is *Homo heidelbergensis* a distinct species? New insight on the Mauer mandible. *Journal of Human Evolution*, 56, 219–46.

Mumby, H. and Vinicius, L. (2008). Primate growth in the slow lane: a study of inter-species variation in the growth constant A. *Evolutionary Biology*, 35, 287–95.

Nadarajah, K., Marlowe, T. J. and Notter, D. R. (1984). Growth patterns of Angus, Charolais, Charolais x Angus and Holstein x Angus cows from birth to maturity. *Journal of Animal Science*, 59, 957–66.

Nahar, B., Hamadani, J. D., Ahmed, T. *et al.* (2008). Effects of psychosocial stimulation on growth and development of severely malnourished children in a nutrition unit in Bangladesh. *European Journal of Clinical Nutrition*, 63, 725–31.

Neiburger, E. J. (1990). Enamel hypoplasias: poor indicators of dietary stress. *American Journal of Physical Anthropology*, 82, 231–2.

Newell, E. A., Guatelli Steinberg, D., Field, M., Cooke, C. and Feeney, R. N. M. (2006). Life history, enamel formation, and linear enamel hypoplasia in the Ceboidea. *American Journal of Physical Anthropology*, 131, 252–60.

Newell-Morris, L. and Sirianni, J. E. (1982). Parameters of bone growth in the fetal and infant macaque (*Macaca nemestrina*) humerus as documented by trichromatic bone labels. In *Factors and Mechanisms Influencing Bone Growth*, ed. A. D. Dixon and B. G. Sarnat. New York: Alan R. Liss, pp. 243–58.

Newell-Morris, L., Tarrant, L. H., Sirianni, J. E. and Munger, R. G. (1980). Trichromatic fluorescent vital labeling of bone in the fetal macaque. *Cellular and Molecular Life Sciences*, 36, 623–4.

Nishida, T., Hamai, M., Hasegawa, T. *et al.* (2003). Demography, female life history, and reproductive profiles among the chimpanzees of Mahale. *American Journal of Primatology*, 59, 99–121.

Nissen, H. W. and Riesen, A. H. (1946). The deciduous dentition of chimpanzee. *Growth*, 9, 265–74.

Nissen, H. W. and Riesen, A. H. (1964). The eruption of the permanent dentition of chimpanzee. *American Journal of Physical Anthropology*, 22, 285–94.

Nomata, N. (1964). Chronological study on the crown formation of the human deciduous dentition. *Bulletin of the Tokyo Medical & Dental University*, 11, 55–76.

Ogilvie, M. D., Curran, B. K. and Trinkaus, E. (1989). Incidence and patterning of dental enamel hypoplasia among the Neandertals. *American Journal of Physical Anthropology*, 79, 25–41.

Ogilvie, M. D. and Trinkaus, E. (1990). Reply to Neiburger. *American Journal of Physical Anthropology*, 82, 232–3.

Ohtsuka, M. and Shinoda, H. (1995). Ontogeny of circadian dentinogenesis in the rat incisor. *Archives of Oral Biology*, 40, 481–5.

Okada, M. and Mimura, T. (1938). Zur Physiologie und Pharmakologie der Hartgewebe. I. Mitteilung: eine Vitalfärbungsmethode mit Bleisalzen und ihre Anwendung bei den Untersuchunger über die rhythmische Streifenbildung der harten Zahngewebe. *Japanese Journal of Medical Sciences Part 4 Pharmacology*, 11, 166–70.

Okada, M. and Mimura, T. (1940). Zur Physiologie und Pharmakologie der Hartgewebe. III. Mitteilung: über die Genese der rhythmischen Streifenbildung der harten Zahngewebe. *Japanese Journal of Medical Sciences Part 4 Pharmacology*, 13, 92–5.

Okada, M. and Mimura, T. (1941). Zur Physiologie und Pharmakologie der Hartgewebe. VII. Mitteilung: über den zeitlichen Verlauf der Schwangerschaft und Entbindung geseher von der Streifenfigur im Dentin des mütterlisches Kaninchens, sowie über eine Blei-Vitalfärbung des fötalen Dentins. *Japanese Journal of Medical Sciences Part 4 Pharmacology*, 14, 6–10.

Okada, M. and Mimura, T. (1942). Zur Physiologie und Pharmakologie der Hartgeweve. VIII. Mitteilung: über den zeitlichen Verlauf der Mobilisierung des subkutanen Bleidepots, betrachtet von der Streifenfigur im Kaninchendentin. *Japanese Journal of Medical Sciences Part 4 Pharmacology*, 15, 8–11.

Okada, M., Mimura, T. and Fuse, S. (1940). Zur Physiologie und Pharmakologie der Hartgewebe. VI. Mitteilung: eine methode der pharmakologischen Untersuchung durch die Anwendung von Streifenfiguren in Kaninchendentin und eine Beobachtung über die Wirkung einiger Pharmaka durch dieselbe Methode. *Japanese Journal of Medical Sciences Part 4 Pharmacology*, 13, 99–101.

Ooë, T. (1979). Development of human first and second permanent molar, with special reference to the distal portion of the dental lamina. *Anatomy & Embryology*, 155, 221–40.

Ooë, T. (1981). *Human Tooth and Dental Arch Development*. Tokyo: Ishiyaku.

Osborn, J. W. and Ten Cate, A. R. (1983). *Advanced Dental Histology*, 4th edn. Dental Practitioner Handbook, Whole No. 6. Bristol: John Wright.

Owen, R. (1845). *Odontography or a Treatise on the Comparative Anatomy of the Teeth: their Physiological Relations, Mode of Development and Microscopic Structure in the Vertebrate Animals*. London: Hyppolyte Baillière.

Parascandola, J. and Ihde, A. J. (1977). Edward Mellanby and the antirachitic factor. *Bulletin of the History of Medicine*, 51, 507–15.

Partridge, T. C. (2005). Dating of the Sterkfontein hominids: progress and possibilities. *Transactions of the Royal Society of South Africa*, 60, 107–9.

Pastore, L., Carroccio, A., Compilato, D., Panzarella, V., Serpico, R. and Muzio, L. L. (2008). Oral manifestations of celiac disease. *Journal of Clinical Gastroenterology*, 42, 224–32.

Patten, B. M. (1968). *Human Embryology*, 3rd edn. New York: McGraw-Hill.

Patten, B. M. (1976). *Patten's Human Embryology: Elements of Clinical Development*, revised edn. New York, London: McGraw-Hill.

Pavelka, M. S. M. and Fedigan, L. M. (1991). Menopause: a comparative life history perspective. *American Journal of Physical Anthropology*, 34, 13–38.

Payton, C. G. (1932). The growth in length of the long bones in the madder-fed pig. *Journal of Anatomy*, 66, 414–25.

Pearson, O. M., Cordero, R. and Busby, A. (2006). How different were Neanderthals' habitual activities? A comparative analysis with diverse groups of recent humans. In *Neanderthals Revisited: New Approaches and Perspectives*, ed. K. Harvati and T. Harrison. Dordrecht: Springer, pp. 135–56.

Peck, A. M. and Vågerö, D. H. (1989). Adult body height, self perceived health and mortality in the Swedish population. *Journal of Epidemiology and Community Health*, 43, 380–4.

Pedersen, P. O. and Scott, D. B. (1951). Replica studies of the surfaces of teeth from Alaskan Eskimo, West Greenland Natives, and American Whites. *Acta Odontologica Scandinavica*, 9, 261–92.

Peppe, D. J., McNulty, K. P., Cote, S. M., Harcourt-Smith, W. E. H., Dunsworth, H. M. and Van Couvering, J. A. (2009). Stratigraphic interpretation of the Kulu Formation (Early Miocene, Rusinga Island, Kenya) and its implications for primate evolution. *Journal of Human Evolution*, 56, 447–61.

Pereira, M. E. and Leigh, S. R. (2003). Modes of primate development. In *Primate Life Histories and Socioecology*, ed. P. M. Kappeler and M. E. Pereira. University of Chicago Press, pp. 149–76.

Pettitt, P. B. (2000). Neanderthal lifecycles: development and social phases in the lives of the last archaics. *World Archaeology*, 31, 351–66.

Pflüger, H. (1924). Eine für Lues congenita charakteristiche Formveränderung (Knospenform) an dem ersten Molaren. *Münchener Medizinsiche Wochenschrift*, 71, 605–7.

Phillips-Conroy, J. E. and Jolly, C. J. (1988). Dental eruption schedules of wild and captive baboons. *American Journal of Primatology*, 15, 17–29.

Piaget, J. and Inhelder, B. (1969). *The Psychology of the Child*. London: Routledge & Kegan Paul.

Pickerill, H. P. (1912). *The Prevention of Dental Caries and Oral Sepsis. Being the Cartwright Prize Essay of the Royal College of Surgeons of England for 1906–1910, with some Additions.* London: Ballière, Tindall & Cox.

Pickerill, H. P. (1913). The structure of the enamel. *Dental Cosmos*, 55, 959–88.

Pickering, T. R., Clarke, R. J. and Moggi Cecchi, J. (2004a). Role of carnivores in the accumulation of the Sterkfontein Member 4 hominid assemblage: a taphonomic reassessment of the complete hominid fossil sample (1936–1999). *American Journal of Physical Anthropology*, 125, 1–15.

Pickering, T. R., Dominguez-Rodrigo, M., Egeland, C. P. and Brain, C. K. (2004b). Beyond leopards: tooth marks and the contribution of multiple carnivore taxa to the accumulation of the Swartkrans Member 3 fossil assemblage. *Journal of Human Evolution*, 46, 595–604.

Pindborg, J. J. (1982). Aetiology of developmental enamel defects not related to fluorosis. *International Dental Journal*, 32, 123–34.

Pounds, J. G., Long, G. J. and Rosen, J. F. (1991). Cellular and molecular toxicity of lead in bone. *Environmental Health Perspectives*, 91, 17–32.

Preiswerk, G. (1895). *Beiträge zur Kentniss der Schmelzstructur bei Säugetieren mit besonderer Berücksichtigung der Ungulaten*. Doctoral dissertation of the University of Basel. Basel: Verlags-Druckerei.

Promislow, D. E. L. and Harvey, P. H. (1990). Living fast and dying young: a comparative analysis of life history variation among mammals. *Journal of Zoology*, 220, 417–37.

Purvis, A., Webster, A. J., Agapow, P. M., Jones, K. E. and Isaac, N. J. B. (2003). Primate life histories and phylogeny. In *Primate Life Histories and Socioecology*, ed. P. M. Kappeler and M. E. Pereira. University of Chicago Press, pp. 25–40.

Purvis, R. J., MacKay, G. S., Cockburn, F. *et al.* (1973). Enamel hypoplasia of the teeth associated with neonatal tetany: a manifestation of maternal vitamin D deficiency. *Lancet*, 2, 811–14.

Putkonen, T. (1962). Dental changes in congenital syphilis. Relationship to other syphilitic stigmata. *Acta Dermato-Venerologica*, 42, 44–62.

Radovcic, J., Smith, F. H., Trinkaus, E. and Wolpoff, M. H. (1988). *The Krapina Hominids. An Illustrated Catalog of Skeletal Collection*. Zagreb: Croatian Natural History Museum.

Rae, T. C., Koppe, T. and Stringer, C. B. (2010). The Neanderthal face is not cold adapted. *Journal of Human Evolution*, 60, 234–9.

Ramirez-Rozzi, F. V. (1995). Time of crown formation in Plio-Pleistocene hominid teeth. In *Structure, Function and Evolution of Teeth. Dental Morphology Meeting, Florence, September 1992*, ed. J. Moggi-Cecchi. Florence: International Institute for the Study of Man, pp. 217–38.

Ramirez-Rozzi, F. V. and Bermudez de Castro, J. M. (2009). Surprisingly rapid growth in Neanderthals. *Nature*, 428, 936–40.

Ramirez-Rozzi, F. V. and Sardi, M. (2007). Crown-formation time in Neandertal anterior teeth revisited. *Journal of Human Evolution*, 53, 108–13.

Ranere, A. J., Piperno, D. R., Holst, I., Dickau, R. and Iriarte, J. (2009). The cultural and chronological context of early Holocene maize and squash domestication in the Central Balsas River Valley, Mexico. *Proceedings of the National Academy of Sciences*, 106, 5014–8.

Rawstron, S. A., Bromberg, K. and Hammerschlag, M. R. (1993a). STD in children: syphilis and gonorrhoea. *Genitourinary Medicine*, 69, 66–75.

Rawstron, S. A., Jenkins, S., Blanchard, S., Ping-Wu, L. and Bromberg, K. (1993b). Maternal and congenital syphilis in Brooklyn, NY. *American Journal of Diseases of Children*, 147, 727–31.

Reeves, J. and Adams, M. (1993). *The Spitalfields Project. Volume 1 – the Archaeology. Across the Styx*. CBA Research Report 85. York: Council for British Archaeology.

Reid, D. J., Beynon, A. D. and Ramirez-Rozzi, F. V. (1998a). Histological reconstruction of dental development in four individuals from a Medieval site in Picardie, France. *Journal of Human Evolution*, 35, 463–77.

Reid, D. J. and Dean, M. C. (2000). The timing of linear hypoplasias on human anterior teeth. *American Journal of Physical Anthropology*, 113, 135–40.

Reid, D. J. and Dean, M. C. (2006). Variation in modern human enamel formation times. *Journal of Human Evolution*, 50, 329–46.

Reid, D. J. and Ferrell, R. J. (2006). The relationship between number of striae of Retzius and their periodicity in imbricational enamel formation. *Journal of Human Evolution*, 50, 195–202.

Reid, D. J., Schwartz, G. T., Dean, M. C. and Chandrasekera, M. S. (1998b). A histological reconstruction of dental development in the common chimpanzee, *Pan troglodytes*. *Journal of Human Evolution*, 35, 427–48.

Retzius, A. (1837). Bemerkungen über den inneren Bau der Zähne, mit besonderer Rücksicht auf den im Zahnknochen vorkommenden Röhrenbau. *Arkiv für Anatomie, Physiologie und Wissenschaftliche Medicin*, 1837, 486–566.

Risnes, S. (1985a). A scanning electron microscope study of the three-dimensional extent of Retzius lines in human dental enamel. *Scandinavian Journal of Dental Research*, 93, 145–52.

Risnes, S. (1985b). Circumferential continuity of perikymata in human dental enamel investigated by scanning electron microscopy. *Scandinavian Journal of Dental Research*, 93, 185–91.

Risnes, S. (1986). Enamel apposition rate and the prism periodicity in human teeth. *Scandinavian Journal of Dental Research*, 94, 394–404.

Risnes, S. (1990). Structural characteristics of staircase-type Retzius lines in human dental enamel analyzed by scanning electron-microscopy. *Anatomical Record*, 226, 135–46.

Risnes, S. (1998). Growth tracks in dental enamel. *Journal of Human Evolution*, 35, 331–50.

Ritzman, T. B., Baker, B. J. and Schwartz, G. T. (2008). A fine line: a comparison of methods for estimating ages of linear enamel hypoplasia formation. *American Journal of Physical Anthropology*, 135, 348–61.

Robbins, A. M., Robbins, M. M., Gerald Steklis, N. and Steklis, H. D. (2006). Age related patterns of reproductive success among female mountain gorillas. *American Journal of Physical Anthropology*, 131, 511–21.

Robine, J. M. and Allard, M. (1998). The oldest human. *Science*, 279, 1831.

Robinson, J. T. (1952). Some hominid features of the ape-man dentition. *Journal of the Dental Association of South Africa*, 7, 1–12.

Robinson, J. T. (1956). *The Dentition of the Australopithecinae*. Transvaal Museum Memoir No 9. Pretoria: Transvaal Museum.

Robson, S. L., van Schaik, C. and Hawkes, K. (2006). The derived features of human life history. In *The Evolution of Human Life History*, ed. K. Hawkes and R. R. Paine. Santa Fe: School of American Research Press, pp. 17–44.

Robson, S. L. and Wood, B. (2008). Hominin life history: reconstruction and evolution. *Journal of Anatomy*, 212, 394–425.

Rolian, C., Lieberman, D. E. and Hallgrímsson, B. (2010). The coevolution of human hands and feet. *Evolution*, 64, 1558–68.

Romero, L. M., Dickens, M. J. and Cyr, N. E. (2009). The reactive scope model – a new model integrating homeostasis, allostasis, and stress. *Hormones and Behavior*, 55, 375–89.

Rose, J. C. (1977). Defective enamel histology of prehistoric teeth from Illinois. *American Journal of Physical Anthropology*, 46, 439–46.

Rose, J. C. (1979). Morphological variations of human enamel prisms within abnormal striae of Retzius. *Human Biology*, 51, 139–51.

Rose, J. C., Armelagos, G. J. and Lallo, J. W. (1978). Histological enamel indicator of childhood stress in prehistoric skeletal samples. *American Journal of Physical Anthropology*, 49, 511–16.

Rosenberg, G. D. and Simmons, D. J. (1980). Rhythmic dentinogenesis in the rabbit incisor: circadian, ultradian, and infradian periods. *Calcified Tissue International*, 32, 29–44.

Rossi, P. F., Bondioli, L., Geusa, G. and Macchiarelli, R. (1999). *Osteodental Biology of the People of Portus Romae (Necropolis of Isola Sacra, 2nd–3rd Cent. AD). I.* Digital Archives of Human Paleobiology. 1. Rome: Soprintendenza Speciale al Museo Nazionale Preistorico Etnografico 'L. Pigorini'. Available as a CD-ROM.

Ruff, C. B. (2003). Long bone articular and diaphyseal structure in old world monkeys and apes. II: estimation of body mass. *American Journal of Physical Anthropology*, 120, 16–37.

Russon, A. E. (2004). Great ape cognitive systems. In *The Evolution of Thought: Evolutionary Origins of Great Ape Intelligence*, ed. A. E. Russon and D. R. Begun. Cambridge University Press, pp. 76–100.

Russon, A. E. and Begun, D. R. (2004). Evolutionary origins of great ape intelligence. In *The Evolution of Thought: Evolutionary Origins of Great Ape Intelligence*, ed. A. E. Russon and D. R. Begun. Cambridge University Press, pp. 353–68.

Sarnat, B. G. and Schour, I. (1941). Enamel hypoplasia (chronologic enamel aplasia) in relation to systemic disease: a chronologic, morphologic and etiologic classification. *Journal of the American Dental Association*, 28, 1989–2000.

Sarnat, B. G. and Schour, I. (1942). Enamel hypoplasia (chronologic enamel aplasia) in relation to systemic disease: a chronologic, morphologic and etiologic classification. *Journal of the American Dental Association*, 29, 397–418.

Scammon, R. E. (1930). The measurement of the body in childhood. In *The Measurement of Man*, ed. J. A. Harris, C. M. Jackson, D. G. Patterson and R. E. Scammon. Minneapolis: University of Minnesota, pp. 171–215.

Scheuer, L. (1998). Age and death and cause of death of the people buried at St Bride's Church, Fleet Street. In *Grave Concerns: Death and Burial in England 1700–1850*, ed. M. Cox. York: Council for British Archaeology.

Scheuer, L. and Black, S. (2004). *The Juvenile Skeleton*. London: Academic Press.

Scheuer, L. and MacLaughlin-Black, S. (1994). Age estimation from the *pars basilaris* of the fetal and juvenile occipital bone. *International Journal of Osteoarchaeology*, 4, 377–82.

Schmidt, W. J. and Keil, A. (1971). *Polarizing Microscopy of Dental Tissues. Theory, Methods and Results from the Structural Analysis of Normal and Diseased Hard Dental Tissues and Tissues Associated with Them in Man and Other Vertebrates*. Oxford: Pergamon Press.

Schneider, B. J. (1968). Lead acetate as a vital marker for the analysis of bone growth. *American Journal of Physical Anthropology*, 29, 197–200.

Schoeninger, M. J. (2009). Stable isotope evidence for the adoption of maize agriculture. *Current Anthropology*, 50, 633–40.

Schour, I. (1936). Neonatal line in enamel and dentin of human deciduous teeth and first permanent molar. *Journal of the American Dental Association*, 23, 1946–55.

Schour, I. and Hoffman, M. M. (1939). Studies in tooth development, II. The rate of apposition of enamel and dentin in man and other animals. *Journal of Dental Research*, 18, 161–75.

Schour, I., Hoffman, M. M., Sarnat, B. G. and Engel, M. B. (1941). Vital staining of growing bones and teeth with Alizarine Red S. *Journal of Dental Research*, 20, 411–18.

Schour, I. and Kronfeld, R. (1938). Tooth ring analysis: IV. Neonatal dental hypoplasia. Analysis of the teeth of an infant with injury of the brain at birth. *Archives of Pathology*, 26, 471–90.

Schour, I. and Massler, M. (1940a). Studies in tooth development: the growth pattern of human teeth. Part I. *Journal of the American Dental Association*, 27, 1778–93.

Schour, I. and Massler, M. (1940b). Studies in tooth development: the growth pattern of human teeth. Part II. *Journal of the American Dental Association*, 27, 1918–31.

Schour, I. and Massler, M. (1941). The development of the human dentition. *Journal of the American Dental Association*, 28, 1153–60.

Schour, I. and Poncher, H. G. (1937). Rate of apposition of human enamel and dentin as measured by the effects of acute fluorosis. *American Journal of Diseases of Children*, 54, 757–76.

Schultz, A. H. (1935). Eruption and decay of the permanent teeth in primates. *American Journal of Physical Anthropology*, 19, 489–581.

Schultz, A. H. (1960). Age changes in primates and their modification in man. In *Human Growth*, ed. J. M. Tanner. Oxford: Pergamon Press, pp. 1–20.

Schultz, P. D. and McHenry, H. M. (1975). Age distribution of enamel hypoplasia in prehistoric California Indians. *Journal of Dental Research*, 54, 913.

Schuman, E. L. and Sognnaes, R. F. (1956). Developmental microscopic defects in the teeth of sub-human primates. *American Journal of Physical Anthropology*, 14, 193–214.

Schwartz, G. T. and Dean, M. C. (2001). The ontogeny of canine dimorphism in extant hominoids. *American Journal of Physical Anthropology*, 115, 269–83.

Schwartz, G. T., Liu, W. and Zheng, L. (2003). Preliminary investigation of dental microstructure in the Yuanmou hominoid (*Lufengpithecus hudienensis*), Yunnan Province, China. *Journal of Human Evolution*, 44, 189–202.

Schwartz, G. T., Reid, D. J. and Dean, C. (2001). Developmental aspects of sexual dimorphism in hominoid canines. *International Journal of Primatology*, 22, 837–60.

Schwartz, G. T., Reid, D. J., Dean, M. C. and Zihlman, A. L. (2006). A faithful record of stressful life events recorded in the dental developmental record of a juvenile gorilla. *International Journal of Primatology*, 27, 1201–19.

Schwartz, J. H. and Tattersall, I. (2002). *Terminology and Craniodental Morphology of Genus Homo (Europe)*. Vol. 1 of The Human Fossil Record. New York: Wiley-Liss.

Schwartz, J. H. and Tattersall, I. (2003). *Terminology and Craniodental Morphology of Genus Homo (Africa and Asia)*. Vol. 2 of The Human Fossil Record. New York: Wiley-Liss.

Selye, H. (1976). Forty years of stress research: principal remaining problems and misconceptions. *Canadian Medical Association Journal*, 115, 53–6.

Semaw, S. (2000). The world's oldest stone artefacts from Gona, Ethiopia: their implications for understanding stone technology and patterns of human evolution between 2.6–1.5 million years ago. *Journal of Archaeological Science*, 27, 1197–214.

Seow, W. K., Brown, J. P., Tudehope, D. A. and O'Callaghan, M. (1984). Dental defects in the deciduous dentition of premature infants with low birth weight and neonatal rickets. *Pediatric Dentistry*, 6, 88–92.

Sheiham, A. (2005). Oral health, general health and quality of life. *Bulletin of the World Health Organization*, 83, 644.

Shellis, R. P. (1984). Variations in growth of the enamel crown in human teeth and a possible relationship between growth and enamel structure. *Archives of Oral Biology*, 29, 697–705.

Shellis, R. P. (1998). Utilization of periodic markings in enamel to obtain information on tooth growth. *Journal of Human Evolution*, 35, 387–400.

Siebert, J. R. and Swindler, D. R. (1991). Perinatal dental development in the chimpanzee (*Pan troglodytes*). *American Journal of Physical Anthropology*, 86, 287–94.

Simpson, M. S. (1981). Effects of demineralizing tetracycline-stained human dentine. *Calcified Tissue International*, 33, 101–4.

Simpson, S. W. (1999). Reconstructing patterns of growth disruption from enamel microstructure. In *Human Growth in the Past: Studies from Bones and Teeth*, ed. R. Hoppa and C. M. FitzGerald. Cambridge University Press, pp. 241–63.

Simpson, S. W. (2001). Patterns of growth disruption in La Florida: evidence from enamel microstructure. In *Bioarchaeology of Spanish Florida*, ed. C. S. Larsen. Gainesville: University Press of Florida, pp. 146–80.

Skinner, M. (1996). Developmental stress in immature hominines from Late Pleistocene Eurasia: evidence from enamel hypoplasia. *Journal of Archaeological Science*, 23, 833–52.

Skinner, M. (1997). Dental wear in immature Late Pleistocene European hominines. *Journal of Archaeological Science*, 24, 677–700.

Skinner, M. and Dupras, T. (1993). Variation in birth timing and location of the neonatal line in human enamel. *Journal of Forensic Sciences*, 38, 1383–90.

Skinner, M. F. (1986). Enamel hypoplasia in sympatric chimpanzee and gorilla. *Journal of Human Evolution*, 1, 289–312.

Skinner, M. F. and Hopwood, D. (2004). Hypothesis for the causes and periodicity of repetitive linear enamel hypoplasia in large, wild African (*Pan troglodytes* and *Gorilla gorilla*) and Asian (*Pongo pygmaeus*) apes. *American Journal of Physical Anthropology*, 123, 216–35.

Skinner, M. F. and Hung, J. T. W. (1986). Localized enamel hypoplasia of the primary canine. *Journal of Dentistry for Children*, 53, 197–200.

Skinner, M. F. and Hung, J. T. W. (1989). Social and biological correlates of localized enamel hypoplasia of the human deciduous canine tooth. *American Journal of Physical Anthropology*, 79, 159–75.

Skinner, M. F. and Newell, E. A. (2003). Localized hypoplasia of the primary canine in bonobos, orangutans, and gibbons. *American Journal of Physical Anthropology*, 120, 61–72.

Skinner, M. F. and Pruetz, J. D. (2012). Reconstruction of periodicity of repetitive linear enamel hypoplasia from perikymata counts on imbricational enamel among dry-adapted chimpanzees (*Pan troglodytes verus*) from Fongoli, Senegal. *American Journal of Physical Anthropology*, 149, 468–82.

Skinner, M. M. and Wood, B. (2006). The evolution of modern human life history: a paleontological perspective. In *The Evolution of Human Life History*, ed. K. Hawkes and R. R. Paine. Santa Fe: School of American Research Press, pp. 331–64.

Smith, B. D. (1997). The initial domestication of *Cucurbita pepo* in the Americas 10,000 years ago. *Science*, 276, 932–4.

Smith, B. H. (1986). Dental development in *Australopithecus* and early *Homo*. *Nature*, 323, 327–30.

Smith, B. H. (1989). Dental development as a measure of life history in primates. *Evolution*, 43, 683–8.

Smith, B. H. (1991a). Age of weaning approximates age of emergence of the first permanent molar in nonhuman primates. *American Journal of Physical Anthropology*, Supplement 12, 163–4.

Smith, B. H. (1991b). Dental development and the evolution of life history in Hominidae. *American Journal of Physical Anthropology*, 86, 157–74.

Smith, B. H. (1991c). Standards of human tooth formation and dental age assessment. In *Advances in Dental Anthropology*, ed. M. A. Kelley and C. S. Larsen. New York: Wiley-Liss, pp. 143–68.

Smith, B. H. (1992). Life history and the evolution of human maturation. *Evolutionary Anthropology: Issues, News, and Reviews*, 1, 134–42.

Smith, B. H. (2000). 'Schultz's rule' and the evolution of tooth emergence and replacement patterns in primates and ungulates. In *Development, Function and Evolution of Teeth*, ed. M. F. Teaford, M. M. Smith and M. W. J. Ferguson. Cambridge University Press, pp. 212–27.

Smith, B. H. and Boesch, C. (2010). Mortality and the magnitude of the "wild effect" in chimpanzee tooth emergence. *Journal of Human Evolution*, 60, 34–46.

Smith, B. H., Crummett, T. L. and Brandt, K. L. (1994). Ages of eruption of primate teeth: a compendium for ageing individuals and comparing life histories. *Yearbook of Physical Anthropology*, 37, 177–232.

Smith, B. H. and Garn, S. M. (1987). Polymorphisms in eruption sequence of permanent teeth in American children. *American Journal of Physical Anthropology*, 74, 289–303.

Smith, M. C., Lantz, E. and Smith, H. V. (1932). The cause of mottled enamel. *Journal of Dental Research*, 12, 149–59.

Smith, M. M. (2003). Vertebrate dentitions at the origin of jaws: when and how pattern evolved. *Evolution & Development*, 5, 394–413.

Smith, T. M. (2006). Experimental determination of the periodicity of incremental features in enamel. *Journal of Anatomy*, 208, 99–113.

Smith, T. M. (2008). Incremental dental development: methods and applications in hominoid evolutionary studies. *Journal of Human Evolution*, 54, 205–24.

Smith, T. M., Martin, L. B. and Leakey, M. G. (2003). Enamel thickness, microstructure and development in *Afropithecus turkanensis*. *Journal of Human Evolution*, 44, 283–306.

Smith, T. M. and Reid, D. J. (2009). Temporal nature of periradicular bands ("Striae Periradicales") on mammalian tooth roots. In *Comparative Dental Morphology*, ed. K. Koppe, G. Meyer and K. W. Alt. Basel: Karger, pp. 85–92.

Smith, T. M., Reid, D. J., Dean, M. C. *et al.* (2007a). New perspectives on chimpanzee and human molar crown development. In *Dental Perspectives on Human Evolution. State-of-the-Art Research in Dental Palaeoanthropology*, ed. S. E. Bailey and J. J. Hublin. Dordrecht: Springer, pp. 177–92.

Smith, T. M., Reid, D. J., Dean, M. C., Olejniczak, A. J. and Martin, L. B. (2007b). Molar development in common chimpanzees (*Pan troglodytes*). *Journal of Human Evolution*, 52, 201–16.

Smith, T. M., Smith, B. H., Reid, D. J. *et al.* (2010a). Dental development of the Taï Forest chimpanzees revisited. *Journal of Human Evolution*, 58, 363–73.

Smith, T. M. and Tafforeau, P. (2008). New visions of dental tissue research: tooth development, chemistry, and structure. *Evolutionary Anthropology*, 17, 213–26.

Smith, T. M., Tafforeau, P., Reid, D. J., Grün, R., Eggins, S. and Boutakiout, M. (2007c). Earliest evidence of modern human life history in North Africa early *Homo sapiens*. *Proceedings of the National Academy of Sciences USA*, 104, 6128–33.

Smith, T. M., Tafforeau, P., Reid, D. J. *et al.* (2010b). Dental evidence for ontogenetic differences between modern humans and Neanderthals. *Proceedings of the National Academy of Sciences*, 107, 20923–8.

Smith, T. M., Toussaint, M., Reid, D. J., Olejniczak, A. J. and Hublin, J. J. (2007d). Rapid dental development in a Middle Paleolithic Belgian Neanderthal. *Proceedings of the National Academy of Sciences USA*, 104, 20220–5.

Song, Y. M., Smith, G. D. and Sung, J. (2003). Adult height and cause-specific mortality: a large prospective study of South Korean men. *American Journal of Epidemiology*, 158, 479–81.

Spocter, M. A. and Manger, P. R. (2007). The use of cranial variables for the estimation of body mass in fossil hominins. *American Journal of Physical Anthropology*, 134, 92–105.

Starling, A. P. and Stock, J. T. (2007). Dental indicators of health and stress in early Egyptian and Nubian agriculturalists: a difficult transition and gradual recovery. *American Journal of Physical Anthropology*, 134, 520–8.

Stimmler, L., Snodgrass, G. J. A. I. and Jaffe, E. (1973). Dental defects associated with neonatal symptomatic hypocalcemia. *Archives of Disease in Childhood*, 48, 217–20.

Streeter, G. L. (1920). *Weight, Sitting Height, Head Size, Foot Length, and Menstrual Age of the Human Embryo*. Washington: Carnegie Institution of Washington.

Suckling, G., Brown, R. and Herbison, G. (1985). The prevalence of developmental defects of enamel in 696 nine-year-old New Zealand children participating in a health development study. *Community Dental Health*, 2, 303–13.

Suckling, G. W., Herbison, G. P. and Brown, R. H. (1987). Etiological factors influencing the prevalence of developmental defects of dental enamel in nine-year-old New Zealand children participating in a health and development study. *Journal of Dental Research*, 66, 1466–9.

Sunderland, E. P., Smith, C. J. and Sunderland, R. (1987). A histological study of the chronology of initial mineralization in the human deciduous dentition. *Archives of Oral Biology*, 32, 167–74.

Swärdstedt, T. (1966). *Odontological Aspects of a Medieval Population in the Province of Jämtland/Mid Sweden*. Doctoral Thesis, University of Lund. Stockholm: Tiden Barnagen.

Sweeney, E. A., Cabrera, J., Urritia, J. and Mata, L. (1969). Factors associated with linear hypoplasia of human deciduous incisors. *Journal of Dental Research*, 48, 1275–9.

Sweeney, E. A. and Guzman, M. (1966). Oral conditions in children from three highland villages in Guatemala. *Archives of Oral Biology*, 11, 687–98.

Sweeney, E. A., Saffir, A. J. and DeLeon, R. (1971). Linear hypoplasia of deciduous incisor teeth in malnourished children. *American Journal of Clinical Nutrition*, 24, 29–31.

Swindler, D. R. (2002). *Primate Dentition. An Introduction to the Teeth of Non-human Primates*. Cambridge University Press.

Swindler, D. R. and McCoy, H. A. (1965). Primate odontogenesis. *Journal of Dental Research*, 44, 283–95.

Swindler, D. R. and Meekins, D. (1991). Dental development of the permanent mandibular teeth in the baboon, *Papio cynocephalus*. *American Journal of Human Biology*, 3, 571–80.

Swindler, D. R., Olshan, A. F. and Sirianni, J. E. (1982). Sex differences in permanent mandibular tooth development in *Macaca nemestrina*. *Human Biology; an International Record of Research*, 54, 45–52.

Swindler, D. R., Orlosky, F. J. and Hendrickx, A. G. (1968). Calcification of the deciduous molars in baboons (*Papio anubis*) and other primates. *Journal of Dental Research*, 47, 167–70.

Tafforeau, P., Boistel, R., Boller, E. *et al.* (2006). Applications of X-ray synchrotron microtomography for non-destructive 3D studies of paleontological specimens. *Applied Physics A Materials Science & Processing*, 83, 195–202.

Tafforeau, P. and Smith, T. M. (2008). Nondestructive imaging of hominoid dental microstructure using phase contrast X-ray synchrotron microtomography. *Journal of Human Evolution*, 54, 272–8.

Taji, S., Hughes, T., Rogers, J. and Townsend, G. (2000). Localised enamel hypoplasia of human deciduous canines: genotype or environment? *Australian Dental Journal*, 45, 83–90.

Talmi, E., Sarnat, B. G. and Eli, I. (1986). The effect of the birth process on the width of the neonatal line (abstract). *Journal of Dental Research*, 65, 576.

Tanner, J. M. (1962). *Growth at Adolescence*, 2nd edn. Oxford: Blackwell Scientific.

Tanner, J. M. (1981). *A History of the Study of Human Growth*. Cambridge University Press.

Tanner, J. M. (1989). *Fetus into Man. Physical Growth from Conception to Maturity*, revised edn. Cambridge, MA: Harvard University Press.

Tanner, J. M. and Cameron, N. (1980). Investigation of the mid-growth spurt in height, weight and limb circumferences in single-year velocity data from the London 1966–67 growth survey. *Annals of Human Biology*, 7, 565–77.

Tanner, J. M. and Whitehouse, R. H. (1976). Clinical longitudinal standards for height, weight, height velocity, weight velocity, and stages of puberty. *Archives of Disease in Childhood*, 51, 170–9.

Tanner, J. M., Whitehouse, R. H. and Takaishi, M. (1966). Standards from birth to maturity for height, weight, height velocity, and weight velocity: British children, 1965. I. *Archives of Disease in Childhood*, 41, 454–71.

Tanner, J. M., Wilson, M. E. and Rudman, C. G. (1990). Pubertal growth spurt in the female rhesus monkey: relation to menarche and skeletal maturation. *American Journal of Human Biology*, 2, 101–6.

Tappen, N. C. and Simmons, D. J. (1975). Correspondence of silver nitrate staining patterns in decalcified bone with the microradiographic image. *The Anatomical Record*, 182, 267–73.

Tattersall, I. (2007). Neanderthals, *Homo sapiens*, and the question of species in paleoanthropology. *Journal of Anthropological Sciences*, 85, 139–46.

Taylor Parker, S. (2002). Evolutionary relationships between molar eruption and cognitive development in anthropoid primates. In *Human Evolution through Developmental Change*, ed. N. Minugh-Purvis and K. J. McNamara. Baltimore & London: Johns Hopkins University Press, pp. 303–16.

Ten Cate, A. R. (1998). *Oral Histology: Development, Structure and Function*, 5th edn. St Louis: C V Mosby.

Thackeray, J. F., Kirschvink, J. L. and Raub, T. D. (2002). Palaeomagnetic analyses of calcified deposits from the Plio-Pleistocene hominid site of Kromdraai, South Africa. *South African Journal of Science*, 98, 537–9.

Thompson, M. E. and Wrangham, R. W. (2008). Diet and reproductive function in wild female chimpanzees (*Pan troglodytes schweinfurthii*) at Kibale National Park, Uganda. *American Journal of Physical Anthropology*, 135, 171–81.

Tonge, C. H. and McCance, R. A. (1973). Normal development of the jaws and teeth in pigs, and the delay and malocclusion produced by calorie deficiencies. *Journal of Anatomy*, 115, 1–22.

Trigger, B. G., Kemp, B. J., O'Connor, D. and Lloyd, A. B. (1983). *Ancient Egypt. A Social History*. Cambridge University Press.

Trinkaus, E. (1995). Neanderthal mortality patterns. *Journal of Archaeological Science*, 22, 121–42.

Trinkaus, E., Churchill, S. E., Ruff, C. B. and Vandermeersch, B. (1999). Long bone shaft robusticity and body proportions of the Saint-Césaire 1 Châtelperronian Neanderthal. *Journal of Archaeological Science*, 26, 753–73.

Ubelaker, D. H. (1978). *Human Skeletal Remains: Excavation, Analysis, Interpretation*. Chicago: Aldine.

van Gerven, D., Beck, R. and Hummert, J. (1990). Patterns of enamel hypoplasia in two Medieval populations from Nubia's Batn al Hajar. *American Journal of Physical Anthropology*, 82, 413–20.

van Gerven, D. P., Sheridan, S. G. and Adams, W. Y. (1995). The health and nutrition of a medieval Nubian population: the impact of political and economic change. *American Anthropologist*, 97, 468–80.

van Schaik, C., Barrickman, N. L., Bastian, M. L., Krakauer, E. B. and van Noordwijk, M. A. (2006). Primate life histories and the role of brains. In *The Evolution of Human Life History*, ed. K. Hawkes and R. R. Paine. Santa Fe: School of American Research Press, pp. 127–54.

van Wagenen, G. and Catchpole, H. R. (1956). Physical growth of the rhesus monkey (*Macaca mulatta*). *American Journal of Physical Anthropology*, 14, 245–73.

van Wagenen, G., Catchpole, H. R., Negri, J. and Butzko, D. (1965). Growth of the fetus and placenta of the monkey (*Macaca mulatta*). *American Journal of Physical Anthropology*, 23, 23–33.

von Ebner, V. (1902). Die Histologie der Zähne mit Einschluß der Histogenes. In *Handbuch der Zahnheilkunde*, ed. J. Scheff. Vienna: A. Holder, pp. 243–302.

Waaler, H. T. (1984). Height, weight and mortality: the Norwegian experience. *Acta Medica Scandinavica*, 215, 1–56.

Walker, M. L. and Herndon, J. G. (2008). Menopause in nonhuman primates? *Biology of Reproduction*, 79, 398–406.

Walker, R., Hill, K., Burger, O. and Hurtado, A. M. (2006). Life in the slow lane revisited: ontogenetic separation between chimpanzees and humans. *American Journal of Physical Anthropology*, 129, 577–83.

Ward, C. V., Flinn, M. and Begun, D. R. (2004). Body size and intelligence in hominoid evolution. In *The Evolution of Thought: Evolutionary Origins of Great Ape Intelligence*, ed. A. E. Russon and D. R. Begun. Cambridge University Press, pp. 335–49.

Waugh, M. A. (1974). Alfred Fournier, 1832–1914. His influence on venereology. *British Journal of Venereology*, 50, 232–6.

Weber, D. F. and Eisenmann, D. (1971). Microscopy of the neonatal line in developing human enamel. *American Journal of Anatomy*, 132, 375–92.

Weinmann, J., Svoboda, J. and Woods, R. (1945). Hereditary disturbances of enamel formation and calcification. *Journal of the American Dental Association*, 32, 397–418.

White, T. D. (1978). Early hominid enamel hypoplasia. *American Journal of Physical Anthropology*, 49, 79–84.

Wich, S. A., Utami-Atmoko, S. S., Setia, T. M. *et al.* (2004). Life history of wild Sumatran orangutans (*Pongo abelii*). *Journal of Human Evolution*, 47, 385–98.

Williams, J. L. (1897). A contribution to the study of pathology of enamel. *Dental Cosmos*, 39, 169–96, 269–301, 353–74.

Willoughby, D. P. (1978). *All About Gorillas*. New Jersey: A.S. Barnes.

Wilson, D. F. and Shroff, F. R. (1970). The nature of the striae of Retzius as seen with the optical microscope. *Australian Dental Journal*, 15, 3–24.

Winkler, L. A. (1995). A comparison of radiographic and anatomical evidence of tooth development in infant apes. *Folia Primatologica*, 65, 1–13.

Winkler, L. A., Schwartz, J. H. and Swindler, D. R. (1991). Aspects of dental development in the orangutan prior to eruption of the permanent dentition. *American Journal of Physical Anthropology*, 86, 255–71.

Winkler, L. A., Schwartz, J. H. and Swindler, D. R. (1996). Development of the orangutan permanent dentition: assessing patterns and variation in tooth development. *American Journal of Physical Anthropology*, 99, 205–20.

Winkler, L. A. and Swindler, D. R. (1990). A comparison of dental development in two neonatal orangutans. *American Journal of Physical Anthropology*, 81, 318.

Witzel, C., Kierdorf, U., Schultz, M. and Kierdorf, H. (2008). Insights from the inside: histological analysis of abnormal enamel microstructure associated with hypoplastic enamel defects in human teeth. *American Journal of Physical Anthropology*, 136, 400–14.

Wood, B. and Lonergan, N. (2008). The hominin fossil record: taxa, grades and clades. *Journal of Anatomy*, 212, 354–76.

World Health Organization (2010). *World Health Statistics 2010*. Geneva: WHO Press.

Worth, J. E. (2001). The ethnohistorical context of bioarchaeology in Spanish Florida. In *Bioarchaeology of Spanish Florida. The Impact of Colonialism*, ed. C. S. Larsen. Gainsville: University Press of Florida, pp. 1–21.

Yen, P. K. J., Shaw, J. H. and Hong, Y. C. (1971). Effects of some staining agents on dentin apposition in young rabbits. *Journal of Dental Research*, 50, 1666–70.

Yilmaz, S., Newman, H. N. and Poole, D. F. G. (1977). Diurnal periodicity of von Ebner growth lines in pig dentine. *Archives of Oral Biology*, 22, 511–13.

Ziegler, A. C. (1971). A theory of the evolution of therian dental formulas and replacement patterns. *Quarterly Review of Biology*, 46, 226–49.

Zihlman, A., Bolter, D. and Boesch, C. (2004). Wild chimpanzee dentition and its implications for assessing life history in immature hominin fossils. *Proceedings of the National Academy of Sciences of the United States of America*, 101, 10541–3.

Zihlman, A. L., Bolter, D. R. and Boesch, C. (2007). Skeletal and dental growth and development in chimpanzees of the Taï National Park, Côte d'Ivoire. *Journal of Zoology*, 273, 63–73.

Zsigmondy, O. (1893). On congenital defects of the enamel. *Dental Cosmos*, 35, 709–17.

Zuckerman, S. (1928). Age changes in the chimpanzee, with special reference to growth of brain, eruption of teeth, and estimation of age; with a note on the Taungs ape. *Proceedings of the Zoological Society of London*, 98, 1–42.

Index

abrasion, 72, 75, 178
accentuated lines, 162, 169, 174–6, 181, 217, 222
additional dentition, 30, 31, 64
ADJ, *see* enamel-dentine junction
adolescence, 15–16
adolescent growth spurt, *see* pubertal growth spurt
African great apes, 232, 251
Afropithecus, 129, 143, 146, 154, 231, 260
age estimation, 15, 53, 58
age for stage, 39
agriculture: adoption or intensification, 215–22
AI, *see* amelogenesis imperfecta
alarm reaction, 202
Alicona InfiniteFocus, 269, 271
Alizarin, 54, 114
alveolar eruption, 42, 43, 144, 146
ameloblast, 32, 90, 97, 98, 165, 168
amelogenesis imperfecta, 162, 163
Andresen's lines, 104, 107
antimere, 119, 167
apatite, 32, 113, 123
ape, 2, 3, 30
apex, 70
apex closure, 32
apical foramen, 70
appositional enamel, *see* cuspal enamel
Ardipithecus, 153, 154, 231, 232
Atapuerca, 131, 132, 209
atlas development charts, 52
attrition, 72, 75, 157, 178
Australia, 219
Australopithecus, 129, 132, 135, 144, 153, 154, 157, 160, 209–11, 231, 232, 251, 258
autofluorescence, 115
Awash valley, 130

baboon, 2, 39, 44, 49, 54, 67, 142, 151, 176, 213, 250
back-scattered electron (BSE) mode, 114, 266
band striae, 174
bell stage, 31
Bielschowsky stain technique, 109

bioarchaeology, 198, 199, 204
birefringence, 174, 267
birth interval, 149, 151, 152, 160
body height or length, 5, 6, 13, 14, 18, 22
body mass, 5, 6, 13, 14, 18, 22, 149, 150, 151, 153, 158
body weight, 5
bonobo, 2, 3, 22, 151, 153, 158, 214
brain growth, 14
brain mass, 150, 154, 156, 158, 160
brochs, 85
brown striae of Retzius, 92–101, 107, 112, 113, 116, 118–23, 124, 127, 143, 166, 167, 169, 170, 176, 222, 224, 268, 270
buccal side, 55
bud stage, 31
bud-form molar, 188
Buffon's *Histoire Naturelle*, 6
Bunon, 184

calcein, 115
calcium metabolism, 13
calcospherites, 88, 102, 108, 115, 128, 267
calculus, 72
canine, 2, 118, 147, 164, 214
Cannon, 201
cap stage, 31
captivity effects, 42
caries, 72
catch-up growth, 16, 200
cattle, 23
CEJ, *see* cement-enamel junction
cement-enamel junction, 70
cervical loop, 31, 32
cervical margin, 70
cervical-type perikymata, 81, 97, 164
cervix, 70
childhood, 14, 24, 152
chimpanzee, 2, 3, 11, 22, 27, 40, 44, 46, 49, 54, 56, 66–7, 68, 76, 127, 128, 129, 132, 135, 136, 138, 140, 146, 147, 150, 151, 152, 153, 154, 156, 157, 158, 159, 211–14, 231, 235–6, 247–8, 252, 257
chronic stress, 202
chronological age, 11

circadian rhythm, 95, 113–18
cleaning teeth, 261
clinical eruption, 42
cluster bands, 174
coeliac disease, 195
cognition, 158–9
collagen, 32, 108
colonisation, 218–20
computed tomography, 50
concrete operations, 158
confocal light microscopy, 92, 95, 119, 127, 132, 143, 175, 182, 268
contour lines of Owen, 104, 107
contrast in light microscopy, 267
cross striation repeat interval, 96, 118–19, 127, 132, 166, 181
cross striations, 90–4, 95, 96, 107, 112–23, 124, 143, 181, 214, 222, 224, 270
cross-sectional growth study, 5, 6, 8, 22
crown, 32, 70, 76
crown completion, 32, 240, 245, 247, 250, 253, 255
crown initiation, 32, 239, 242, 244, 247, 250, 252, 254
crown-heel length, 5, 11, 13
crown-rump length, 5, 9, 13, 19
crypt, 42, 43
CT, see computed tomography
cumulated EDJ extension time, 124
cumulative EDJ extension rate, 137
cumulative percentage graph, 35
cumulative prism lengths, 124
cusp development in molars, 54, 95
cuspal enamel, 123–6, 136, 139, 147, 168, 172
cuspal-type perikymata, 81, 96, 164
cusps, 70

daily secretion rate, see mean enamel formation rate
dark field microscopy, 174, 268
DDE, see developmental defects of dental enamel
death assemblage, 199
deciduous dentition, 28, 43, 44, 47, 48, 54, 57, 58, 63, 64, 66, 75, 100, 136, 177, 180, 186, 191, 192, 214, 222–3, 232, 233, 235, 237, 239–41
decussation, 90, 113, 123
defect width, 167
demineralised preparations, 51, 114
Demirjian system, 61
dental caries, 199
dental development, 28–69, 154–5
dental endowment at weaning, 156
dental follicle, 31, 54
dental impressions, 262
dental lamina, 30
dental papilla, 31, 32
dental precocity, 156
dentinal tubules, 102
dentine, 70, 72, 101–4

dentine incremental lines, 104–10
dentine matrix, 32, 87, 102, 108, 115
dentition, 28
depth of field, 268
developing edge of tooth, 70
development stages, 32
developmental defects of dental enamel index, 163, 220, 221
diabetes, 194
diagenesis, 101
Dickson Mounds, 215
diet, 16
distal side, 55
Dryopithecini, 231
Dryopithecus, 129, 143, 146, 154, 231, 232, 260
DSR, see mean enamel formation rate
duration of hypoplasia, 167

ectomesenchyme, 30
EDJ, see enamel-dentine junction
Egypt, 220
embryo, 11
enamel, 2, 70, 75–101, 112
enamel matrix, 32, 54, 71, 87, 95, 127
enamel organ, 31
enamel-dentine junction, 31, 90, 95, 110, 121, 124, 127, 136, 162, 168, 217, 219
energy-dispersive X-ray spectroscopy, 114
engineer's measuring microscope, 271
entoconid, 55
environmental effects on growth, 16–18, 39, 41
environmental secondary electron detector, 265
epiphyseal fusion, 15
epithelium, 30
erosions, 162, 185
eruption, 28, 30, 32, 33, 40, 42–9, 144–7, 155, 157, 233–8
Everhart Thornley detector, 266
exfoliation, 45, 72
extension rate, 124, 136–9
extinction (polarising microscopy), 267

Family, 231, 232
Fauchard, 184
Fels growth study, 61
femur length, 9
fertilisation, 10
fever, 194
flight or fight response, 201
Florida, 218
fluorescence microscopy, 115
fluorochromes, 115
fluorosis, 189
foetus, 11, 13
formal operations, 158
formation schedule of teeth, 49–67, 139–44, 239–60
fossil assemblage, 41

fossil primates, 128
fossils, 1
Fournier, 186
Fournier's molar, 186
fractures of crown, 72
full-term, 11, 56, 64
furcation, 70
furrow-form defects, 86, 163–7, 175, 181, 224

GAS, *see* general adaptation syndrome
gastrointestinal conditions, 191
general adaptation syndrome, 202
genotype, 16
gestation period, 149, 151, 156
gestational age, 9, 10, 18, 30, 64
gibbon, 3, 22, 150, 211, 212, 214, 231
Gigantopithecus, 129, 143, 231, 232, 260
gingival emergence, 42, 43, 44–9, 144, 146
gnarled enamel, 90, 123
gorilla, 2, 3, 11, 22, 48, 68, 128, 136, 139, 146, 147, 150, 151, 152, 153, 156, 157, 158, 176, 211–14, 231, 235, 237, 252, 257
Gorilla, 231, 232
Gorillina, 231
Gran Dolina, 131, 132
granular layer of Tomes, 104, 108
graticule, 270
ground substance, 32, 108
growth attainment curve, 6
growth rate, 6
growth velocity, 6, 11
growth velocity curve, 6, 13, 14, 18, 22, 23

Haavikko dental development tables, 63, 244–6
haematoxylin stain, 51, 108, 114, 194
head circumference, 9
health, 198–200
Heliopithecus, 231
Hertwig's epithelial root sheath, 32
Holocene, 129
homeostasis, 201, 202
hominid, 3, 129
Hominidae, 3, 231, 232
Hominidea, 205
hominin, 1, 3
Hominina, 231
Homininae, 3, 231, 232
hominine, 3
Hominini, 3, 231, 232
hominoid, 3, 129
Hominoidea, 3, 231, 232
Homo, 3, 130, 145, 153, 154, 231, 232, 251, 256
human, 2, 3, 6–18, 25, 26, 30, 41, 43, 44–6, 49, 64–6, 68, 76, 116, 122, 127, 128, 131, 132, 135, 136, 138, 141, 143, 146, 147, 150, 151, 152, 153, 156, 158, 159, 214, 231, 233–4, 239–46, 251, 252, 254, 256, 260

human life history, 150–2
Hunter-Schreger bands, 90
Hutchinson, 186
Hutchinson's incisor, 186
Hylobatidae, 231, 232
hylobatids, 3
hypocalcaemia, 13, 192
hypocalcification, 71, 162, 189
hypocone, 55
hypoconid, 55
hypoconulid, 55
hypoplasia, 42, 56, 76, 86–7, 96, 104, 121, 122, 162–97, 200, 201, 203, 204–27
hypoplasia recording, 176–84
hypoplastic area method, 180

imbrication lines, 75
imbricational enamel, *see* lateral enamel
implantation, 10
in utero development, 9–13, 18, 30, 54, 121, 192, 239
incisor, 29, 164, 186
increment margin, 78
infancy, 14
infant, 11
interglobular spaces, 104, 162, 190
internal enamel epithelium, 31, 32, 54, 194
intertubular dentine, 102
intradian lines, 94, 113
Irhoud, 131, 132, 155
Isola Sacra, 222–3
isolated deep pits, 85

juvenile, 14

Krapina, 206, 207, 208
Kraus and Jordan dental development charts, 54
Kreshover, 194–5
Kromdraai, 129, 209

labial side, 55
laser scanning confocal microscope, 269
lateral enamel, 123, 126–7, 136, 140
lead acetate, 113
LEH, *see* linear enamel hypoplasia
LHPC, *see* localized hypoplasia of the primary canine
life history, 149–61
life history related variables, 153
lifespan, 149, 150
light microscopy, 95, 112, 114, 123, 127, 166, 174, 181, 215, 217, 219, 222, 223–5, 266–9
light scattering, 70, 71, 93, 95, 113, 120, 122, 175, 189, 268
line striae, 174
linear enamel hypoplasia, 163
lingual side, 55

localised hypoplasia of the primary canine, 163, 214–15
Logan and Kronfeld development tables, 51
logarithmic transformation, 38
London Atlas dental development chart, 53
longitudinal growth study, 5, 8, 18, 22
long-period lines, 107, 110, 118–23, 126
Lufengpithecini, 231
Lufengpithecus, 129, 143, 231, 232, 260

macaque, 2, 18–22, 27, 33, 48, 54, 56, 67, 116, 122, 151, 212, 237–8, 250
macrophotography, 265
Makapansgat, 129
malnutrition, 17, 192, 193
mamelons, 55, 57, 70, 186
mangabey, 212
markers for dental tissues, 113
Massler, Schour and Poncher diagram, 179, 191
matrix secretion phase, 32, 97
maturation, 32, 50, 71, 115, 128
maturity, 149, 151, 152
mean age, 38
mean age at attainment, 38
mean enamel formation rate, 124, 127, 133–6, 165, 251
measurements, 270–1
measurements of developing tooth height, 58, 59–60, 241
median age, 37, 38
medical histories, 191
Mellanby, 189
menarche, 15, 19, 27, 152
menopause, 151
mesial side, 55
metacone, 55
metaconid, 55
MFH system, *see* Moorrees, Fanning and Hunt
micro-CT, 50, 127, 142, 269–70
micro-radiography, 92, 120
microtome sectioning, 51
mid-crown type perikymata, 78, 97, 167
Middle Palaeolithic, 131, 207
Middle Stone Age, 155, 161
mid-growth spurt, 6, 14
midpoint age of attainment, 38
Miocene, 129
Mississippian culture, 216
mixed dentition, 29
mixed-longitudinal study, 5
molar, 2, 118, 121, 124, 144, 164
molar cusps, *see* cusp development in molars
monkeys, 2, 22, 30, 128, 150, 211, 212
Montbeillard's son, 6
Moon's molar, 188
Moorrees, Fanning and Hunt dental development charts, 60, 61, 246

morbidity, 198
mortality, 198, 199
Mousterian, 131, 155, 161
mulberry molar, 168, 173, 186

Nariokotome, 130, 145, 155, 160
Neanderthal, 130, 132–3, 135, 137, 138, 142, 144, 154, 155, 157, 161, 205–9, 254
neonatal, 64
neonatal body mass, 151
neonatal hypoplasia, 56, 192–4
neonatal line, 100, 112, 121–3, 139, 174, 175, 192, 222, 224
neonatal period, 13
neonatal tetany, 192
neonate, 11, 13, 55, 56
neural growth pattern, 25
Nomata deciduous development table, 57
Nubia, 220, 221

occlusal eruption, 43, 144, 146
odontoblasts, 32, 102
ogive, 35
Omo basin, 130
opacities, 163
orangutan, 2, 3, 22, 48, 50, 56, 67, 68, 128, 129, 135, 139, 146, 147, 150, 151, 152, 153, 156, 157, 158, 211–14, 231, 235, 237, 249, 252, 257
ovulation, 10

palatal side, 55
Pan, 231, 232
Papio, 254
paracone, 55
Paranthropus, 129, 132, 135, 137, 144, 145, 153, 154, 157, 160, 209–11, 231, 232, 258, 259
pathologic Retzius lines, 174
pathological bands, 174
pathological striae, 174–6, 219
peak height velocity, 138
perikyma groove, 79, 95
perikyma ridge, 79
perikymata, 75, 77–86, 96–8, 100, 121, 123, 127, 143, 145, 164, 167, 169, 175, 181, 182, 224, 270
perikymata packing pattern, 132
perikymata spacing, 225
perikymata spacing measurements, 184
periodicity, *see* cross striation repeat interval
periradicular bands, 76, 110
permanent dentition, 28, 33, 43, 45, 48, 49, 55, 58, 61, 64–6, 67, 118, 136, 177, 179, 186, 191, 232, 234, 236–7, 238, 242–60
phenotype, 16
PHV, *see* peak height velocity
Piaget, 158
pit-form defects, 86, 168, 173, 181, 224

Index

plane-form defects, 86, 167–70, 175, 181, 186
Pleistocene, 129, 130
Pliocene, 129
polarising microscopy, 90, 107, 174, 267
pongid, 3
Pongidae, 3, 231
Ponginae, 3, 232
pongine, 3
Pongo, 231, 232
portable scanning confocal optical microscope, 269
post-canine, 2
post-natal growth, 13–15
post-term, 11
predentine, 128
premolars, 2, 29
prenatal development, *see in utero* development
preoperations, 158
preterm, 11
prevalence, 204–5
primary epithelial band, 30
primate life history, 149
primates, 1, 23
prism boundaries, 90, 95, 98, 112, 123, 124, 127, 174
prism-free true surface zone enamel, 98
probits, 38
Proconsul, 128, 132, 135, 137, 142, 143, 231, 254, 259
Proconsuloidea, 231
prosimians, 2, 29
protocone, 55
protoconid, 55
PSCOM, 269
pubertal growth spurt, 6, 8, 15, 18, 21, 22, 23
puberty, 15–16
pulp, 70
pulp chamber, 32, 72, 101

Qafzeh, 131, 132, 155
quadrants of dentition, 29
quartiles, 37

radiography, 50, 60, 64, 66
radiolucency, 50
radio-opaque, 50
registration lines, 96
regression, 126, 135, 251
Reid and Dean crown surface charts, 76, 179, 192
replicas for microscopy, 76, 262
reproductive growth pattern, 25
rhythmic incremental lines, 174
rickets, 185, 189–91
rLEH, 213
Rome, 222–3
root, 32, 43, 70, 76, 101
root resorption, 72

rubella, 185
Rusinga Island, 128

Salpêtriere, 184
Sangiran, 137
Scammon's growth curves, 24
scanning electron microscopy, 76, 92, 114, 181, 182, 219, 265–6
Schour and Massler dental development chart, 51
Schultz's rule, 157
scurvy, 185
SE mode, 266
secondary dentine, 72
secondary electron mode, 77, 266
secretion rate, 113–18
section plane, 112, 121, 123
sectioning teeth, 262–4
secular trend, 16
Selye, 201
sensorimotor, 158
serial sections, 51
severity of hypoplasia, 167
sexual dimorphism, 2, 7, 15, 16, 18, 22, 27, 67, 147, 153
short-period lines, 107, 112–23, 126
siamang, 3, 22, 150, 231
Sima de los Huesos, 131, 132, 209
sitting height, 5
Sivapithecini, 231
Sivapithecus, 129, 143, 146, 154, 231, 232, 260
skew, 37
Skhūl, 131, 132, 155
smallpox, 185
sodium fluoride, 114
somatic growth pattern, 25
Spanish missions, 218
Spearman-Kärber method, 38
Spitalfields, 58, 63, 117, 223–5
SR-μCT, 270
St Brides, 223–5
stage of exhaustion, 202
stage of resistance, 202
statistical methods, 35–9
stature, 5, 16, 200
stereomicroscope, 265
Sterkfontein, 129, 209
storing teeth, 261
stress, 17, 167, 176, 201–4, 216
stress response, 202
stressor, 201, 202
successional dentition, 30, 31, 43, 65
supine length, 5
surface overlapping projections, 85
Swärdstedt chart, 179, 206, 216
Swartkrans, 129, 209

Index

synchrotron radiation micro-CT, 92, 95, 119, 120, 127, 132, 138, 143, 270
syphilis, 185–8

tandem scanning microscope, 269
tempo of growth, 17, 33
tetracyclines, 115, 139, 141
Theropithecus, 213
tissue turnover, 2
Tomes' process, 90, 98
Tomes' process pits, 79, 98, 128, 165, 168, 172
tooth germs, 30
tooth ring analysis, 122, 191–2
transmitted light microscopy, 266
Tribe, 231, 232
trunk height, 5
trunk length, 5
TSM, 269
tuberculosis, 194
Turkana Boy, 130

Ubelaker dental development chart, 53

Upper Palaeolithic, 131, 132, 155, 157, 161, 207

vaccinia, 195
variation in growth rate, 7, 15, 17, 19, 33, 44, 64, 179
velocity, 6
velocity curve, 8, 19
venereal disease, 185–8
vitamin D, 189–91, 193
von Ebner's lines, 104, 107

weaning, 14, 149, 152, 155–7, 159
wear facets, 72
wear of teeth, 72, 123, 127, 157, 178, 186, 217
wild versus captive, 42
Wilson bands, 162, 174–6, 218, 219, 222
Woodland culture, 216

X-ray, *see* radiography
X-ray diffraction, 123

zahnreihe, 30

For EU product safety concerns, contact us at Calle de José Abascal, 56–1°,
28003 Madrid, Spain or eugpsr@cambridge.org.

www.ingramcontent.com/pod-product-compliance
Ingram Content Group UK Ltd.
Pitfield, Milton Keynes, MK11 3LW, UK
UKHW050727110725
460672UK00006B/56